New Challenges in Wood and Wood-Based Materials II

New Challenges in Wood and Wood-Based Materials II

Editors

Lubos Kristak
Roman Réh
Ivan Kubovský

MDPI • Basel • Beijing • Wuhan • Barcelona • Belgrade • Manchester • Tokyo • Cluj • Tianjin

Editors

Lubos Kristak
Technical University in Zvolen
Slovakia

Roman Réh
Technical University in Zvolen
Slovakia

Ivan Kubovský
Technical University in Zvolen
Slovakia

Editorial Office
MDPI
St. Alban-Anlage 66
4052 Basel, Switzerland

This is a reprint of articles from the Special Issue published online in the open access journal *Polymers* (ISSN 2073-4360) (available at: https://www.mdpi.com/journal/polymers/special_issues/Wood_Based_Mater_2022).

For citation purposes, cite each article independently as indicated on the article page online and as indicated below:

LastName, A.A.; LastName, B.B.; LastName, C.C. Article Title. *Journal Name* **Year**, *Volume Number*, Page Range.

ISBN 978-3-0365-7160-7 (Hbk)
ISBN 978-3-0365-7161-4 (PDF)

© 2023 by the authors. Articles in this book are Open Access and distributed under the Creative Commons Attribution (CC BY) license, which allows users to download, copy and build upon published articles, as long as the author and publisher are properly credited, which ensures maximum dissemination and a wider impact of our publications.

The book as a whole is distributed by MDPI under the terms and conditions of the Creative Commons license CC BY-NC-ND.

Contents

Preface to "New Challenges in Wood and Wood-Based Materials II" vii

Lubos Kristak, Roman Réh and Ivan Kubovský
New Challenges in Wood and Wood-Based Materials II
Reprinted from: *Polymers* 2023, 15, 1409, doi:10.3390/polym15061409 1

Johannes Jorda, Emanuele Cesprini, Marius-Cătălin Barbu, Gianluca Tondi, Michela Zanetti and Pavel Král
Quebracho Tannin Bio-Based Adhesives for Plywood
Reprinted from: *Polymers* 2022, 14, 2257, doi:10.3390/polym14112257 9

Johannes Jorda, Günther Kain, Marius-Catalin Barbu, Alexander Petutschnigg and Pavel Král
Influence of Adhesive Systems on the Mechanical and Physical Properties of Flax Fiber Reinforced Beech Plywood
Reprinted from: *Polymers* 2021, 13, 3086, doi:10.3390/polym13183086 23

Ivana Tureková, Martina Ivanovičová, Jozef Harangózo, Stanislava Gašpercová and Iveta Marková
Experimental Study of the Influence of Selected Factors on the Particle Board Ignition by Radiant Heat Flux
Reprinted from: *Polymers* 2022, 14, 1648, doi:10.3390/polym14091648 39

Nikola Perković, Jure Barbalić, Vlatka Rajčić and Ivan Duvnjak
Compressive Strength Properties Perpendicular to the Grain of Hollow Glue-Laminated Timber Elements
Reprinted from: *Polymers* 2022, 14, 3403, doi:10.3390/polym14163403 53

Farshid Abdoli, Maria Rashidi, Akbar Rostampour-Haftkhani, Mohammad Layeghi and Ghanbar Ebrahimi
Withdrawal Performance of Nails and Screws in Cross-Laminated Timber (CLT) Made of Poplar (*Populus alba*) and Fir (*Abies alba*)
Reprinted from: *Polymers* 2022, 14, 3129, doi:10.3390/polym14153129 73

Marcin Szczepanski, Ahmed Manguri, Najmadeen Saeed and Daniel Chuchala
The Effect of Openings' Size and Location on Selected Dynamical Properties of Typical Wood Frame Walls
Reprinted from: *Polymers* 2022, 14, 497, doi:10.3390/polym14030497 93

Anna Danihelová, Zuzana Vidholdová, Tomáš Gergeľ, Lucia Spišiaková Kružlicová and Michal Pástor
Thermal Modification of Spruce and Maple Wood for Special Wood Products
Reprinted from: *Polymers* 2022, 14, 2813, doi:10.3390/polym14142813 107

Milan Gaff, Hana Čekovská, Jiří Bouček, Danica Kačíková, Ivan Kubovský, Tereza Tribulová, et al.
Flammability Characteristics of Thermally Modified Meranti Wood Treated with Natural and Synthetic Fire Retardants
Reprinted from: *Polymers* 2021, 13, 2160, doi:10.3390/polym13132160 121

Martin Zachar, Iveta Čabalová, Danica Kačíková and Tereza Jurczyková
Effect of Natural Aging on Oak Wood Fire Resistance
Reprinted from: *Polymers* 2021, 13, 2059, doi:10.3390/polym13132059 135

Michal Dudiak, Ladislav Dzurenda and Viera Kučerová
Effect of Sunlight on the Change in Color of Unsteamed and Steamed Beech Wood with Water Steam
Reprinted from: *Polymers* **2022**, *14*, 1697, doi:10.3390/polym14091697 . . . . **145**

Ladislav Dzurenda, Michal Dudiak and Eva Výbohová
Influence of UV Radiation on the Color Change of the Surface of Steamed Maple Wood with Saturated Water Steam
Reprinted from: *Polymers* **2022**, *14*, 217, doi:10.3390/polym14010217 . . . . **157**

Daniel Filgueira, Cristian Bolaño, Susana Gouveia and Diego Moldes
Enzymatic Functionalization of Wood as an Antifouling Strategy against the Marine Bacterium *Cobetia marina*
Reprinted from: *Polymers* **2021**, *13*, 3795, doi:10.3390/polym13213795 . . . . **167**

Igor Wachter, Tomáš Štefko, Peter Rantuch, Jozef Martinka and Alica Pastierová
Effect of UV Radiation on Optical Properties and Hardness of Transparent Wood
Reprinted from: *Polymers* **2021**, *13*, 2067, doi:10.3390/polym13132067 . . . . **181**

Preface to "New Challenges in Wood and Wood-Based Materials II"

Wood is a natural material that is available in large quantities and is easy to produce, making it the perfect material to consider for the circular economy. Its importance has dramatically increased in recent years. This increase is accompanied by the development of new research methods that open new possibilities in the areas related to wood and wood products in the process of their production, processing, and final use. The main topics of this Special Issue were: the knowledge of the quality of wood and other lignocellulose materials; the processes of their effective utilization and processing for more efficient processing; the adoption of techniques and research around using wood for environmentally friendly composite production and the positive impact of this on the environment; wood's interaction with solid substances and with different mechanical loads, chemicals, and other substances; and the different forms of energy and surface modification of wood and wood composites.

Lubos Kristak, Roman Réh, and Ivan Kubovský
Editors

Editorial

New Challenges in Wood and Wood-Based Materials II

Lubos Kristak *, Roman Réh and Ivan Kubovský

Faculty of Wood Sciences and Technology, Technical University in Zvolen, 96001 Zvolen, Slovakia
* Correspondence: kristak@tuzvo.sk

Wood is a natural material that is available in large quantities and is easy to produce, making it the perfect material to consider for the circular economy. Its importance has dramatically increased in recent years. This increase is accompanied by the development of new research methods that open new possibilities in the areas related to wood and wood products in the process of their production, processing, and final use. The main topics of this Special Issue were knowledge of the quality of wood and other lignocellulose materials; the processes of their effective utilization and processing for more efficient processing; the adoption of techniques and research around using wood for environmentally friendly composite production and the positive impact of this on the environment; wood's interaction with solid substances and with different mechanical loads, chemicals and other substances; and the different forms of energy and surface modification of wood and wood composites.

Section 1 *Wood-Based Panels*

Over the last 50 years, the use of wood adhesives in the manufacturing of wood-based panel goods has increased the efficiency of wood resources. Wood adhesives are becoming more popular as the need for wood-based panels grows [1]. Traditional wood-based panels are produced with synthetic, formaldehyde-based adhesives that are commonly made from fossil-derived constituents, such as urea, phenol, melamine, etc. Along with their numerous advantages, such as chemical versatility, high reactivity, and excellent adhesive performance, these adhesives are characterized by certain problems, and connected with hazardous volatile organic compounds (VOCs) [2,3]. Recently, increased amounts of biobased adhesives have been used in the production of wood composites to meet the current need for the development of sustainable and innovative materials which will make the wood-based panel industry more sustainable and lower its dependence on fossil fuels [4]. Increased interest in developing sustainable adhesives from different renewable biomass feedstock focuses mostly on lignin, tannin, starch, and proteins [5].

Authors Jorda et al. [6] focused their research on bio-based adhesive formulation consisting of Quebracho tannin extract and furfural. They investigated its usability as an adhesive for five-layered beech (Fagus sylvatica L.) plywood, detailing its press parameters and their influence on its physical and mechanical properties. The prepared fully bio-based adhesive formulation showed good viscosity and curing behavior at a relatively low temperature (100 °C), good bending, and acceptable tensile shear strength in a dry environment, which was comparable to industrial applied PF adhesive. The authors have concluded that the proposed Quebracho tannin and furfural formulation can improve and contribute to recyclability for specific interior plywood applications as a key element of the bio-based circular economy. Another piece of research conducted by Jorda et al. [7] focused on using adhesive systems for flax fiber-reinforced beech plywood. Epoxy resin, urea-formaldehyde, melamine-urea formaldehyde, isocyanate MDI prepolymer, and polyurethane had a different role in improving its mechanical properties. Epoxy resin showed excellent properties in the case of flax fiber reinforcement, while urea-formaldehyde, melamine urea-formaldehyde, and isocyanate MDI prepolymer helped improve the modulus of elasticity, modulus of rupture, shear strength, and screw withdrawal resistance. On the other hand, the tensile strength was lowered in the case of

Citation: Kristak, L.; Réh, R.; Kubovský, I. New Challenges in Wood and Wood-Based Materials II. *Polymers* **2023**, *15*, 1409. https://doi.org/10.3390/polym15061409

Received: 21 February 2023
Accepted: 9 March 2023
Published: 12 March 2023

Copyright: © 2023 by the authors. Licensee MDPI, Basel, Switzerland. This article is an open access article distributed under the terms and conditions of the Creative Commons Attribution (CC BY) license (https://creativecommons.org/licenses/by/4.0/).

these adhesive systems. Polyurethane was the last tested adhesive system that showed the lowered mechanical properties of flax fiber-reinforced plywood. As the authors concluded, further research is mandatory to determine the influence of press parameters, in addition to factors such as veneer and flax fiber manufacturing, and pre-treatment of the flax fiber [8,9]. Research in the area of fiber reinforcement used for veneer-based products has become recently very popular. Alongside flax fibers, several natural fibers such as hemp or kenaf, or synthetic fibers such as carbon have shown very good potential in increasing the mechanical properties of veneer-based products [10–12].

Alongside plywood, particleboards are another major wood-based product in global trade. Their global popularity and production have seen an upward trend in recent years [13–15]. In terms of quality assessment, particleboards have only a few disadvantages, and flammability is among them [16]. Several researchers have evaluated the flammability of wood-based composite materials [17–19]. The assumption of a fire hazard requires an appropriate description of the fire ignition and fire development [20,21]. The authors Turekova et al. [22] focused in their research on the significance of the influence of heat flux density and particleboards' properties on their thermal resistance (time to ignition). The obtained results showed that the time to ignition depended significantly on the thickness of the particleboard. Additionally, the linear dependence of the distance of the sample from the heat source was confirmed.

Glued-laminated timber (Glulam) is a wood-based product preferred for wide-span timber structures due to its lower variability in strength properties, and the possibility of producing it in almost any length and shape [23]. Hollow glulam beam has some advantages compared to traditional solid glulam beams, such as the convenience for wiring construction, comparably lightweight, and higher strength-to-weight ratio [24]. Perkovic et al. in their paper [25] presented innovative hollow Glulam elements intended for log-house construction. The research also involved testing the compression strength of elliptical hollow cross-section Glulam specimens made of softwood and hardwood, as well as full cross-section glue-laminated softwood timber specimens. The results showed that the compression strength perpendicular to the grain of hollow specimens decreased by about 55% compared to the full cross-section. By removing the holes in the central part of the cross-section, the stress was reduced. Additionally, the distance of the holes from the edges defined the local cracking. This research followed the long-term research focused on innovative hollow Glulam elements [26,27].

Cross-laminated timber (CLT) as a structural, plate-like timber product gained global recognition in the last three decades. The cross-laminated configuration improves the stability, rigidity, and mechanical properties of the product [28]. CLT is commonly utilized as a wall, roof, floor diaphragm, and other structural components. The connections between structural components (wall to wall, wall to floor, floor to floor, wall to foundation, and wall to concrete cores) must be designed for adequate strength, stiffness, and ductility when CLT panels are used in structures [29,30]. Research conducted by Abdoli et al. [31] was focused on the withdrawal resistance of nine types of conventional fasteners (stainless steel nails, concrete nails and screws, drywall screws, three types of partially and fully threaded wood screws, and two types of lag screws) with three loading directions and two-layer arrangements in 3-ply CLTs made of poplar and fir. Research results confirmed that the fastener type had the most significant impact on the withdrawal resistance, so changing the fastener type from nails to screws increased it by about 5–11 times, which is in accordance with other studies.

Wood frame walls have also gained popularity in recent decades. The frames are usually made from spruce or pine timber surrounded (mostly) by OSB casing and filled with thermal insulation material [32,33]. Szcepanski et al. [34] tested three walls of different widths, with and without two different sizes of openings, in several locations, to monitor the change in their natural frequencies under dynamic loads. The results showed that the effect of the size and location of the openings on the natural frequency is significant (and more sensitive for the location than for the sizes). The relative change in the natural frequencies of a wall without and with an opening in a specific place could be up to 30%. The authors concluded that the appropriate size and place were found to be small openings at the top of the walls.

Section 2 *Wood*

Wood is a valuable industrial and building material with many excellent properties in its natural state. Modification of wood is applied to overcome weak points that are mainly related to hygroscopicity, low dimensional stability, hardness, wear resistance, low resistance to different forms of radiation (UV, IR, etc.), and low resistance to bio-deterioration against fungi and termites [35–37]. Though many aspects of wood modification treatments are known, the fundamental influence of the process on product performance, the environment, and end-of-life scenarios remains relatively unknown [38,39]. A very popular and ecological method is the thermal modification of wood. Thermal modification is a well-established commercial technology for improving the durability and dimensional stability of wood [40–42].

The unique mechanical and acoustical properties of wood and its aesthetic appeal still make it the material of choice for musical instruments. The desire of musical instrument manufacturers to reduce the negative characteristics of raw material wood through special modification processes has existed for a century now [43,44]. In research by Danihelova et al. [45], the authors investigated the suitability of thermally modified (ThermoWood process) Norway spruce and sycamore maple for special wood products, mainly for musical instruments. Selected physical and acoustical characteristics (PACHs) including density, dynamic modulus of elasticity along the wood grain, specific modulus, speed of sound along the wood grain, resonant frequency, acoustic constant, logarithmic decrement, loss coefficient, acoustic conversion efficiency, sound quality factor and the timbre of sound were evaluated. Fast Fourier transform was used to analyze the sound produced. Based on the results, the mild thermal modification resulted in a decrease in mass and density, and in an increase in the speed of sound and dynamic modulus of elasticity at all temperatures of modification. The thermally modified wood showed higher sound radiation and lower loss coefficients than unmodified wood, and influenced the timbre of sound of both wood species.

Wood is one of the most sustainable, aesthetically pleasing, and environmentally friendly materials; however, the hazards of fire make wood a very desirable material for further investigation. As well as having ignition resistance and a low heat release rate, wood products are required to resist burn-through and maintain structural integrity when exposed to fire or heat. The changes in the residential fire environment, and the lack of fire behavior training are significant factors that are contributing to the continued climb in traumatic deaths and injuries to firefighters [46–50]. In research by Gaff et al. [51], the authors investigated the effect of synthetic and natural flame retardants on the flammability characteristics and chemical changes in thermally modified meranti wood. The basic chemical composition was evaluated to clarify the relationships between temperature modifications and incineration. Weight loss, burning speed, the maximum burning rate, and the time taken to reach the maximum burning rate were evaluated. The thermal modification did not confirm a positive contribution to the flammability and combustion properties of meranti wood. The effect of synthetic retardant on all combustion properties was significantly higher compared to that of natural retardant.

Some significant factors affecting the behavior of wood during exposure to fire are its chemical composition, wood species, density, moisture content, permeability, anatomy, and aging process [52–54]. Authors Zachar et al. [55] provided research dealing with the effect of the age of oak wood (0,10,40,80, and 120 years) on its fire resistance. The authors determined the chemical composition of wood by the wet chemistry method, and elementary analysis was performed according to ISO standards. From the fire's technical properties, the flame ignition, spontaneous ignition temperature, and mass burning rate were evaluated. Results showed that lignin content did not change, the content of extractives and cellulose increased and the content of holocellulose decreased with the higher age of the wood. The elementary analysis confirmed the lowest proportional content of nitrogen, sulfur, and phosphor, and the highest content of carbon in the oldest wood. The difference among the values of spontaneous ignition activation energy is clear evidence of higher resistance to the initiation of older wood in comparison with younger wood. The research confirmed that the oldest sample is the least thermally resistant, due to the different chemical compositions.

Steam treatment is often applied to wood to improve the stability and permeability of the wood, obtain a desirable color and soften the wood [56–58]. In the paper by authors Dudiak et al. [59], the differences in the color changes of unsteamed and steamed beech wood were caused by long-term exposure to sunlight on the surface of the wood in interiors for 36 months. The lower value of the total color difference of steamed beech wood indicated the fact that steaming of beech wood with saturated water steam had a positive effect on the color stability and partial resistance of steamed beech wood to the initiation of photochemical reactions induced by UV–VIS wavelengths of solar radiation. Spectra ATR-FTIR analyses declared the influence of UV–VIS components of solar radiation on unsteamed and steamed beech wood, and confirmed the higher color stability of the steamed beech wood. Another study, conducted by Dzurenda et al. [60], dealt with maple wood steamed with a saturated steam-air mixture to give a pale pink-brown, pale brown, and brown-red color. Subsequently, samples of unsteamed and steamed maple wood were irradiated with a UV lamp in a Xenotest after drying to test the color stability of the steamed maple wood. The results showed that the surface of unsteamed maple wood changes color far more markedly under the influence of UV radiation than the surface of steamed maple wood. Differential ATR-FRIT spectra confirmed the effect of UV radiation on unsteamed and steamed maple wood and confirmed the higher color stability of the steamed maple wood.

The protection of wood is particularly demanding in adverse environments. The use of wood in marine environments is a major challenge due to the high sensitivity of wood to both water and marine microorganisms [61,62]. Another interesting study by Filgueira et al. [63] focused on the development of a new green methodology based on the laccase-assisted grafting of lauryl gallate onto wood veneers, to improve their marine antifouling properties. Different wood species (beech, pine, and eucalyptus) were effectively hydrophobized through the enzymatic treatment. The reaction condition played an important role in the extent of hydrophobization; the treated wood species were also a major factor. The results observed in the study confirmed the potential efficiency of laccase-assisted treatments to improve the marine antifouling properties of wood.

With increasing energy demand and requirements for environmental conservation, the replacement of petroleum-based materials with bio-based materials is an interesting opportunity. Transparent wood is an example of a multifunctional wood composite. Optically transparent wood integrates mechanical and optical properties, and is a promising contender for intelligent buildings, structural optics, and photonics applications [64–68]. In research conducted by Wachter et al. [69], the effect of monochromatic UV-C radiation on transparent wood was evaluated. Samples of basswood were treated using a lignin modification method to preserve most of the lignin, and subsequently impregnated with refractive-index-matched types of acrylic polymers. Optical and mechanical properties were measured to describe the degradation process over 35 days. Results confirmed that exposure to UV-C radiation has a significant effect on the color of transparent wood. Samples became darker with increasing exposure time, and their color shifted towards shades of red and yellow; this can be possibly explained by the reactivation of chromophores. Additionally, the transmittance of light was significantly affected by UV-C radiation. The influence of UV-C radiation on hardness was significantly lower than in the case of optical properties. The measured values showed that the resulting hardness of transparent wood depends mainly on the hardness of the acrylic polymer used.

We would like to thank our Section Managing Editor, Chris Chen for his professional attitude and assistance with publishing.

Author Contributions: Conceptualization, L.K., R.R. and I.K. All authors have read and agreed to the published version of the manuscript.

Acknowledgments: This work was supported by the Slovak Research and Development Agency under contracts No. APVV-20-004, No. APVV-20-0159, No. APVV-19-0269, and No. SK-CZ-RD-21-0100.

Conflicts of Interest: The authors declare no conflict of interest.

References

1. Hussin, M.H.; Latif, N.H.A.; Hamidon, T.S.; Idris, N.N.; Hashim, R.; Appaturi, J.N.; Brosse, N.; Ziegler-Devin, I.; Chrusiel, L.; Fatriasari, W.; et al. Latest advancements in high-performance bio-based wood adhesives: A critical review. *J. Mater. Res. Technol.* **2022**, *21*, 3909–3946.
2. Kristak, L.; Antov, P.; Bekhta, P.; Lubis, M.A.R.; Iswanto, A.H.; Reh, R.; Sedliacik, J.; Savov, V.; Taghiyari, H.R.; Papadopoulos, A.N.; et al. Recent progress in ultra-low formaldehyde emitting adhesive systems and formaldehyde scavengers in wood-based panels: A review. *Wood Mater. Sci. Eng.* **2022**, 1–20. [CrossRef]
3. Antov, P.; Lee, S.H.; Lubis MA, R.; Yadav, S.M. Potential of Nanomaterials in Bio-Based Wood Adhesives: An Overview. *Emerg. Nanomater. Oppor. Chall. For. Sect.* **2022**, 25–63.
4. Antov, P.; Savov, V.; Neykov, N. Sustainable bio-based adhesives for eco-friendly wood composites. A review. *Wood Res.* **2020**, *65*, 51–62. [CrossRef]
5. Aristri, M.A.; Lubis, M.A.R.; Yadav, S.M.; Antov, P.; Papadopoulos, A.N.; Pizzi, A.; Fatriasari, W.; Ismayati, M.; Iswanto, A.H. Recent Developments in Lignin- and Tannin-Based Non-Isocyanate Polyurethane Resins for Wood Adhesives—A Review. *Appl. Sci.* **2021**, *11*, 4242. [CrossRef]
6. Jorda, J.; Cesprini, E.; Barbu, M.C.; Tondi, G.; Zanetti, M.; Král, P. Quebracho Tannin Bio-Based Adhesives for Plywood. *Polymers* **2022**, *14*, 2257.
7. Jorda, J.; Kain, G.; Barbu, M.-C.; Petutschnigg, A.; Král, P. Influence of Adhesive Systems on the Mechanical and Physical Properties of Flax Fiber Reinforced Beech Plywood. *Polymers* **2021**, *13*, 3086. [CrossRef] [PubMed]
8. Jorda, J.S.; Barbu, M.C.; Kral, P. Natural fibre-reinforced veneer-based products. *Pro Ligno* **2019**, *15*, 206–211.
9. Jorda, J.; Kain, G.; Barbu, M.-C.; Köll, B.; Petutschnigg, A.; Král, P. Mechanical Properties of Cellulose and Flax Fiber Unidirectional Reinforced Plywood. *Polymers* **2022**, *14*, 843. [CrossRef]
10. Auriga, R.; Gumowska, A.; Szymanowski, K.; Wronka, A.; Robles, E.; Ocipka, P.; Kowaluk, G. Performance properties of plywood composites reinforced with carbon fibers. *Compos. Struct.* **2020**, *248*, 112533. [CrossRef]
11. Kowaluk, G. Properties of lignocellulosic composites containing regenerated cellulose fibers. *BioResources* **2014**, *9*, 5339–5348. [CrossRef]
12. Bazli, M.; Heitzmann, M.; Hernandez, B.V. Durability of fibre-reinforced polymer-wood composite members: An overview. *Compos. Struct.* **2022**, *295*, 115827.
13. Mantanis, G.I.; Athanassiadou, E.T.; Barbu, M.C.; Wijnendaele, K. Adhesive systems used in the European particleboard, MDF and OSB industries. *Wood Mater. Sci. Eng.* **2018**, *13*, 104–116. [CrossRef]
14. Hua, L.S.; Chen, L.W.; Geng, B.J.; Kristak, L.; Antov, P.; Pędzik, M.; Rogoziński, T.; Taghiyari, H.R.; Lubis, M.A.R.; Fatriasari, W.; et al. Particleboard from agricultural biomass and recycled wood waste: A review. *J. Mater. Res. Technol.* **2022**, *20*, 4630–4658.
15. Pędzik, M.; Janiszewska, D.; Rogoziński, T. Alternative lignocellulosic raw materials in particleboard production: A review. *Ind. Crops Prod.* **2021**, *174*, 114162.
16. Taib, M.N.A.M.; Antov, P.; Savov, V.; Fatriasari, W.; Madyaratri, E.W.; Wirawan, R.; Osvaldová, L.M.; Hua, L.S.; Ghani, M.A.A.; Al Edrus, S.S.A.O.; et al. Current progress of biopolymer-based flame retardant. *Polym. Degrad. Stab.* **2022**, *205*, 110153.
17. Pedieu, R.; Koubaa, A.; Riedl, B.; Wang, X.-M.; Deng, J. Fire-retardant properties of wood particleboards treated with boric acid. *Eur. J. Wood Prod.* **2012**, *70*, 191–197. [CrossRef]
18. Tudor, E.M.; Scheriau, C.; Barbu, M.C.; Réh, R.; Krišťák, Ľ.; Schnabel, T. Enhanced Resistance to Fire of the Bark-Based Panels Bonded with Clay. *Appl. Sci.* **2020**, *10*, 5594. [CrossRef]
19. Madyaratri, E.W.; Ridho, M.R.; Aristri, M.A.; Lubis, M.A.R.; Iswanto, A.H.; Nawawi, D.S.; Antov, P.; Kristak, L.; Majlingová, A.; Fatriasari, W. Recent advances in the development of fire-resistant biocomposites—A review. *Polymers* **2022**, *14*, 362. [CrossRef]
20. Palumbo, M.; Formosa, J.; Lacasta, A.M. Thermal degradation and fire behaviour of thermal insulation materials based on food crop by-products. *Constr. Build. Mater.* **2015**, *79*, 34–39.
21. Marková, I.; Hroncova, E.; Tomaskin, J.; Turekova, I. Thermal analysis of granulometry selected wood dust particles. *BioResources* **2018**, *13*, 8041–8060. [CrossRef]
22. Tureková, I.; Ivanovičová, M.; Harangózo, J.; Gašpercová, S.; Marková, I. Experimental Study of the Influence of Selected Factors on the Particle Board Ignition by Radiant Heat Flux. *Polymers* **2022**, *14*, 1648. [CrossRef]
23. Tannert, T.; Vallée, T.; Müller, A. Critical review on the assessment of glulam structures using shear core samples. *J. Civ. Struct. Health Monit.* **2012**, *2*, 65–72.
24. Shang, P.; Sun, Y.; Zhou, D.; Qin, K.; Yang, X. Experimental study of the bending performance of hollow glulam beams. *Wood Fiber Sci.* **2018**, *50*, 3–19.

25. Perković, N.; Barbalić, J.; Rajčić, V.; Duvnjak, I. Compressive Strength Properties Perpendicular to the Grain of Hollow Glue-Laminated Timber Elements. *Polymers* **2022**, *14*, 3403. [CrossRef]
26. Perković, N.; Rajčić, V.; Pranjić, M. Behavioral Assessment and Evaluation of Innovative Hollow Glue-Laminated Timber Elements. *Materials* **2021**, *14*, 6911. [CrossRef]
27. Perković, N.; Rajčić, V. Mechanical and Fire Performance of Innovative Hollow Glue-Laminated Timber Beams. *Polymers* **2022**, *14*, 3381. [CrossRef]
28. Espinoza, O.; Buehlmann, U. Cross-laminated timber in the USA: Opportunity for hardwoods? *Curr. For. Rep.* **2018**, *4*, 1–12.
29. Hossain, A.; Popovski, M.; Tannert, T. Cross-laminated timber connections assembled with a combination of screws in withdrawal and screws in shear. *Eng. Struct.* **2018**, *168*, 1–11.
30. Abdoli, F.; Rashidi, M.; Rostampour-Haftkhani, A.; Layeghi, M.; Ebrahimi, G. Effects of fastener type, end distance, layer arrangement, and panel strength direction on lateral resistance of single shear lap joints in cross-laminated timber (CLT). *Case Stud. Constr. Mater.* **2023**, *18*, e01727.
31. Abdoli, F.; Rashidi, M.; Rostampour-Haftkhani, A.; Layeghi, M.; Ebrahimi, G. Withdrawal Performance of Nails and Screws in Cross-Laminated Timber (CLT) Made of Poplar (Populus alba) and Fir (Abies alba). *Polymers* **2022**, *14*, 3129. [CrossRef] [PubMed]
32. Burawska-Kupniewska, I.; Krzosek, S.; Mańkowski, P. Efficiency of visual and machine strength grading of sawn timber with respect to log type. *Forests* **2021**, *12*, 1467. [CrossRef]
33. Szczepański, M.; Migda, W.; Jankowski, R. Experimental study on dynamics of wooden house wall panels with different thermal isolation. *Appl. Sci.* **2019**, *9*, 4387. [CrossRef]
34. Szczepanski, M.; Manguri, A.; Saeed, N.; Chuchala, D. The Effect of Openings' Size and Location on Selected Dynamical Properties of Typical Wood Frame Walls. *Polymers* **2022**, *14*, 497. [CrossRef] [PubMed]
35. Sandberg, D.; Kutnar, A.; Mantanis, G. Wood modification technologies-a review. *Iforest-Biogeosciences For.* **2017**, *10*, 895. [CrossRef]
36. Todaro, L.; Liuzzi, S.; Pantaleo, A.M.; Lo Giudice, V.; Moretti, N.; Stefanizzi, P. Thermo-modified native black poplar (Populus nigra L.) wood as an insulation material. *iForest-Biogeosciences For.* **2021**, *14*, 268. [CrossRef]
37. Lovaglio, T.; D'Auria, M.; Gindl-Altmutter, W.; Lo Giudice, V.; Langerame, F.; Salvi, A.M.; Todaro, L. Thermal Modification and Alkyl Ketene Dimer Effects on the Surface Protection of Deodar Cedar (Cedrus deodara Roxb.) Wood. *Forests* **2022**, *13*, 1551. [CrossRef]
38. Jones, D.; Sandberg, D.; Goli, G.; Todaro, L. *Wood Modification in Europe: A State-of-the-Art about Processes, Products and Applications*; Firenze University Press: Florence, Italy, 2019; p. 123.
39. Čabalová, I.; Výbohová, E.; Igaz, R.; Kristak, L.; Kačík, F.; Antov, P.; Papadopoulos, A.N. Effect of oxidizing thermal modification on the chemical properties and thermal conductivity of Norway spruce (Picea abies L.) wood. *Wood Mater. Sci. Eng.* **2022**, *17*, 366–375. [CrossRef]
40. Umar, I.; Zaidon, A.; Lee, S.H.; Halis, R. Oil-heat treatment of rubberwood for optimum changes in chemical constituents and decay resistance. *J. Trop. For. Sci.* **2016**, *28*, 88–96.
41. Ali, M.R.; Abdullah, U.H.; Ashaari, Z.; Hamid, N.H.; Hua, L.S. Hydrothermal modification of wood: A review. *Polymers* **2021**, *13*, 2612. [CrossRef]
42. Lee, S.H.; Lum, W.C.; Zaidon, A.; Maminski, M. Microstructural, mechanical and physical properties of post heat-treated melamine-fortified urea formaldehyde-bonded particleboard. *Eur. J. Wood Wood Prod.* **2015**, *73*, 607–616. [CrossRef]
43. Pfriem, A. Thermally modified wood for use in musical instruments. *Drv. Ind.* **2015**, *66*, 251–253. [CrossRef]
44. Danihelová, A.; Spišiak, D.; Reinprecht, L.; Gergeľ, T.; Vidholdová, Z.; Ondrejka, V. Acoustic Properties of Norway spruce wood modified with staining fungus (Sydowia polyspora). *BioResources* **2019**, *14*, 3432–3444. [CrossRef]
45. Danihelová, A.; Vidholdová, Z.; Gergeľ, T.; Spišiaková Kružlicová, L.; Pástor, M. Thermal Modification of Spruce and Maple Wood for Special Wood Products. *Polymers* **2022**, *14*, 2813. [CrossRef] [PubMed]
46. Lowden, L.A.; Hull, T.R. Flammability behaviour of wood and a review of the methods for its reduction. *Fire Sci. Rev.* **2013**, *2*, 1–19. [CrossRef]
47. Kerber, S. Analysis of Changing Residential Fire Dynamics and Its Implications on Firefighter Operational Timeframes. *Fire Technol.* **2012**, *48*, 865–891. [CrossRef]
48. Osvaldova, L.M.; Petho, M. Occupational Safety and Health During Rescue Activities. *Procedia Manuf.* **2015**, *3*, 4287–4293. [CrossRef]
49. Gašparík, M.; Osvaldová, L.M.; Čekovská, H.; Potůček, D. Flammability characteristics of thermally modified oak wood treated with a fire retardant. *BioResources* **2017**, *12*, 8451–8467.
50. Gaff, M.; Kačík, F.; Gašparík, M.; Todaro, L.; Jones, D.; Corleto, R.; Osvaldová, L.M.; Čekovská, H. The effect of synthetic and natural fire-retardants on burning and chemical characteristics of thermally modified teak (Tectona grandis L. f.) wood. *Constr. Build. Mater.* **2019**, *200*, 551–558. [CrossRef]
51. Gaff, M.; Čekovská, H.; Bouček, J.; Kačíková, D.; Kubovský, I.; Tribulová, T.; Zhang, L.; Marino, S.; Kačík, F. Flammability Characteristics of Thermally Modified Meranti Wood Treated with Natural and Synthetic Fire Retardants. *Polymers* **2021**, *13*, 2160. [CrossRef]
52. Dietenberger, M.A.; Hasburgh, L.E. Wood products: Thermal degradation and fire. *Ref. Modul. Mater. Sci. Mater. Eng.* **2016**, 1–8.

53. Osvaldova, L.M.; Kosutova, K.; Lee, S.H.; Fatriasari, W. Ignition and burning of selected tree species from tropical and northern temperate zones. *Adv. Ind. Eng. Polym. Res.* **2023**. [CrossRef]
54. Topaloglu, E.; Ustaomer, D.; Ozturk, M.; Pesman, E. Changes in wood properties of chestnut wood structural elements with natural aging. *Maderas. Cienc. Tecnol.* **2021**, *23*. [CrossRef]
55. Zachar, M.; Čabalová, I.; Kačíková, D.; Jurczyková, T. Effect of Natural Aging on Oak Wood Fire Resistance. *Polymers* **2021**, *13*, 2059. [CrossRef] [PubMed]
56. Yilgor, N.; Unsal, O.; Kartal, S.N. Physical, mechanical, and chemical properties of steamed beech wood. *For. Prod. J.* **2001**, *51*, 89.
57. Dzurenda, L. Mode for hot air drying of steamed beech blanks while keeping the colours acquired in the steaming process. *Acta Fac. Xylologiae Zvolen* **2022**, *64*, 81–88.
58. Adamčík, L.; Kminiak, R.; Banski, A. The effect of thermal modification of beech wood on the quality of milled surface. *Acta Fac. Xylologiae Zvolen* **2022**, *64*, 57–67.
59. Dudiak, M.; Dzurenda, L.; Kučerová, V. Effect of Sunlight on the Change in Color of Unsteamed and Steamed Beech Wood with Water Steam. *Polymers* **2022**, *14*, 1697. [CrossRef]
60. Dzurenda, L.; Dudiak, M.; Výbohová, E. Influence of UV Radiation on the Color Change of the Surface of Steamed Maple Wood with Saturated Water Steam. *Polymers* **2022**, *14*, 217. [CrossRef]
61. Borges, L.M.S. Biodegradation of wood exposed in the marine environment: Evaluation of the hazard posed by marine wood-borers in fifteen European sites. *Int. Biodeterior. Biodegrad.* **2014**, *96*, 97–104. [CrossRef]
62. Marais, B.N.; Brischke, C.; Militz, H. Wood durability in terrestrial and aquatic environments–A review of biotic and abiotic influence factors. *Wood Mater. Sci. Eng.* **2022**, *17*, 82–105. [CrossRef]
63. Filgueira, D.; Bolaño, C.; Gouveia, S.; Moldes, D. Enzymatic Functionalization of Wood as an Antifouling Strategy against the Marine Bacterium Cobetia marina. *Polymers* **2021**, *13*, 3795. [CrossRef] [PubMed]
64. Chutturi, M.; Gillela, S.; Yadav, S.M.; Wibowo, E.S.; Sihag, K.; Rangppa, S.M.; Bhuyar, P.; Siengchin, S.; Antov, P.; Kristak, L.; et al. A comprehensive review of the synthesis strategies, properties, and applications of transparent wood as a renewable and sustainable resource. *Sci. Total Environ.* **2022**, *864*, 161067. [CrossRef] [PubMed]
65. Li, Y.; Vasileva, E.; Sychugov, I.; Popov, S.; Berglund, L. Optically transparent wood: Recent progress, opportunities, and challenges. *Adv. Opt. Mater.* **2018**, *6*, 1800059. [CrossRef]
66. Wachter, I.; Rantuch, P.; Štefko, T. Properties of Transparent Wood. In *Transparent Wood Materials: Properties, Applications, and Fire Behaviour*; Springer Nature Switzerland: Cham, Switzerland, 2023; pp. 1–13.
67. Li, Y.; Fu, Q.; Yang, X.; Berglund, L. Transparent wood for functional and structural applications. *Philos. Trans. R. Soc. A Math. Phys. Eng. Sci.* **2018**, *376*, 20170182. [CrossRef] [PubMed]
68. Vandličková, M.; Markova, I.; Osvaldová, L.M.; Gašpercová, S.; Svetlík, J. Evaluation of African padauk (*Pterocarpus soyauxii*) explosion dust. *BioResources* **2020**, *15*, 401–414. [CrossRef]
69. Wachter, I.; Štefko, T.; Rantuch, P.; Martinka, J.; Pastierová, A. Effect of UV Radiation on Optical Properties and Hardness of Transparent Wood. *Polymers* **2021**, *13*, 2067. [CrossRef]

Disclaimer/Publisher's Note: The statements, opinions and data contained in all publications are solely those of the individual author(s) and contributor(s) and not of MDPI and/or the editor(s). MDPI and/or the editor(s) disclaim responsibility for any injury to people or property resulting from any ideas, methods, instructions or products referred to in the content.

Article

Quebracho Tannin Bio-Based Adhesives for Plywood

Johannes Jorda [1,2,*], Emanuele Cesprini [3], Marius-Cătălin Barbu [1,4], Gianluca Tondi [3], Michela Zanetti [3] and Pavel Král [2]

[1] Forest Products Technology and Timber Construction Department, Salzburg University of Applied Sciences, Markt 136a, 5431 Kuchl, Austria; marius.barbu@fh-salzburg.ac.at
[2] Department of Wood Science and Technology, Mendel University, Zemědělská 3, 61300 Brno, Czech Republic; kral@mendelu.cz
[3] Land Environmental Agriculture & Forestry Department, University of Padua, Viale dell'Università 16, 3520 Legnaro, Italy; emanuele.cesprini@studenti.unipd.it (E.C.); gianluca.tondi@unipd.it (G.T.); michela.zanetti@unipd.it (M.Z.)
[4] Faculty for Furniture Design and Wood Engineering, Transilvania University of Brasov, B-Dul. Eroilor Nr. 29, 500036 Brasov, Romania
* Correspondence: johannes.jorda@fh-salzburg.ac.at

Abstract: Wood-based products are traditionally bonded with synthetic adhesives. Resources availability and ecological concerns have drawn attention to bio-based sources. The use of tannin-based adhesives for engineered wood products has been known for decades, however, these formulations were hardly used for the gluing of solid wood because their rigidity involved low performance. In this work, a completely bio-based formulation consisting of Quebracho (*Schinopsis balancae*) extract and furfural is characterized in terms of viscosity, gel time, and FT-IR spectroscopy. Further, the usability as an adhesive for beech (*Fagus sylvatica*) plywood with regard to press parameters (time and temperature) and its influence on physical (density and thickness) and mechanical properties (modulus of elasticity, modulus of rupture and tensile shear strength) were determined. These polyphenolic adhesives presented non-Newtonian behavior but still good spreading at room temperature as well as evident signs of crosslinking when exposed to 100 °C. Within the press temperature, a range of 125 °C to 140 °C gained suitable results with regard to mechanical properties. The modulus of elasticity of five layered 10 mm beech plywood ranged between 9600 N/mm^2 and 11,600 N/mm^2, respectively, with 66 N/mm^2 to 100 N/mm^2 for the modulus of rupture. The dry state tensile shear strength of ~2.2 N/mm^2 matched with other tannin-based formulations, but showed delamination after 24 h of water storage. The proposed quebracho tannin-furfural formulation can be a bio-based alternative adhesive for industrial applicability for special plywood products in a dry environment, and it offers new possibilities in terms of recyclability.

Keywords: plywood; quebracho; tannin furfural; biogenic adhesives

Citation: Jorda, J.; Cesprini, E.; Barbu, M.-C.; Tondi, G.; Zanetti, M.; Král, P. Quebracho Tannin Bio-Based Adhesives for Plywood. *Polymers* **2022**, *14*, 2257. https://doi.org/10.3390/polym14112257

Academic Editor: Ivan Kubovský

Received: 10 May 2022
Accepted: 29 May 2022
Published: 31 May 2022

Publisher's Note: MDPI stays neutral with regard to jurisdictional claims in published maps and institutional affiliations.

Copyright: © 2022 by the authors. Licensee MDPI, Basel, Switzerland. This article is an open access article distributed under the terms and conditions of the Creative Commons Attribution (CC BY) license (https://creativecommons.org/licenses/by/4.0/).

1. Introduction

Lignocellulosics are abundant bio-resources, nowadays perceived as a gamechanger in the scope of the climate crisis. Wood products contribute as carbon dioxide (CO_2) storage sinks due to increasing the time that CO_2 captured in forests is kept out of the atmosphere. Encouraging more forest growth, wood products enhance the efficiency of forest sinks by acting as carbon stores [1]. Indeed, numerous studies have emphasized the environmental benefits of wood-based materials compared to mineral-based compounds [2,3], due to the low embodied emissions and the lower material intensity of wood [4]. Within wood-based products, wood panels cover a major assortment of applications in the construction, packaging, and furniture sectors [5]. In order to achieve well-distinct properties, wood panel manufacturing adapts the dimensions of engineered wood products (EWPs) through the intelligent (re)assembly of wooden parts. Assembly that is regularly done with the application of adhesive resins. The high market that EWPs are gaining have caused environmental concerns related to the emissions from formaldehyde and other volatile compound that underlie the main adhesives used [6]. Consequently, workable alternatives are required in accordance with environmental standards and safety and market demands to direct future development through a sustainable use of wood. For instance, the generation of adhesives from bio-resources enables both a reduction in the use of chemical reagents harmful to health and a further move away from petroleum derivatives, thus decreasing the carbon footprint of the final product [7]. Actually, recent market forecasts highlight the importance of the bio-adhesives field and a growth from between USD 3.7–6.0 billion in 2020 to USD 5.2–9.7 billion by 2025–2028 is expected [8,9]. Different renewable substances have been proposed as a building block to manufacture bio-resins, from plant protein such as soy, starch based polysaccharides, and lignocellulosic molecules such as lignin and tannins [10,11]. Additionally, to overcome issues related to toxic reagents such as the formaldehyde traditionally used to manufacture wood-based products [12], different bio-based formaldehyde free formulations have been developed. Oktay et al. (2021) used bio-based corn-starch Mimosa tannin sugar adhesives for panels to meet the EN 312:2010 particleboard (P2) standard requirements for interior fittings in a dry state [13]. Similar results are given by Paul et al. (2021) for particleboard bonded with lignin-based adhesives [14]. Plywood, for example, assembled with PVOH–lignin–hexamine showed a dry tensile shear strength of 0.95 N/mm^2 [15] or with a soybean meal-based adhesive, which displayed excellent water resistance with a tensile shear strength exceeding 1 N/mm^2 [16]. Ghahri et al. (2022) reported a wet state tensile shear strength of ~0.8 N/mm^2 for a Quebracho tannin and isolate soy protein adhesive without hardener [17]. According to the mentioned research, tannins are of particular interest due to their chemical structure and good reactivity [18,19], which make these compounds great candidates. Tannins are classified into hydrolysable and condensed, the former class are mixtures of simple phenols, such as pyrogallol and ellagic acid, and esters of glucose, with gallic and digallic acids [20]. The latter, also known as proanthocyanins or flavanol, constitutes more than 90% of world production [20], which due to its reactivity is more suitable for industrial application. Condensed tannins are polyhydroxy-flavan-3-ol oligomers bonded together mostly by C-C bonds between the A rings of the flavanol units and the pyran rings of other flavanol units [19]. Particularly, the polyphenolic structure suggests the comparison and the possible replacement of phenol-formaldehyde (PF) synthetic resins used for gluing EWPs, whose production has seen a sharp increase in the last decade [21]. Moreover, during processing, PF resins have the highest environmental impact of all major synthetic resins [22], ranking tannins as a potential prime substitute.

Deep research has been carried out on the application of tannin adhesives [23–26]. However, it is useful to mention that a synthetic crosslinker is almost always required to form the three-dimensional polymeric structure. In the current study, an entirely renewable tannin-based adhesive is proposed, using furfural as hardener. Furfural, belonging to the furan compounds, is produced through the acid hydrolysis of biomass [27], and agricultural residues can be used, too [28]. The renewability and the abundance of lignocellulosic biomass make it a viable resource.

This study proposes a new Quebracho tannin-furfural adhesive formulation to be used for the production of plywood. The aim of our investigation is to characterize the Quebracho tannin-furfural adhesive, previously studied and compared with the main synthetic and non-synthetic hardeners [29], in terms of gel time, viscosity, and FT-IR spectroscopy as well as to determine the mechanical performance of five layered beech (*Fagus sylvatica*) plywood with regard to the press parameter of time and temperature, which consequently contribute to the production and the use of bio-based adhesives.

2. Materials and Methods

2.1. Materials

The tannin-based adhesives were prepared using Quebracho (*Schinopsis balancae*) tannin extract (Fintan 737B), kindly provided by the company Silvateam (S. Michele Mondovì,

Cuneo, Italy) and furfural (99%) obtained by International Furan Chemical IFC (Rotterdam, The Netherlands). Sodium hydroxide was purchased from Alfa Aesar (Thermo Fisher, Waltham, MA, USA) and it was applied to change the pH of the formulation.

Pre-conditioned (20 °C, 65% relative air humidity) rotary cut defect free beech (*Fagus sylvatica*) veneers, purchased from Europlac (Topolčany, Slovakia,), with a nominal thickness of 2.2 mm, an average density of 0.72 g/cm^3, and an average moisture content of 12% were used to prepare the plywood for this study.

2.2. Methods

2.2.1. Adhesive Preparation

The tannin-furfural formulation was prepared by mixing under vigorous stirring the commercial extract with water to obtain a 65% homogeneous suspension. The starting pH of 6.7 was adjusted to 8 by adding a 33% sodium hydroxide solution and finally 10% of furfural calculated on solid tannin was added.

2.2.2. Adhesive Characterization

Gel time: 5 g of the formulation were inserted into a glass test tube which was immersed in an oil bath at 100 °C. The transition time to obtain a solid was recorded using a stopwatch. The tests were repeated three times.

Viscosity: A freshly prepared formulation was analyzed with Rheometer Kinexus Lab from Malvern Panalytical (Malvern, UK). The measurement was conducted at 25 °C using cone-shaped geometry spindles with a diameter of 4 cm and a gap between the plates of 0.15 mm. The rotational speed was set from 10 s^{-1} to 300 s^{-1}.

FT-IR: A Frontier ATR-FT-MIR from Perkin Elmer (Waltham, MA, USA) was used for scanning the industrial Quebracho powder, Quebracho furfural formulation dried at room temperature for 24 h and the same formulation cross-linked at 100 °C for 24 h. Every spectrum was acquired with the ATR diamond device with 32 scans from 4000 to 600 cm^{-1} and the fingerprint spectral region between 1800 and 600 cm^{-1} was considered after normalization and baseline correction.

2.2.3. Plywood Preparation

The plywood consisted of five layered 90° crosswise oriented 2.2 mm thick beech veneer plies. Adhesive application was carried out manually by weighing the required adhesive amount of 150 g/m^2 per glue line with a KERN ITB 35K1IP device (Balingen-Frommern, Germany). Pressing was conducted using a Höfler HLOP 280 (Taiskirchen, Austria). Pressure was set to 3 N/mm^2; press-time was 10 min, 15 min, respectively 20 min and press-temperature was 110 °C, 125 °C and 140 °C.

A pretest to determine the time depended temperature behavior within the glue line during hot pressing as well as in order to check temperature difference between press and glue line was conducted using a Lutron electronic enterprise BTM 4208SD (Taipei City, Taiwan) datalogger with K-couple thermo-wired sensors. The sensors were placed on the outer plies surfaces and within the glue lines between the singular plies. The temperature at the press control unit was adjusted according to the pretest results.

After pressing, the boards were stored until mass constancy under a climate of 20 °C and 65% relative humidity. Test specimen were cut from the plywood boards for the determination of density, bending strength (MOR), stiffness (MOE), and tensile shear strength (TSS).

2.2.4. Plywood Characterization

The density was determined according to EN 323:2005, and it was obtained from the bending test specimen [30].

The density profile was measured with a DENSE-LAB X (EWS, Hammeln, Germany) and the specimen dimensions 50 mm × 50 mm. The thickness was obtained from the bending test specimens. The "Degree of compression" (DoC) was calculated by the percentage based difference between the theoretical thickness of 11 mm of non-compressed veneer ply stack before pressing and the actual thickness of the bending test specimens according to Spulle et al. (2021) [31].

Dry state tensile shear strength (TSS) and 24 h water soaking TSS was determined according to EN 314:2005 with specimen dimensions 100 mm × 25 mm [32]. Modulus of rupture (MOR) and modulus of elasticity (MOE) were determined by a three-point bending test according to EN 310:2005 with specimen dimensions 250 mm × 50 mm [33].

All mechanical properties (SS, MOE, and MOR) were determined using a Zwick/Roell 250 8497.04.00 test device (Ulm, Germany) under constant climatic conditions (rel. humidity 65%, ambient temperature 20 °C). The set-up and the number of specimens of the conducted tests is given in Table 1.

Table 1. Number of test specimen for the physical and mechanical properties testing of Quebracho tannin-furfural bonded five layered beech plywood.

Temperature [°C]	Time [min]								
	10			15			200		
	110	125	140	110	125	140	110	125	140
	Number of Test Specimens N								
Density	5	5	5	5	5	5	5	5	5
Density profile	5	5	5	5	5	5	5	5	5
Thickness	5	5	5	5	5	5	5	5	5
MOE/MOR	5	5	5	5	5	5	5	5	5
TSS dry state and 24 h	5	5	5	5	5	5	5	5	5

2.2.5. Data Analysis

For statistical evaluation IBM SPSS (Armonk, NY, USA) was used for descriptive data exploration and univariate and multivariate methods for the evaluation of the different Quebracho tannin-furfural bonded plywood test specimen. To determine differences between the press parameters, an ANOVA at a significance level of 95% was used. Multivariate ANOVA was used to determine the influence of "Temperature" and "Press-time" with the "Density" as covariant. The significance of correlations (Pearson) were evaluated using two-sided confidence intervals of 95%.

3. Results & Discussion

3.1. Adhesive Characterization

Tannin-furfural adhesives showed the most favorable hardening conditions at pH 8 [29]. Due to the limited viscosity of the adhesive at 50% solid content (s.c.), in this work tannin formulations with 65% s.c. were tested for their viscosity, gel time, and hardening. It was observed that concentrated tannin-furfural formulation presents a non-Newtonian pseudoplastic behavior (Figure 1), described as an increase of shear rate leading to a decrease of viscosity.

In these conditions, the formulation easily resulted in being homogeneously spread on wood. The curing behavior of the formulation was measured through gel time at 100 °C, which was 238 (+/−10) seconds that is slightly slower than commercial urea-formaldehyde's (UF) as it hardens after 127 s [34] but rather faster than phenol-formaldehyde's (PF) with a gelation time ranging within 10 min [35].

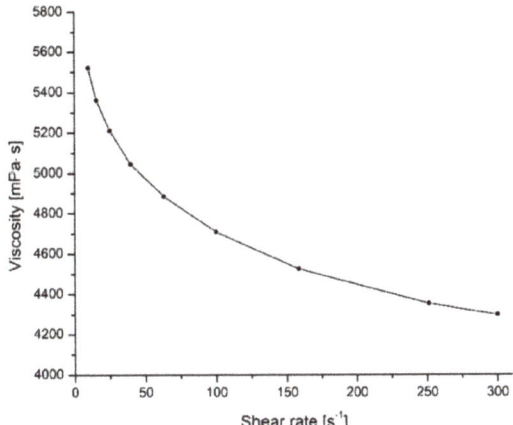

Figure 1. Viscosity of a 65% tannin furfural formulation.

From the chemical point of view, the curing process was observed comparing the spectra of the resin exposed 24 h at 25 and 100 °C. Figure 2 reports the spectra of the dry resin before and after curing. Comparing the two spectra, the most evident difference is that after curing the bands become broader suggesting the formation of polymeric structures, in particular the region at lower wavenumber become almost flat due to the steric hindrance for out of plane C-OH wagging and C-H bending vibrations [36,37]. Further major observations are the decreasing/disappearing of some signals such as those at 1670, 1392, 1018, 929, and 758 cm^{-1}, which are related to furfural compounds [38,39]. According to these observations, the crosslinking process could be similar to that observed for the polymer with Mimosa extract, involving the bridging through methylene–furanic units [40].

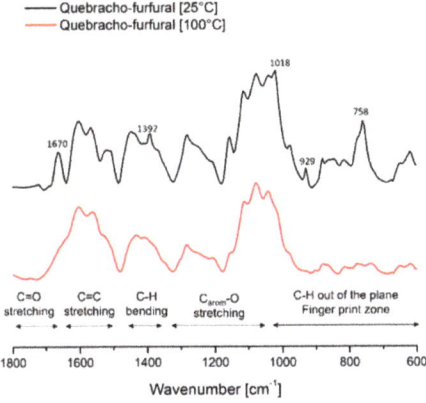

Figure 2. ATR-FTIR of Quebracho tannin furfural formulations at room temperature (black) and cured at 100 °C (red).

The main differences between Mimosa and Quebracho tannin is related to the nature of the B ring where the former bonds three hydroxyl groups (pyrogallic unit) [40]. Conversely, the B ring of Quebracho bonds principally two hydroxyl groups (catechol unit) decreasing the reactivity due to the chemistry of phenol [29]. Thus concluding, the reaction between Quebracho and furfural mainly involves the benzene ring A, as reported in Figure 3.

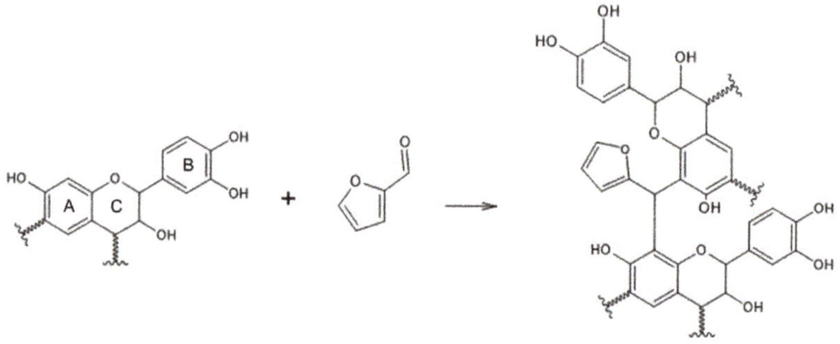

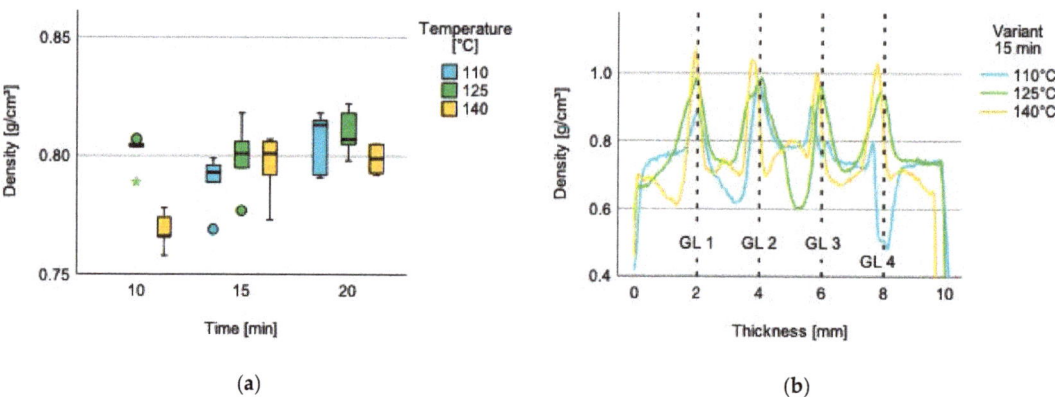

Figure 3. Possible product from Quebracho tannin and furfural reaction.

3.2. Plywood Characterization

3.2.1. Density

Density is one of the major physical parameters influencing the mechanical properties of plywood while enhancing MOE and tensile strength (TS) [41]. The mean density of the tested groups range between 0.768 g/cm^3 (press temperature 140 °C; press time 10 min) and 0.810 g/cm^3 (press temperature 125 °C; press time 20 min) (Figure 4a). The gained results are within the range compared to the values mentioned in the literature for identical five-layered beech plywood set-ups [42,43]. Testing of the density specimen for 110 °C press temperature and 10 min press time was not possible due to delamination after pressing and conditioning.

Figure 4. (a) Density grouped by time and temperature. Dots and stars within the box plot indicate outliers and (b) density profile for the 5-layers plywood glued for 15 min with Quebracho tannin-furfural adhesives at different temperature.

The density profile, plotting density against thickness, displays a method to gain information of the bonding performance within the adhesive layer [44,45]. The selected density profiles (Figure 4b) of specimen from the test set 15 min and three different temperatures, demonstrating differing bonding behavior. The specimen for the press temperature of 110 °C reveals delamination within the glue line (GL) 4 due to a significant sharp declined density gap and wider thickness. Further, a double peaking at glue line 3 indicates inappropriate bonding behavior. Test specimen for the press temperature of 125 °C illustrates a deeper adhesive penetration into the plies adjacent to the glue line due to wider and slightly lower density peaks than the selected specimen of the press temperature 140 °C. Compared to the previous described samples, the specimen for 140 °C has a sharper curvature of the density peaks indicating a reduced adhesive penetration into the adjacent wood layers, and a higher degree of compression is visible due to the lower thickness (<10 mm) compared to the other test specimen (>10 mm) with a lower temperature (Figure 5a).

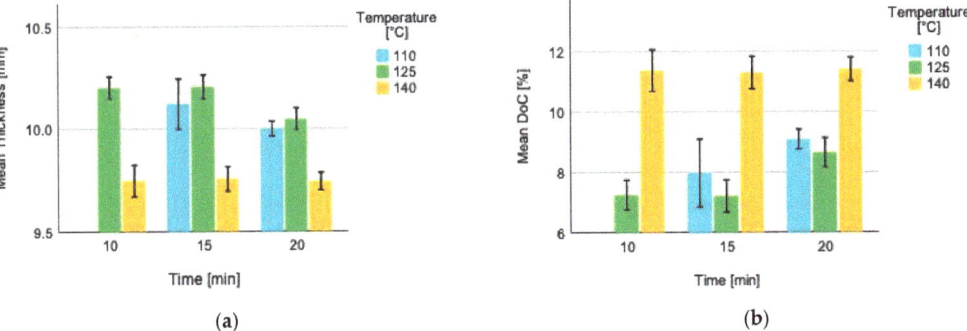

Figure 5. (a) Mean thickness and (b) mean degree of compression (DoC) of five layered Quebracho tannin furfural bonded plywood. The brackets within the figure indicate the interval ± 1 standard deviation (SD).

The thickness of the Quebracho tannin-furfural bonded five layered plywood ranges between 9.74 mm (press temperature 140 °C; press time 20 min) and 10.21 mm (press temperature 125 °C; press time 15 min) (Figure 5a), respectively, and between 11.45% and 7.18% for the degree of compression (DoC) (Figure 5b). This is according to Bekhta et al. (2009), stating a compression of ~10% for plywood manufacturing [46].

Thickness and therefore the degree of compression is influenced by the moisture content of veneers, press time and temperature. An elongated press time with a higher temperature influences the chemical wood structure due to a shift toward the glass transition of the singular chemical wood constituents while softening the natural polymeric cellular fiber composite character of wood [47].

3.2.2. Bending Properties

Modulus of elasticity (MOE) ranges between 448 (SD = 34) N/mm^2 (press-time 15 min/-temp. 110 °C) and 11,628 (SD = 592) N/mm^2 (press-time 10 min/-temperature 140 °C) (Figure 6a). Modulus of rupture (MOR) ranges between 18.73 (SD = 2.65) N/mm^2 (press-time 15 min/-temperature 110 °C) and 104.61 (SD = 20.67) N/mm^2 (press-time 15 min/-temperature 140 °C) (Figure 6b). Testing of specimen of test-group press time 10 min and press temperature 110 °C could not be carried out due to delamination after pressing and within conditioning. All tested specimen regardless of the test group failed within the adhesive layers, indicating low cohesive strength. Notable is the shift of the failure pattern from the pressure zone to the tension zone of the three-point bending test specimens, with increasing press-temperature and time. This fact reveals an improved adhesive performance with increasing press-time and temperature (Figure 7).

The modulus of elasticity is clearly affected by the combination of temperature and time. At a higher temperature, similar MOE are achieved independently of the pressing time.

Applying 20 min curing time, 110 °C is already sufficient to exceed 9000 N/mm^2, while at 140 °C with 10 min already a modulus of elasticity exceeding 11,000 N/mm^2 is reached. Hence, an increase of the temperature above 15 min does not further influence the final MOE. Additionally, the modulus of rupture (MOR) is dependent on the combined effect between temperature and time, where temperature is still crucial (Figure 6b). It can be observed that, the overall preferable pressing conditions for the bending properties require higher temperature (140 °C) and a pressing time of 10 to 15 min.

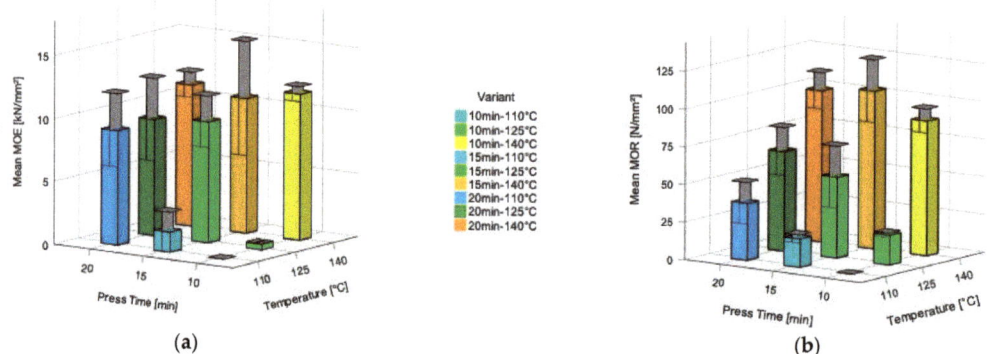

Figure 6. Influence of press time and press temperature of five layered Quebracho Tannin furfural bonded beech plywood on (**a**) Modulus of elasticity (MOE) and (**b**) Modulus of rupture (MOR). The top of the column indicates the means and the bars within the figure represent the standard deviation (SD).

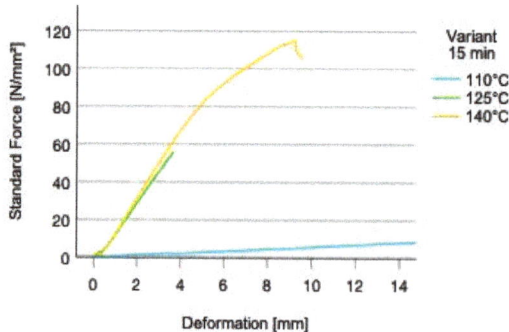

Figure 7. Stress-deformation behavior of selected samples for 15 min press time and three different temperatures.

Comparing the presented MOE and MOR values to the literature, there is a general divergent picture. Niemz (1993) stated general values for MOE between 1500 and 7000 N/mm² for plywood without regard to adhesives [48]. Values for MOE according to DIN 68 705-5 range between 5900 and 9600 N/mm² [49]. Hrazsky and Kral (2005) stated a mean MOE of 12,493 N/mm² and a mean MOR of 77.50 N/mm² for seven layered foiled 10 mm thick beech plywood [50]. Biadala et al. (2020) obtained a mean MOE of 13,720 N/mm² for three layered phenol-formaldehyde bonded beech plywood with a nominal veneer thickness of 1.7 mm and a MOR of 158.4 N/mm² [51]. Lower values are given by Dieste et al. (2008) for MOE with a mean of 9369 N/mm² of *Fagus sylvatica* five layered phenolic resin (150 g/m²) bonded plywood at 140 °C press temperature, 10 min of pressing and a pressure of 1 N/mm² [52]. This is 20% lower compared to the presented mean MOE of 11,628 N/mm² for 10 min and 140 °C of the current study. The variation within the numbers can be explained by the natural variation of native wood and its anisotropic behavior. Lohmann (2008) stated for the MOE of *Fagus sylvatica* a range between 10,000 to 18,000 N/mm² and for MOR 74 to 210 N/mm² [53]. Additionally, the mechanical performance of wood-based materials is influenced by the press parameters, according to Réh et al. (2021), as well as the specific lay-up of laminar wood-based products [54]. Further, the type of adhesive has a significant influence on MOE and MOR [55], concluding that the presented adhesive formulation can compete with synthetic phenolic resins in terms of MOE and MOR.

3.2.3. Tensile Shear Strength

Tensile shear strength had been tested in the dry state and after 24 h water storage. The results for the dry state tensile shear strength range between 0.00 N/mm² (press-time 10 min; press-temperature 110 °C), respectively, 1.74 (SD = 0.32) N/mm² (press-time 10 min; press-temperature 140 °C) and 2.29 (SD = 0.69) N/mm² (press-time 15 min; and press-temperature 125 °C) (Figure 8). It has to be noted that specimens of test group 110 °C/10 min failed subsequently before testing due to delamination and only two specimen per test group 125 °C/10 min and test group 110 °C/15 min due to delamination during specimen cutting could be tested. This indicates a poor bonding behavior within the glue line. Tensile shear strength testing at dry state revealed excellent results even at moderate curing temperature (125 °C) with limited influence of the press time.

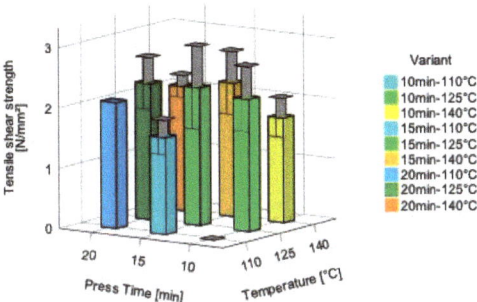

Figure 8. Dry state tensile shear strength for the 5-layered Quebracho tannin bonded plywood. The brackets within the columns of the figure indicate the standard deviation.

All tested specimens regardless of the test group failed within the glue line without displaying a wood fracture pattern according to EN 314 [56].

Testing of tensile shear strength for class 1 plywood applications according to EN 314 with 24 h water storage could not be carried out due to delamination failure of all test specimens within the 24 h of immersion into water [32].

Compared to the literature, Xi et al. (2020) gained values for tensile shear strength at a dry state between 0.98 and 1.99 N/mm² for three layered poplar (*Populus tremuloides*) plywood bonded with different Mimosa tannin glucose mixtures [57]. Similar results are stated by Hafiz et al. (2020) for tannin phenol-formaldehyde (TPF) co-polymer bonded rubber wood (*Hevea brasiliensis*) plywood in a range between 1.71 and 2.58 N/mm² and 3.41 N/mm² for the phenol-formaldehyde (PF) bonded reference [58]. Compared to industrial applicated adhesives, Jorda et al. (2021) stated for five layered beech (*Fagus sylvatica*) urea-formaldehyde (UF) bonded plywood a mean tensile shear strength in a dry state of 5.47 N/mm², for melamine-urea formaldehyde (MUF) 6.29 N/mm² and polyurethane (PUR) of 6.74 N/mm² [42]. Biadala et al. (2020) obtained a tensile shear strength value for phenol formaldehyde resin bonded beech plywood after 24 h water soaking of 2.99 N/mm², respectively 2.44 N/mm² after the boiling test [51]. Concluding that the presented Quebracho tannin-furfural adhesive formulation is capable of preserving with other mentioned tannin adhesives formulations for dry state tensile shear strength. Compared to industrial applicated adhesives, the dry-state performance is significantly lower and after 24 h water exposure incapable in terms of water resistance. This could be related to tensions induced by swelling of the singular veneer plies, especially beech (*Fagus sylvatica* L.) reacts sensitive to moisture induced swelling and shrinkage, resulting in low stress transfer capability within the glue line due to the brittle structure of the hardened Quebracho tannin-furfural adhesive. This is in line with several studies mentioning the brittle behavior of tannin-based adhesive formed glue lines [59–61].

3.2.4. Statistical Considerations

Significant correlations between thickness and MOE (R = −0.609; p-value 0.001), thickness and MOR (R = −0.823; p-value 0.001) and the correlation between MOE and MOR (R = 0.831; p-value 0.001) could be stated. The correlation of MOE versus density (R = 0.098; p-value 0.546) and MOR versus density (R = −0.025; p-value 0.876) is not detected. No correlation between tensile shear strength versus MOE (R = −0.147; p-value 0.456) and tensile shear strength versus MOR (R = −0.105; p-value 0.596) are detected.

The selected press parameters "time" and "temperature" have been accessed by uni- and multivariate methods to determine the influence on density, thickness, modulus of elasticity, modulus of rupture, and tensile shear strength (Table 2).

Table 2. Results of statistical significance for one-way ANOVA.

Properties	Variable	Mean Square	F-Value	p-Value
Density	Temperature	0.001	4.671	0.016 *
	Time	0.001	6.021	0.005 *
Thickness	Temperature	0.654	90.577	<0.001 *
	Time	0.036	0.876	0.425
MOE	Temperature	130,074.609	7.985	0.001 *
	Time	489,227,828.8	2.360	0.108
MOR	Temperature	17,277.397	47.926	<0.001 *
	Time	609.421	0.483	0.621
TSS	Temperature	0.360	1.850	0.178
	Time	0.206	0.994	0.384

* The p-value lower than $\alpha = 0.05$ displays significant influence on the physical and mechanical plywood properties.

The one-way ANOVA for the factor "temperature" reveals the influence on density (p-value = 0.016; η^2 0.202), thickness (p-value < 0.001; η^2 0.830), modulus of elasticity (p-value = 0.001; η^2 0.301) and modulus of rupture (p-value < 0.001; η^2 0.721). It does not influence tensile shear strength (p-value = 0.178; η^2 0.129). A significant influence can be stated for the factor "time" on density (p-value = 0.005; η^2 0.246) but not for thickness (p-value = 0.425; η^2 0.045), tensile shear strength (p-value = 0.384; η^2 0.074), modulus of elasticity (p-value = 0.108; η^2 0.113) and modulus of rupture (p-value = 0.621; η^2 0.025).

The multivariate test conducted for the factors "time" and "temperature" with the covariant "density" displays a similar picture for the factor "temperature" significantly influencing thickness (p-value < 0.001), modulus of elasticity (p-value < 0.001), modulus of rupture (p-value < 0.001) and tensile shear strength (p-value 0.048). The factor "time" does significantly influence the thickness (p-value < 0.001) but not tensile shear strength (p-value 0.127), modulus of elasticity (p-value 0.428) and modulus of rupture (p-value 0.271).

Comparing the trend of the estimated marginal means trends for temperature, increasing the temperature between 110 °C to 125 °C increases thickness. A further increase in temperature significantly decreases the thickness. This can be explained by the glass transition of the singular chemical constituents of wood resulting in a shape change of the cellular structure [45]. The factor press time displays a similar trend.

Interaction effects between the factors "time" and "temperature" are given for thickness and the mechanical properties of modulus of elasticity and modulus of rupture with a p-value < 0.001 but not for tensile shear strength with a p-value of 0.303.

For MOE, the "time" has a great influence at a low temperature; but, reaching a temperature between 125 and 140 °C, the increase of pressing time does not lead to improving properties. Similar behavior is found for MOR, but the temperature must reach 140 °C to achieve best features. Tensile shear strength is influenced by time only at 110 °C. With increasing temperature no similar trends are observed, as stated for MOE and MOR.

Concluding the importance of the factor "temperature" on the performance of the mechanical properties led to suggest a temperature range between 125 °C and 140 °C in order to gain sufficient bonding quality. It can be explained by the phenolic character of tannin. The industrial applicated temperature for hot pressing of plywood with PF adhesives is ~130 °C [62].

4. Conclusions

The aim of the study was to determine the adhesive characteristics gel time and viscosity as well as the influence of the press parameters, time and temperature, on the selected physical and mechanical properties—density, thickness, modulus of elasticity, modulus of rupture and tensile shear strength—of a totally bio-based sustainable Quebracho tannin-furfural bonded, five-layered beech plywood.

The presented adhesive formulation has shown good viscosity and curing behavior at a relatively low temperature (100 °C), producing polymers after curing. The non-reactivity at room temperature has to be highlighted as a clear advantage in terms of industrial application due to a prolonged open-time and storage duration. Their use as a fully bio-based sustainable adhesive for plywood displayed good bending (modulus of elasticity range ~9600 to ~11,600 N/mm^2; modulus of rupture range 66 to 100 N/mm^2) and acceptable tensile shear strength (~2.2 N/mm^2) in a dry environment, especially for the test specimens in the temperature range 125–140 °C, concluding that the presented formulation is comparable to industrial applicated PF adhesives. Depending on the field of application, as a negative drawback, the low water-resistance due to the brittle character of the adhesive layer structure has to be mentioned as it limits the use of the proposed Quebracho tannin-furfural formulation. On the other hand, it can improve and contribute to recyclability for specific interior plywood applications, as a key element of the bio-based circular economy.

Further research should focus on improving the elastic character of the glue line and enhancing the water resistance of the adhesive, likewise by adding some proportion of isocyanate or epoxy resins in order to further improve the mechanical properties of the adhesive. Additionally, the usability of different wood species, due to the fact that beech (*Fagus sylvatica*) reacts sensitively to moisture induced swelling and shrinkage. Further investigation of press parameters such as pressure and adhesive amount per layer should be taken into consideration. This study used 3 N/mm^2 as press pressure whereas other studies about tannin-based adhesives range between 1.2 N/mm^2 [63] and 1.6 N/mm^2 [64] as well as ~1.4 N/mm^2 [51] for phenol formaldehyde plywood. For industrial application, the adhesive amount per layer could be further optimized.

Author Contributions: Conceptualization and methodology, J.J. and E.C.; validation and formal analysis, J.J., E.C., G.T. and M.-C.B.; investigation, J.J. and E.C.; resources and funding acquisition, M.-C.B. and M.Z.; writing—original draft preparation, review, and editing, J.J., E.C., G.T., M.-C.B. and P.K.; supervision and project administration, G.T. and M.-C.B. All authors have read and agreed to the published version of the manuscript.

Funding: This research received no external funding.

Institutional Review Board Statement: Not applicable.

Informed Consent Statement: Not applicable.

Data Availability Statement: The data presented in this study are available on request from the corresponding author.

Conflicts of Interest: The authors declare no conflict of interest.

References

1. Beyer, G.; Defays, M.; Fischer, M.; Fletcher, J.; de Munck, E.; de Jaeger, F.; Van Riet, C.; Vandeweghe, K.; Wijnendaele, K. *Tackle Clim. Change—Use Wood*; CEI-Bois: Brussels, Belgium, 2011; p. 84.
2. Oliver, C.D.; Nassar, N.T.; Lippke, B.R.; McCarter, J.B. Carbon, Fossil Fuel, and Biodiversity Mitigation with Wood and Forests. *J. Sustain. For.* **2014**, *33*, 248–275. [CrossRef]
3. Moncaster, A.M.; Pomponi, F.; Symons, K.E.; Guthrie, P.M. Why Method Matters: Temporal, Spatial and Physical Variations in LCA and Their Impact on Choice of Structural System. *Energy Build.* **2018**, *173*, 389–398. [CrossRef]
4. Churkina, G.; Organschi, A.; Reyer, C.P.O.; Ruff, A.; Vinke, K.; Liu, Z.; Reck, B.K.; Graedel, T.E.; Schellnhuber, H.J. Buildings as a Global Carbon Sink. *Nat. Sustain.* **2020**, *3*, 269–276. [CrossRef]
5. Irle, M.; Barbu, M.C. *Wood-Based Panels: An Introduction for Specialists*; Thoemen, M., Irle, M., Sernek, M., Eds.; Cost Action E49; Brunel University Press: London, UK, 2010; pp. 1–94.
6. Klarić, S.; Obučina, M. New Trends in Engineering Wood Technologies. *Lect. Notes Netw. Syst.* **2020**, *76*, 712–727. [CrossRef]
7. Heinrich, L.A. Future Opportunities for Bio-Based Adhesives-Advantages beyond Renewability. *Green Chem.* **2019**, *21*, 1866–1888. [CrossRef]
8. Alliedmarketresearch. Available online: https://www.alliedmarketresearch.com/bioadhesives-market-A11324 (accessed on 27 May 2022).
9. Marketsandmarkets. Available online: https://www.marketsandmarkets.com/Market-Reports/bioadhesive-market-16386893.html (accessed on 27 May 2022).
10. Hemmilä, V.; Adamopoulos, S.; Karlsson, O.; Kumar, A. Development of Sustainable Bio-Adhesives for Engineered Wood Panels-A Review. *RSC Adv.* **2017**, *7*, 38604–38630. [CrossRef]
11. Ferdosian, F.; Pan, Z.; Gao, G.; Zhao, B. Bio-Based Adhesives and Evaluation for Wood Composites Application. *Polymers* **2017**, *9*, 70. [CrossRef]
12. Kristak, L.; Antov, P.; Bekhta, P.; Lubis, M.A.R.; Iswanto, A.H.; Reh, R.; Sedliacik, J.; Savov, V.; Taghiyari, H.R.; Papadopoulos, A.N.; et al. Recent Progress in Ultra-Low Formaldehyde Emitting Adhesive Systems and Formaldehyde Scavengers in Wood-Based Panels: A Review. *Wood Mater. Sci. Eng.* **2022**. [CrossRef]
13. Oktay, S.; Kızılcan, N.; Bengü, B. Development of Bio-Based Cornstarch—Mimosa Tannin—Sugar Adhesive for Interior Particleboard Production. *Ind. Crops Prod.* **2021**, *170*, 113689. [CrossRef]
14. Paul, G.B.; Timar, M.C.; Zeleniuc, O.; Lunguleasa, A.; Coșereanu, C. Mechanical Properties and Formaldehyde Release of Particleboard Made with Lignin-Based Adhesives. *Appl. Sci.* **2021**, *11*, 8720. [CrossRef]
15. Lubis, M.A.R.; Labib, A.; Sudarmanto; Akbar, F.; Nuryawan, A.; Antov, P.; Kristak, L.; Papadopoulos, A.N.; Pizzi, A. Influence of Lignin Content and Pressing Time on Plywood Properties Bonded with Cold-Setting Adhesive Based on Poly (Vinyl Alcohol), Lignin, and Hexamine. *Polymers* **2022**, *14*, 2111. [CrossRef] [PubMed]
16. Zhang, Y.; Shi, R.; Xu, Y.; Chen, M.; Zhang, J.; Gao, Q.; Li, J. Developing a Stable High-Performance Soybean Meal-Based Adhesive Using a Simple High-Pressure Homogenization Technology. *J. Clean. Prod.* **2020**, *256*, 120336. [CrossRef]
17. Ghahri, S.; Pizzi, A.; Hajihassani, R. A Study of Concept to Prepare Totally Biosourced Wood Adhesives from Only Soy Protein and Tannin. *Polymers* **2022**, *14*, 1150. [CrossRef]
18. Pizzi, A. Recent Developments in Eco-Efficient Bio-Based Adhesives for Wood Bonding: Opportunities and Issues. *J. Adhes. Sci. Technol.* **2006**, *20*, 829–846. [CrossRef]
19. Shirmohammadli, Y.; Efhamisisi, D.; Pizzi, A. Tannins as a Sustainable Raw Material for Green Chemistry: A Review. *Ind. Crops Prod.* **2018**, *126*, 316–332. [CrossRef]
20. Pizzi, A.; Mittal, K.L. *Handbook of Adhesive Technology*, 3rd ed.; CRC Press: Boca Raton, FL, USA, 2017; pp. 223–262. [CrossRef]
21. Xu, Y.; Guo, L.; Zhang, H.; Zhai, H.; Ren, H. Research Status, Industrial Application Demand and Prospects of Phenolic Resin. *RSC Adv.* **2019**, *9*, 28924–28935. [CrossRef] [PubMed]
22. Arias, A.; González-García, S.; Feijoo, G.; Moreira, M.T. Tannin-Based Bio-Adhesives for the Wood Panel Industry as Sustainable Alternatives to Petrochemical Resins. *J. Ind. Ecol.* **2021**, *26*, 627–642. [CrossRef]
23. Pizzi, A. The Chemistry and Development of Tannin/Urea–Formaldehyde Condensates for Exterior Wood Adhesives. *J. Appl. Polym. Sci.* **1979**, *23*, 2777–2792. [CrossRef]
24. Navarrete, P.; Pizzi, A.; Pasch, H.; Rode, K.; Delmotte, L. Characterization of Two Maritime Pine Tannins as Wood Adhesives. *J. Adhes. Sci. Technol.* **2013**, *27*, 2462–2479. [CrossRef]
25. Engozogho Anris, S.P.; Bikoro Bi Athomo, A.; Safou-Tchiama, R.; Leroyer, L.; Vidal, M.; Charrier, B. Development of Green Adhesives for Fiberboard Manufacturing, Using Okoume Bark Tannins and Hexamine–Characterization by 1H NMR, TMA, TGA and DSC Analysis. *J. Adhes. Sci. Technol.* **2021**, *35*, 436–449. [CrossRef]
26. Ballerini, A.; Despres, A.; Pizzi, A. Non-Toxic, Zero Emission Tannin-Glyoxal Adhesives for Wood Panels. *Holz Roh-Werkst.* **2005**, *63*, 477–478. [CrossRef]
27. Kabbour, M.; Luque, R. Furfural as a Platform Chemical: From Production to Applications. In *Biomass, Biofuels, Biochemicals—Recent Advances in Development of Platform Chemicals*; Saravanamurugan, S., Pandey, A., Riisager, A., Eds.; Elsevier B.V.: Amsterdam, The Netherlands, 2020. [CrossRef]
28. Bozell, J.J.; Petersen, G.R. Technology Development for the Production of Biobased Products from Biorefinery Carbohydrates—The US Department of Energy's "Top 10" Revisited. *Green Chem.* **2010**, *12*, 539–555. [CrossRef]

29. Cesprini, E.; Šket, P.; Causin, V.; Zanetti, M. Development of Quebracho (*Schinopsis balansae*) Tannin-Based Thermoset Resins. *Polymers* **2021**, *13*, 4412. [CrossRef] [PubMed]
30. EN 323; Wood-Based Panels—Determination of Density. European Committee for Standardization: Brussels, Belgium, 2005.
31. Spulle, U.; Meija, A.; Kūliņš, L.; Kopeika, E.; Liepa, K.H.; Šillers, H.; Zudrags, K. Influence of Hot Pressing Technological Parameters on Plywood Bending Properties. *BioResources* **2021**, *16*, 7550–7561. [CrossRef]
32. EN 314-1; Plywood—Bonding Quality—Test Methods. European Committee for Standardization: Brussels, Belgium, 2005.
33. EN 310; Wood-Based Panels—Determination of Modulus of Elasticity in Bending and of Bending Strength. European Committee for Standardization: Brussels, Belgium, 2005.
34. Navarrete, P.; Pizzi, A.; Tapin-Lingua, S.; Benjelloun-Mlayah, B.; Pasch, H.; Rode, K.; Delmotte, L.; Rigolet, S. Low Formaldehyde Emitting Biobased Wood Adhesives Manufactured from Mixtures of Tannin and Glyoxylated Lignin. *J. Adhes. Sci. Technol.* **2012**, *26*, 1667–1684. [CrossRef]
35. Hauptt, R.A.; Sellers, T. Characterizations of Phenol-Formaldehyde Resol Resins. *Ind. Eng. Chem. Res.* **1994**, *33*, 693–697. [CrossRef]
36. Ricci, A.; Olejar, K.J.; Parpinello, G.P.; Kilmartin, P.A.; Versari, A. Application of Fourier Transform Infrared (FTIR) Spectroscopy in the Characterization of Tannins. *Appl. Spectrosc. Rev.* **2015**, *50*, 407–442. [CrossRef]
37. Tondi, G.; Petutschnigg, A. Middle Infrared (ATR FT-MIR) Characterization of Industrial Tannin Extracts. *Ind. Crops Prod.* **2015**, *65*, 422–428. [CrossRef]
38. Mohamad, N.; Abd-Talib, N.; Kelly Yong, T.L. Furfural Production from Oil Palm Frond (OPF) under Subcritical Ethanol Conditions. *Mater. Today Proc.* **2020**, *31*, 116–121. [CrossRef]
39. Kane, S.N.; Mishra, A.; Dutta, A.K. Synthesis of Furfural from Water Hyacinth (*Eichornia croassipes*) This. *Mater. Sci. Eng.* **2017**, *172*, 012027. [CrossRef]
40. Tondi, G. Tannin-Based Copolymer Resins: Synthesis and Characterization by Solid State 13C NMR and FT-IR Spectroscopy. *Polymers* **2017**, *9*, 223. [CrossRef] [PubMed]
41. Wagenführ, A.; Scholz, F. *Taschenbuch der Holztechnik*; Carl Hanser Verlag: München, Germany, 2008.
42. Jorda, J.; Kain, G.; Barbu, M.-C.; Petutschnigg, A.; Král, P. Influence of Adhesive Systems on the Mechanical and Physical Properties of Flax Fiber Reinforced Beech Plywood. *Polymers* **2021**, *13*, 3086. [CrossRef] [PubMed]
43. Jorda, J.; Kain, G.; Barbu, M.-C.; Köll, B.; Petutschnigg, A.; Král, P. Mechanical Properties of Cellulose and Flax Fiber Unidirectional Reinforced Plywood. *Polymers* **2022**, *14*, 843. [CrossRef] [PubMed]
44. Mansouri, H.R.; Pizzi, A.; Leban, J.M. Improved Water Resistance of UF Adhesives for Plywood by Small PMDI Additions. *Holz Roh-Werkst.* **2006**, *64*, 218–220. [CrossRef]
45. Luo, J.; Luo, J.; Gao, Q.; Li, J. Effects of Heat Treatment on Wet Shear Strength of Plywood Bonded with Soybean Meal-Based Adhesive. *Ind. Crops Prod.* **2015**, *63*, 281–286. [CrossRef]
46. Bekhta, P.; Hiziroglu, S.; Shepelyuk, O. Properties of Plywood Manufactured from Compressed Veneer as Building Material. *Mater. Des.* **2009**, *30*, 947–953. [CrossRef]
47. Cabral, J.P.; Kafle, B.; Subhani, M.; Reiner, J.; Ashraf, M. Densification of Timber: A Review on the Process, Material Properties, and Application. *J. Wood Sci.* **2022**, *68*, 20. [CrossRef]
48. Niemz, P. *Physik des Holzes und der Holzwerkstoffe*; DRW Verlag Weinbrenner: Leinfelden-Echterdingen, Germany, 1993.
49. DIN 68705-5; Sperrholz Teil 5—Bau-Furniersperrholz aus Buche. Deutsches Institut für Normung: Berlin, Germany, 1980.
50. Hrázský, J.; Král, P. Assessing the Bending Strength and Modulus of Elasticity in Bending of Exterior Foiled Plywoods in Relation to Their Construction. *J. For. Sci.* **2005**, *51*, 77–94. [CrossRef]
51. Biadała, T.; Czarnecki, R.; Dukarska, D. Water Resistant Plywood of Increased Elasticity Produced from European Wood Species. *Wood Res.* **2020**, *65*, 111–124. [CrossRef]
52. Dieste, A.; Krause, A.; Bollmus, S.; Militz, H. Physical and Mechanical Properties of Plywood Produced with 1.3-Dimethylol-4.5-Dihydroxyethyleneurea (DMDHEU)-Modified Veneers of *Betula* sp. and *Fagus Sylvatica*. *Holz Roh-Werkst.* **2008**, *66*, 281–287. [CrossRef]
53. Lohmann, U. *Holz Handbuch*; DRW: Echterdingen-Leinenfelden, Germany, 2010; pp. 46–47.
54. Réh, R.; Krišťák, Ľ.; Sedliačik, J.; Bekhta, P.; Božiková, M.; Kunecová, D.; Vozárová, V.; Tudor, E.M.; Antov, P.; Savov, V. Utilization of Birch Bark as an Eco-Friendly Filler in Urea-Formaldehyde Adhesives for Plywood Manufacturing. *Polymers* **2021**, *13*, 511. [CrossRef] [PubMed]
55. Bal, B.C.; Bektaþ, Ý. Some Mechanical Properties of Plywood Produced from Eucalyptus, Beech, and Poplar Veneer. *Maderas. Cienc. Tecnol.* **2014**, *16*, 99–108. [CrossRef]
56. EN 314-2; Plywood—Bonding Quality—Part 2 Requirements. European Committee for Standardization: Brussels, Belgium, 2005.
57. Xi, X.; Pizzi, A.; Frihart, C.R.; Lorenz, L.; Gerardin, C. Tannin Plywood Bioadhesives with Non-Volatile Aldehydes Generation by Specific Oxidation of Mono- and Disaccharides. *Int. J. Adhes. Adhes.* **2020**, *98*, 102499. [CrossRef]
58. Hafiz, N.L.M.; Tahir, P.M.D.; Hua, L.S.; Abidin, Z.Z.; Sabaruddin, F.A.; Yunus, N.M.; Abdullah, U.H.; Abdul Khalil, H.P.S. Curing and Thermal Properties of Co-Polymerized Tannin Phenol-Formaldehyde Resin for Bonding Wood Veneers. *J. Mater. Res. Technol.* **2020**, *9*, 6994–7001. [CrossRef]
59. Pizzi, A.; Scharfetter, H.O. The Chemistry and Development of Tannin-based Adhesives for Exterior Plywood. *J. Appl. Polym. Sci.* **1978**, *22*, 1745–1761. [CrossRef]

60. Ayla, C.; Parameswaran, N. Macro- and Microtechnological Studies on Beechwood Panels Bonded with *Pinus Brutia* Bark Tannin. *Holz Roh-Werkst.* **1980**, *38*, 449–459. [CrossRef]
61. Ferreira, É.D.S.; Lelis, R.C.C.; Brito, E.D.O.; Iwakiri, S. Use of Tannin from *Pinus oocarpa* Bark for Manufacture of Plywood. In Proceedings of the LI International Convention of Society of Wood Science and Technology, Concepción, Chile, 10–12 November 2008; pp. 10–12.
62. Sedliačik, J.; Bekhta, P.; Potapova, O. Technology of Low-Temperature Production of Plywood Bonded with Modified Phenol-Formaldehyde Resin. *Wood Res.* **2010**, *55*, 123–130.
63. Moubarik, A.; Pizzi, A.; Allal, A.; Charrier, F.; Charrier, B. Cornstarch and Tannin in Phenol-Formaldehyde Resins for Plywood Production. *Ind. Crops Prod.* **2009**, *30*, 188–193. [CrossRef]
64. Stefani, P.M.; Peña, C.; Ruseckaite, R.A.; Piter, J.C.; Mondragon, I. Processing Conditions Analysis of *Eucalyptus Globulus* Plywood Bonded with Resol-Tannin Adhesives. *Bioresour. Technol.* **2008**, *99*, 5977–5980. [CrossRef]

Article

Influence of Adhesive Systems on the Mechanical and Physical Properties of Flax Fiber Reinforced Beech Plywood

Johannes Jorda [1], Günther Kain [1,2,*], Marius-Catalin Barbu [1,3], Alexander Petutschnigg [1] and Pavel Král [4]

[1] Forest Products Technology and Timber Construction Department, Salzburg University of Applied Sciences, Markt 136a, 5431 Kuchl, Austria; jjorda.lba@fh-salzburg.ac.at (J.J.); marius.barbu@fh-salzburg.ac.at (M.-C.B.); alexander.petutschnigg@fh-salzburg.ac.at (A.P.)

[2] Department for Furniture and Interior Design, Higher Technical College Hallstatt, Lahnstraße 69, 4830 Hallstatt, Austria

[3] Faculty for Furniture Design and Wood Engineering, Transilvania University of Brasov, B-dul. Eroilor nr. 29, 500036 Brasov, Romania

[4] Department of Wood Science and Technology, Mendel University, Zemědělská 3, 61300 Brno, Czech Republic; kral@mendelu.cz

* Correspondence: gkain.lba@fh-salzburg.ac.at; Tel.: +43-699-819-764-42

Abstract: In order to improve the acceptance of broader industrial application of flax fiber reinforced beech (*Fagus sylvatica* L.) plywood, five different industrial applicated adhesive systems were tested. Epoxy resin, urea-formaldehyde, melamine-urea formaldehyde, isocyanate MDI prepolymer, and polyurethane displayed a divergent picture in improving the mechanical properties—modulus of elasticity, modulus of rupture, tensile strength, shear strength and screw withdrawal resistance—of flax fiber-reinforced plywood. Epoxy resin is well suited for flax fiber reinforcement, whereas urea-formaldehyde, melamine urea-formaldehyde, and isocyanate prepolymer improved modulus of elasticity, modulus of rupture, shear strength, and screw withdrawal resistance, but lowered tensile strength. Polyurethane lowered the mechanical properties of flax fiber reinforced plywood. Flax fiber reinforced epoxy resin bonded plywood exceeded glass fiber reinforced plywood in terms of shear strength, modulus of elasticity, and modulus of rupture.

Keywords: wood-based composite; fiber reinforced plywood; flax fiber

Citation: Jorda, J.; Kain, G.; Barbu, M.-C.; Petutschnigg, A.; Král, P. Influence of Adhesive Systems on the Mechanical and Physical Properties of Flax Fiber Reinforced Beech Plywood. *Polymers* **2021**, *13*, 3086. https://doi.org/10.3390/polym13183086

Academic Editor: Ivan Kubovský

Received: 26 August 2021
Accepted: 6 September 2021
Published: 13 September 2021

Publisher's Note: MDPI stays neutral with regard to jurisdictional claims in published maps and institutional affiliations.

Copyright: © 2021 by the authors. Licensee MDPI, Basel, Switzerland. This article is an open access article distributed under the terms and conditions of the Creative Commons Attribution (CC BY) license (https://creativecommons.org/licenses/by/4.0/).

1. Introduction

Wood is a natural, polymeric, cellular fiber composite that is broadly available and has been used for all kinds of application purposes throughout the history of mankind [1]. To overcome solid wood disadvantages of anisotropy, biodegradability, and dimensional limitations, respectively, various wood-based products such as cross laminated timber (CLT), plywood, oriented strand board (OSB), particleboard (PB), or medium/high density fiberboard (MDF/HDF) have been developed. Natural caused solid wood inhomogeneity are thereby reduced by downsizing raw material geometry [2] and creating homogeneous composite material products with the support of joining materials [3].

Plywood is considered to be the oldest wood-based composite material based on a laminar structure with two distinct fields of application for structural construction purposes and furniture/interior design products [3], and also for several applications for niche market products like transportation, construction, sport equipment, etc. [4]. To enhance the mechanical properties of wood-based products such as plywood and laminated veneer lumber (LVL), fiber reinforcement is well discussed and several experimental studies have been conducted, primarily focusing on synthetic glass and carbon fiber reinforcement, dating back to the 1960s [5].

Bal et al. (2015) reinforced phenol-formaldehyde (PF) bonded poplar (Samsun I-77/51 clone) plywood with woven glass fiber (GF) fabric, significantly improving the modulus of elasticity and modulus of rupture for perpendicular samples, and noted a decreasing factor for inequalities between parallel and perpendicular specimens. In addition, density increased whereas thickness swelling and water absorption decreased [6]. Furthermore, screw withdrawal resistance, screw-head pull-through, and lateral nail resistance of glass-fiber reinforced phenol-formaldehyde bonded plywood improved significantly aside from increasing maximum load capacity [7]. Liu et al. (2019) conducted research on different experimental plies of poplar (Populus euramenicana), eucalyptus (Eucalyptus grandis), poplar/eucalyptus, and carbon fiber reinforced plywood for construction formwork [8]. Veneers were bonded with PF resin, whereas carbon fiber fabric was impregnated with epoxy resin and used for bonding carbon fiber to veneer. The combination of different wood species improved flexural plywood performance, which was surpassed by carbon fiber reinforcement. The position of the fiber reinforcement over the plywood cross section is significant for its performance. Surface fiber reinforcement increases the longitudinal modulus of elasticity and the modulus of rupture. In addition, it improves the ultimate load carrying capacity of plywood and influences the failure mode to shear delamination failure caused by the strengthened surface layer. Auriga et al. (2020) studied the effect of randomly unidirectional parallel and perpendicular orientated carbon fiber (CF) reinforcement located internal of the melamine-formaldehyde (MUF) glue line of veneer plies. CF reinforcement was located at two different positions: external at the outer glue line and internal surrounding the core veneer ply. The results displayed increasing modulus of rupture (MOR) and modulus of elasticity (MOE) and the influence of fiber reinforcement location on MOR and MOE [9]. Guan et al. (2020) evaluated the three point bending performance of unidirectional CF (EL 203,631 N/mm^2) reinforced eucalyptus (EL 11,619 N/mm^2)/poplar (EL 6751 N/mm^2) epoxy resin bonded plywood (thickness 17.5 mm, 17.8 mm, 17.65 mm, 18.0 mm) by digital image correlation (DIC) and finite element analysis (FEA), concluding the usability of FEA for the prediction of material failure behavior [10].

The studies display the effectivity of fiber reinforcement in order to improve physical and mechanical properties of laminar structured wood-based products. Due to rising consumer awareness and resource scarcity, fiber reinforcement based on natural fibers such as flax, hemp, ramie, or basalt can be used in this multilayer laminar composite structure to overcome negative impacts on environmental and resource availability issues. While the concept of natural fiber reinforcement such as flax is not new, research efforts of the last decades have been dominated by synthetic based fiber reinforcement [11].

Several studies have been conducted to investigate the influence of different natural non-/lignocellulose-based fibers in improving the mechanical and physical properties of solid wood and wood-based products. Speranzini and Tralascia (2010) reinforced LVL and solid wood with (FRP) fiber reinforced plastic glass and carbon fibers and natural fibers such as basalt, flax, and hemp. The four point bending test revealed a lower MOR and MOE for natural fibers compared to FRP reinforcement, but still significant improvements compared to the non-reinforced samples [12]. Moezzipour et al. (2017) studied the effect of kenaf and date palm fiber reinforcement on the mechanical and physical properties of horn beam plywood bonded with urea-formaldehyde (UF). Concluding the effectivity of utilizing natural fibers for reinforced plywood products to enhance mechanical performance [13]. Kramár et al. (2020) used non-/impregnated basalt scrim with an area weight of 360 g/m^2 to enhance the mechanical properties of MUF bonded PB. The effects of different fiber reinforcement positions within the structure on MOR, MOE, internal bond (IB) strength, screw withdrawal resistance (SWR), and thickness swelling (TS) were examined. The study revealed that basalt fiber scrim located at the outer positions significantly improved the strength-to-weight-ratio of particleboards [14]. Jorda et al. (2020) investigated the influence of flax-fiber-reinforcement bonded with epoxy-resin on three-dimensional molded plywood. Improved load capacity and stiffness of flax-fiber reinforced molded plywood structures could be measured [15]. Valdes et al. (2020) reinforced CLT with flax fiber fabrics bonded with bicomponent thixotropic epoxy resin [16]. The study showed that reinforcement of three-layered solid wood panels (SWP) significantly improved load-carrying capacity and stiffness, while the effect for five-layered panels was negligible.

Concluding from the studies, epoxy resin is the main source for bonding fiber reinforced wood-based composites with synthetic and natural fibers. Some attempts had been made to use PF, UF, and MUF [6–9,13,14]. For broader industrial production applications, the use of different adhesives is desirable due to the high production costs of epoxy resin systems. Based on its lignocellulosic origin, flax fiber may be bonded with standard industrial plywood adhesives to improve mechanical properties and contribute to broader industrial applications due to limited production process changes and investment costs.

The aim of this study was to investigate the influence of standard industrial adhesives such as UF, MUF, polyurethane (PU-K), isocyanate MDI based pre-polymer (PU-AN), and epoxy resin on the mechanical properties (modulus of elasticity, modulus of rupture, bending strength, tensile strength shear strength, screw withdrawal resistance) of woven flax fiber fabric reinforced beech (*Fagus sylvatica* L.) plywood. The effect of resin type and fiber reinforcement on the panel characteristics named before were evaluated using multivariate statistics.

2. Materials and Methods

2.1. Materials and Sample Preparation

Pre-conditioned (20 °C, 65% relative air humidity) zero defect rotary cut beech (*Fagus sylvatica* L.) veneers (distributed by Europlac, Topolčany, Slovakia) with the dimensions of 0.75 m × 0.75 m, a thickness of 2.2 mm, an average density of 0.72 g/cm^3, and an average moisture content of 12% were used in this study as wooden raw material.

Twill woven flax fabric LINEO FlaxPly Balanced Fabric 200 (Ecotechnilin, Valliquerville, France) with a thickness of 0.4 mm, a density of 1.27 g/cm^3, and a grammage of 200 g/m^2 acted as fiber reinforcement. For epoxy resin synthetic fiber reinforcement, textile reference samples of a twill woven e-glass fabric (distributed by DD Composite, Bad Liebenwerda, Germany) with a thickness of 0.5 mm and a grammage of 200 g/m^2 was used (Figure 1a).

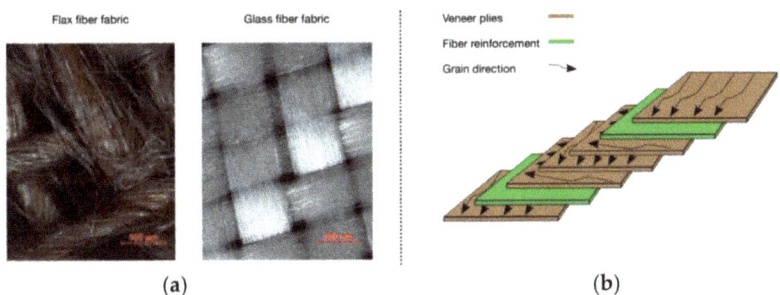

Figure 1. (a) Flax fiber vs. glass fiber fabric (b) Lay up of 5 × 90° reinforced plywood.

Five different commercially available and industrially applied adhesives were used to bond the veneer plies and the flax and glass fiber fabric. Epoxy resin SR GreenPoxy 56 (Sicomin, Chateneuf les Matigues, France) with hardener SD 7561 was used. Specifications were: density of 1.198 g/cm^3 and 0.971 g/cm^3, respectively, initial viscosity of 0.7 Pa*s and a resin/hardener ratio of 100:36 g were used (press time 13 h, temperature 20 °C). The adhesive application per glue line was set to 200 g/m^2. Polyurethane adhesive (PUR) Polyurethan 501 Kleiberit (Kleiberit, Weingarten, Germany) with a density of 1.13 g/cm^3 and viscosity of 7.50 Pa*s (press time 1 h, temperature 20 °C, pressure 0.6 N/mm^2). The adhesive application per glue line was set to 150 g/m^2. Isocyanate MDI based prepolymer adhesive PUR system 2010 AkzoNobel (Akzo Nobel, Stockholm, Sweden) (PU-AN) was used with a density of 1.160 g/cm^3 and viscosity of 6.0 to 19.0 Pa*s (press time 22 min, temperature 20 °C, pressure 2 N/mm^2). Adhesive application per glue line was set to 200 g/m^2. Urea-formaldehyde (UF) 1274 Akzo Nobel (Akzo Nobel, Stockholm, Sweden) with hardener 2545 Akzo Nobel with a density of 1.300 g/cm³ and 1.450 g/cm³, respectively, was used with a viscosity of 1.5 to 3.5 Pa*s/2.0 to 10.0 Pa*s and a resin/hardener ratio 100:20 g (press time 10 min, temperature 90 °C, pressure 1.8 N/mm²). Glue amount per glue line was set to 160 g/m². Melamine urea-formaldehyde (MUF) 1247 Akzo Nobel (Akzo Nobel, Stockholm, Sweden) was also used with hardener 2526 Akzo Nobel, a density of 1.270 g/cm³, respectively, 1.070 g/cm³, a viscosity of 10 to 25 Pa*s/1.7 to 2.7 Pa*s, and a resin/hardener ratio 100:50 g (press time 12 min, temperature 65 °C, pressure 2 N/mm²). Glue amount per glue line was set to 300 g/m².

Two lay-ups of plywood were introduced (Figure 1b). The reference samples consisted of five 90° cross laid veneers layers, whereas the fiber reinforced samples consisted of the identical five 90° cross laid veneers layers with two layers of flax or glass fabric (Figure 1b), respectively. These were located in the first glue line on each side in order to improve tensile strength under bending and to minimize the effect of shear stress.

Based on the lay-up, boards with dimensions of 600 mm × 600 mm and a thickness of 10 mm for the non-fiber reinforced reference samples and respectively 11.2 mm for the flax and glass fiber reinforced were produced using a Höfler HLOP 280 press (Taiskirchen, Austria). Lay-up and adhesive application were carried out manually. Adhesive application was controlled by weighing with a KERN ITB 35K1IP device (Baligen-Frommern, Austria). The boards were pressed according to the specific parameters given for each singular adhesive type. Before further testing, the boards where stored for conditioning until mass constancy under constant climate conditions (relative humidity 65%, 20 °C) was achieved. Test specimens were cut from these boards for density, moisture content (MC), tensile- (TS), shear strength (SS), and screw withdrawal resistance (SWR) (Table 1).

Table 1. Design of the experiment for the influence of adhesive systems on flax fiber reinforced plywood.

Adhesive	Type of Reinforcement	Adhesive Applic. (g/m²)	Board Thickn. (mm)	Density	MC	TS	SS	MOE	MOR	SWR
				\multicolumn{7}{c}{Number of Specimens (N)}					
Epoxy	non		10	10	3	5	9	10	10	9
	flax	200	11.2	10	3	5	9	10	10	9
	glass		11.2	10	3	5	9	10	10	9
UF	non		10	10	3	5	9	10	10	9
	flax	160	11.2	10	3	5	9	10	10	9
MUF	non		10	9	3	5	9	10	10	9
	flax	300	11.2	9	3	5	9	10	10	9
PU-AN	non		10	10	3	5	9	10	10	9
	flax	200	11.2	10	3	5	9	10	10	9
PUR	non		10	10	3	5	9	10	10	9
	flax	150	11.2	10	3	5	9	10	10	9

2.2. Testing

The density was determined according to EN 323:2005 [17], the moisture content according to EN 322:2005 [18] with the specimen size 50 mm × 50 mm, and obtained from bending test specimens after testing. The tensile strength (TS) was measured according to DIN 52377 [19] with specimen dimensions given in Figure 2.

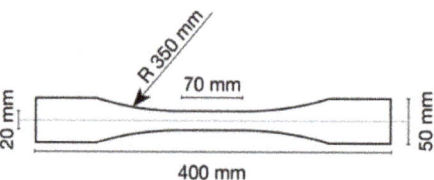

Figure 2. Tensile strength specimen dimensions.

The shear strength (SS) was determined based on EN 314:2005 [20,21] with specimen dimensions of 100 mm × 25 mm (Figure 3).

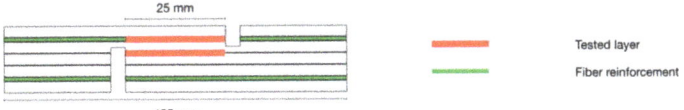

Figure 3. Shear test specimen made of five veneer plies both sides fiber reinforced.

Modulus of rupture (MOR) and modulus of elasticity (MOE) were determined by a three-point bending test according to EN 310:2005 [22] with specimen dimensions of 250 mm × 50 mm for reference samples and 274 mm × 50 mm for fiber reinforced samples. The screw withdrawal resistance (SWR) was measured according to EN 320:2011 [23] with specimen dimensions of 50 mm × 50 mm and thread screws of ST 4.2 mm. TS, SS, MOR, MOE, and SWR was determined using a Zwick/Roell 250 8497.04.00 test device (Zwick/Roell, Ulm, Germany) and constant climatic conditions (relative humidity 65%, ambient temperature 20 °C). For the statistical analysis, IBM SPSS was used for the descriptive statistics, correlation, and two-way ANOVAs with the consideration of the first order interaction effects for determining the influence of the factors "type of fiber reinforcement" and "adhesive type".

3. Results and Discussion

3.1. Density

The results (Table 2) displayed a low effect Pearson correlation between density and applied amount of glue (*p*-value 0.01; R^2 = 0.065) as well as the significant influence of the factors "type of adhesive" (*p*-value 0.00) and the influence of "fiber reinforcement" (*p*-value 0.00).

Table 2. Density of different specimens.

Adhesive	Reinforcement	N	Density (g/cm³)			
			Min	Mean	Max	SD
Epoxy	non	10	0.79	0.80	0.81	0.01
	flax	10	0.83	0.84	0.86	0.01
	glass	10	0.85	0.87	0.89	0.02
UF	non	10	0.77	0.78	0.79	0.01
	flax	10	0.79	0.81	0.83	0.01
MUF	non	9	0.77	0.79	0.81	0.01
	flax	9	0.82	0.84	0.86	0.01
PU-AN	non	10	0.75	0.77	0.78	0.01
	flax	10	0.80	0.82	0.83	0.01
PUR	non	10	0.77	0.79	0.80	0.01
	flax	10	0.77	0.79	0.81	0.01

Epoxy resin bonded flax fiber reinforced samples with a mean density of 0.843 (standard deviation (SD) = 0.009) g/cm³ increased up to 5.5%, respectively 8.8% for glass fiber reinforcement with a mean density of 0.870 (SD = 0.015) g/cm³ compared to the reference with a mean density of 0.799 (SD = 0.007) g/cm³. The urea-formaldehyde flax reinforcement with a mean density of 0.808 (SD = 0.013) g/cm³ increased up to 4.1% compared to the reference mean density of 0.776 g/cm³. The melamine urea-formaldehyde bonded flax reinforced sample mean density 0.844 (SD = 0.012) g/cm³ increased by 7.0% compared to the reference mean density of 0.789 (SD = 0.012) g/cm³. The isocyanate MDI based prepolymer adhesive (PU-AN) flax reinforcement mean density 0.816 (SD = 0.009) g/cm³ increased by 6.7% compared to the reference mean density with 0.765 (SD = 0.008) g/cm³. PUR bonded flax fiber reinforcement increased by 0.5% with a mean density of 0.794 (SD = 0.014) g/cm³ in comparison to the reference mean density of 0.790 (SD = 0.012) g/cm³.

The mean density of flax fiber reinforcement increased between 0.032 g/cm³ (4.1%) and 0.055 g/cm³ (7.0%) with the exception of PUR with 0.004 g/cm³ (0.5%). Enhanced mean density for the fiber reinforcement specimen can be explained by the additional amount of adhesive for the supplementary glue lines and the layers of woven fiber fabric. In addition, density of the different test groups was influenced by the specific adhesive density. The resin density range was 1.13 g/cm^3 to 1.3 g/cm^3. In detail, the density for epoxy resin was 1.198 g/cm^3, UF 1.3 g/cm^3, MUF 1.270 g/cm^3, PU-AN 1.16 g/cm^3, and PUR 1.130 g/cm^3.

According to Wagenführ and Scholz (2008), density is one of the main influencing parameters for plywood properties, besides the veneer thickness and the solid resin content. Increasing board density correlates with increasing compression strength, enhancing MOE and TS [24].

The low standard deviation values for each test group indicates an even glue application and plywood board production process.

3.2. Moisture Content

Epoxy resin bonded flax fiber reinforced MC mean of 9.46 (SD = 0.14)% decreased by 3.76%, respectively 18.31% for the fiber reinforced specimen, compared to the reference mean of 9.83 (SD = 0.14)%. UF bonded flax fiber reinforced MC mean of 10.27 (SD = 0.14)% increased slightly by 0.58% in contrast to the reference mean of 10.27 (SD = 0.09)%. MUF bonded flax fiber reinforced MC mean of 11.42 (SD = 0.04)% decreased by 3.63% compared to the reference MC mean of 11.85 (SD = 2.00)%. Comparability is to be questioned by the standard deviation of SD = 2.00. Isocyanate MDI based prepolymer adhesive (PU-AN) flax fiber reinforced MC mean of 9.75 (SD = 0.32)% decreased by 6.7% in contrast to the reference MC mean of 10.45 (SD = 0.15)%. The PUR bonded flax fiber reinforced MC mean of 8.84 (SD = 0.04)% decreased by 15% compared to the reference MC mean of 10.04 (SD = 0.05)% (Table 3).

Table 3. Postproduction moisture content.

Adhesive	Reinforcement	N	Moisture Content (%)	
			Mean	SD
Epoxy	non	5	9.83	0.14
	flax	5	9.46	0.23
	glass	5	8.03	0.15
UF	non	5	10.27	0.09
	flax	5	10.33	0.14
MUF	non	5	11.85	2.00
	flax	5	11.42	0.04
PU-AN	non	5	10.45	0.15
	flax	5	9.75	0.32
PUR	non	5	10.04	0.05
	flax	5	8.84	0.04

It was concluded that for epoxy resin, PU-AN adhesive, and PUR that flax fiber reinforcement reduced the moisture content by 3.76%, 6.7%, and 15%. MUF bonded plywood displayed the highest moisture content of 11.85% and also a high standard deviation of 2.00.

Moisture content (MC) is influenced by the type of adhesive with a p-value 0.000 and fiber reinforcement (p-value 0.001). The influence of fiber reinforcement on the moisture content was questioned for urea formaldehyde (p-value 0.568) and melamine urea formaldehyde (p-value 0.731).

Moisture content of wood and wood based products influence several mechanical properties such as MOE, MOR, compression-, and TS within the hygroscopic region [25]. Aydin et al. (2006) displayed the influence of veneer MC on the mechanical properties of UF and MUF bonded poplar and spruce plywood. Increased veneer MC lowered the MOR, SS, and MOE with the positive effect of decreasing formaldehyde emissions [26]. The effect of decreasing equilibrium moisture content is stated by Bal et al. (2015) for the PF adhesive bonded GF reinforced poplar plywood compared to the control group specimens [6].

3.3. Tensile Strength

The results for ultimate tensile strength f_t and maximum tensile force F_{max} were evaluated with an ANOVA including the factors adhesive type and fiber reinforcement. The influence of the adhesive type was slight given the f_t with a *p*-value of 0.057 and R^2 of 0.165. The maximum tensile force F_{max} was independent of the applied type of adhesive (*p*-value 0.303; R^2 = 0.091). The factor "fiber reinforcement" slightly influenced f_t due to a *p*-value of 0.054 with R^2 0.106, whereas F_{max} was independent (*p*-value 0.788; R^2 = 0.009). The interaction between adhesive type and fiber reinforcement was significant (*p* = 0.001) (Table 4).

Table 4. Tensile strength f_t.

Adhesive	Reinforcement	N	Tensile Strength (N/mm^2)			
			Min	Mean	Max	SD
Epoxy	non	5	69.94	76.47	82.46	4.68
	flax	5	76.20	83.50	88.71	4.92
	glass	5	73.76	88.81	93.65	8.48
UF	non	5	88.57	93.16	98.44	3.56
	flax	5	78.04	90.51	94.28	7.00
MUF	non	5	70.70	86.78	96.72	10.00
	flax	5	72.93	79.64	90.74	7.15
PU-AN	non	5	82.13	94.47	99.62	7.28
	flax	5	74.02	85.32	91.92	7.37
PUR	non	5	91.66	95.46	97.24	2.20
	flax	5	65.38	78.25	89.56	8.64

For epoxy resin bonded plywood with a mean f_t of 76.47 (SD = 4.68) N/mm^2, the flax fiber reinforcement increased by 9.18% with a mean of f_t 83.50 (SD = 4.93) N/mm^2 and 16.14% for the mean f_t of 76.47 (SD = 4.68) N/mm^2 for glass fiber reinforced specimens. Excluding one outlier and comparing the median f_t 93.11 (SD = 7.00) N/mm^2 for UF flax fiber reinforced with median reference sample with 92.94, there was a slight increase of 0.18%.

In contrast to MUF, the isocyanate MDI based prepolymer adhesive and PUR revealed a negative influence of flax fiber reinforcement on tensile strength f_t. In detail, MUF flax fiber reinforced samples with a mean f_t of 79.64 (SD = 7.15) N/mm^2 reached 91.77% of the reference sample with a mean f_t of 86.78 (SD = 9.99) N/mm^2, displaying a decline of 8.23%. The isocyanate MDI based prepolymer adhesive bonded flax fiber reinforced plywood mean f_t declined by 9.62% compared to the reference mean f_t of 94.47 (SD = 7.28) N/mm^2. PUR flax fiber reinforced mean f_t of 78.25 (SD = 8.643) N/mm^2 decreased by 18.03% compared to the reference mean f_t of 95.47 (SD = 8.64) N/mm^2.

The maximum tensile force F_{max} was reached at approximately 20.00 kN for all test groups excluding the epoxy references with a F_{max} of 17.42 kN.

The Pearson correlation between moisture content and tensile strength f_t was significant for the PUR flax fiber reinforced specimen (*p*-value 0.01, R^2 = 1.00). No correlation was found for epoxy (*p*-value 0.645; R^2 = 0.281), UF (*p*-value 0.850; R^2 = 0.054), MUF (*p*-value 0.183; R^2 = 0.910), and the isocyanate MDI based prepolymer adhesive (PU-AN) flax fiber reinforced specimen (*p*-value 0.889; R^2 = 0.030).

The general stated correlation between increasing density and tensile strength according to Niemz (1993) could not be manifested due to the Pearson correlation with R^2 = 0.070 and a *p*-value of 0.051 (Figure 4a). Comparing the TS to the range given by Niemz (1993) for plywood between 30 to 60 N/mm^2, the results exceeded the range by 16.47 (76.47) to 35.46 (95.36) N/mm^2 [27].

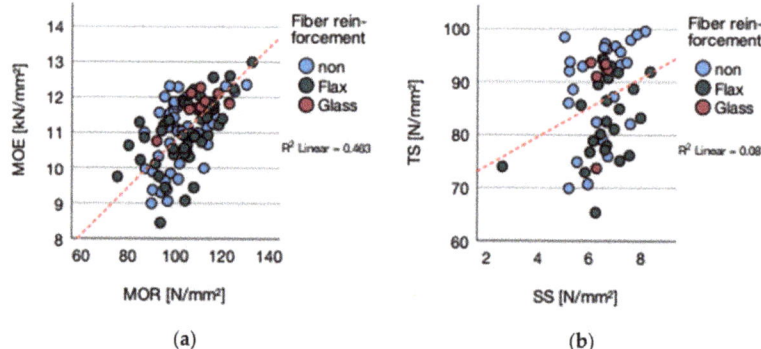

Figure 4. (a) Tensile strength vs. density and (b) tensile vs. shear strength.

Further research is mandatory for a better understanding of the decline of the tensile strength for MUF, PU-AN, and PUR bonded flax fiber reinforced plywood. The decline could not be explained by bonding performance if compared to the values of the tensile shear strength. Nevertheless, there was a correlation between tensile strength f_t and shear strength f_t (p-value 0.029; R^2 = 0.087) (Figure 4b). In addition, influences of high moisture content on tensile strength, and maximum tensile strength in the range 5 to 10% MC [27] have to be neglected if compared to the measured moisture content.

3.4. Tensile Shear Strength

Shear strength f_t (Table 5) was significantly influenced by the type of the adhesive (p-value 0.000). Fiber reinforcement had no significant influence on shear strength f_t (p-value 0.561). Interactions between fiber reinforcement and the applied adhesive type were slight for shear strength f_t (p-value 0.045).

Table 5. Shear strength.

Adhesive	Reinforcement	N	Shear Strength (N/mm^2)			
			Min	Mean	Max	SD
Epoxy	non	9	5.20	6.28	6.92	0.58
	flax	9	4.80	6.81	7.70	0.86
	glass	9	6.03	6.51	6.88	0.28
UF	non	9	5.01	5.47	5.83	0.29
	flax	9	5.29	6.40	6.83	0.44
MUF	non	9	5.03	6.29	7.18	0.78
	flax	9	5.82	6.45	7.17	0.50
PU-AN	non	9	6.99	7.61	8.14	0.32
	flax	9	2.59	7.20	8.82	1.96
PUR	non	9	6.49	6.74	7.19	0.23
	flax	9	6.08	6.39	6.84	0.28

A general correlation between moisture content, respectively density and shear strength f_t was not detected (MC vs. f_t: R^2 0.070; p-value 0.697/density vs. f_t: R^2 0.036; p-value 0.722).

Comparing the SS means of the applied adhesives, a divergent picture is given (Table 5). Epoxy flax fiber reinforced specimens increased f_t by 8.4%, respectively by 3.7% for glass fiber reinforcement in comparison with the mean of the reference samples. UF glued flax fiber reinforced plywood enhanced shear strength ft by 17.0%. MUF bonded flax fiber reinforcement increased by 2.5%. The isocyanate MDI based prepolymer adhesive (PU-AN) bonded flax fiber reinforced plywood mean of 7.20 (SD = 1.96) N/mm^2 decreased by 5.4% in comparison to the non-reinforced reference mean of 7.61 (SD = 7.61) N/mm^2. A different picture is seen in the comparison of the median enhancing fiber reinforcement (median 8.07 N/mm^2) by 6.5%, in contrast to the reference median of 7.58 N/mm^2. Due to eliminating the influence of the outlier and standard deviation of 1.96, PUR flax fiber reinforced plywood lowered shear strength f_t by 5.2%, indicating complications with the glue line.

Bonding strength between veneers is mainly determined by the properties of the adhesives. All specimens exceeded the limit value of 1 N/mm^2 for the shear strength mean indicated in EN 314-2 [21]. For example, the mean value of UF bonded non fiber reinforced beech plywood with a shear strength of 5.47 N/mm^2 was three times higher than the findings of Bekhta et al. (2020), with a shear strength mean of 1.51 N/mm^2 [28]. UF proved to be satisfactory with flax fiber and the adhesive matrix. MUF and PU-AN displayed acceptable improvements. UF is widely used for plywood production due to low price, high bonding strength, and desirable water resistance [29]. The difference between flax and glass fiber for epoxy resin indicates that flax is well suitable. One method to improve the bonding performance between flax fiber and epoxy resin is given by Sbardella et al. (2021), suggesting the use of zinc oxide (ZnO) nanorods [30]. The decrease in PUR indicates adhesive application problems during the manufacturing process. Further testing is mandatory to make a final statement on the suitability of PUR and flax fiber reinforcement. This is due to the fact that polyurethane based adhesives are commonly used for all kinds of applications because of their self-supporting excellent bond strength, fast curing, and environmental influence resistance [31]. In addition, according to Somarathna et al. (2018), several studies have proven the suitability of polyurethane adhesives surpassing the performance of epoxy resins in terms of quasi-static, dynamic, impact, and cyclic loading. Furthermore, the lower costs of polyurethane adhesives compared to epoxy resins [32] should be mentioned. One explanation for the weak bonding performance could be that binding of natural fibers is strongly influenced by their lignocellulosic origin and inherent hydrophilic character, causing weak binding between the fiber and the polymeric adhesive [33]. This is in line with the decreasing post-production moisture content of 15% for PUR compared to the non-reinforced reference, displaying the lowest moisture content value with 8.84% for references and flax fiber reinforced samples. In addition, Lavalette et al. (2016) mentioned an optimum wood moisture content between 30 and 60% for efficient bonding of veneer plies with polyurethane adhesives [34]. To improve the understanding of interaction effects of bonding performance, Li et al. (2020) suggested the combination of lap-shear tests with digital image correlation (DIC) as a valuable investigation method to determine the bonding strength of plywood [29].

3.5. Modulus of Elasticity and Modulus of Rupture

The MOE and MOR (Table 6) were significantly influenced by the applied type of adhesive (*p*-value 0.00), whereas fiber reinforcement in general had no significant influence on MOE (*p*-value 0.219) and MOR (*p*-value 0.253). Interaction effects between the factors "type of adhesive" and "fiber reinforcement" are given with a *p*-value of 0.00.

Based on the different adhesives, the influence of the factor fiber reinforcement tested by ANOVA varies. For the epoxy resin bonded plywood, no significant influence of fiber reinforcement on MOE (*p*-value 0.198; R^2 = 0.113) and MOR (*p*-value 0.008; R^2 = 0.304) could be stated. MOE (*p*-value 0.480; R^2 = 0.028) and MOR (*p*-value 0.151; R^2 = 0.111) of the UF bonded plywood was not influenced by the fiber reinforcement. MOE of MUF bonded plywood was slightly influenced by fiber reinforcement (*p*-value 0.003; R^2 = 0.431) and for MOR (*p*-value 0.107; R^2 = 0.154), the effect could not be stated. Fiber reinforcement did not affect MOE (*p*-value 0.829; R^2 = 0.003) and MOR (*p*-value 0.747; R^2 = 0.006) of the isocyanate MDI based prepolymer adhesive (PU-AN) bonded plywood. Fiber reinforcement significantly influenced MOE (*p*-value 0.000; R^2 = 0.769) and MOR (*p*-value 0.001; R^2 = 0.461) of the polyurethane bonded plywood.

Table 6. MOE and MOR.

Adhesive	Reinforcement	N	MOE (N/mm^2)				MOR (N/mm^2)			
			Min	Mean	Max	SD	Min	Mean	Max	SD
Epoxy	non	10	11,011	11,738	12,099	323	87.00	99.30	109.59	6.41
	flax	10	11,297	12,071	13,000	555	84.84	113.57	137.57	13.24
	glass	10	10,762	11,772	12,277	416	92.31	109.99	122.92	8.04
UF	non	10	9995	11,259	12,317	623	109.63	114.84	121.24	3.05
	flax	10	9448	11,056	11,663	638	106.39	112.15	119.93	4.78
MUF	non	9	9976	10,917	11,700	508	87.43	99.44	109.87	6.63
	flax	9	11,055	11,620	12035	330	94.25	105.58	115.29	8.52
PU-AN	non	10	8997	9530	10,130	366	90.19	95.65	101.92	4.26
	flax	10	3885	9384	10,742	2075	50.31	97.50	111.09	17.33
PUR	non	10	11,427	11,962	12,360	319	97.77	110.87	130.24	12.16
	flax	10	9745	10,477	11,355	517	75.33	91.64	103.76	9.63

Comparing the median for MOE, flax fiber reinforcement increased by 2.6%, respectively by 0.3% for the glass fiber reinforced sample. The UF bonded flax fiber reinforced plywood MOE mean decreased by 2.3%. The MUF bonded flax fiber reinforced specimen increased by 6.5%. The MOE of the isocyanate MDI based prepolymer (PU-AN) bonded flax fiber reinforced plywood increased by 8.9% and the PUR flax fiber reinforced specimen decreased by 11.2%.

The epoxy resin bonded flax fiber reinforced MOR compared by the median increased by 16.65%, respectively by 11.09% for glass fiber reinforcement. The median of MOR for the UF bonded flax fiber reinforced sample lowered by 2.91%. The MUF based flax fiber reinforced plywood increased MOR by 4.65%. The isocyanate MDI based prepolymer flax fiber reinforced specimen improved MOR by 8.20% and for the PUR samples, it decreased by 14.72%.

The correlation between density and MOE is given (p-value 0.000; $R^2 = 0.125$) (Figure 5a). A correlation between moisture content and MOE was not detected (p-value 0.203; $R^2 = 0.052$) (Figure 5b). Within the singular types of adhesives, a correlation for MOE and density was only significant for MUF bonded plywood (p-value 0.001; $R^2 = 0.917$) and for moisture content (p-value = 0.001; $R^2 = 0.854$). The general correlation between MOR and MOE is given by a p-value 0.000; $R^2 = 0.463$ (Figure 6).

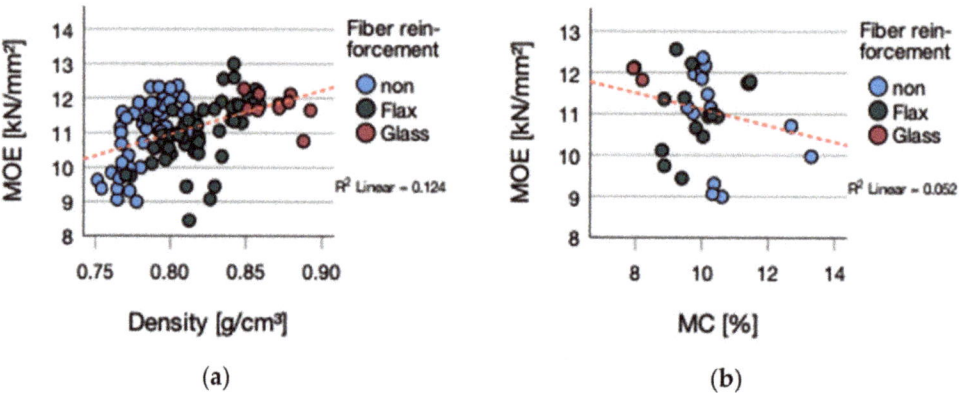

Figure 5. (a) MOE vs. density (b) MOE vs. MC.

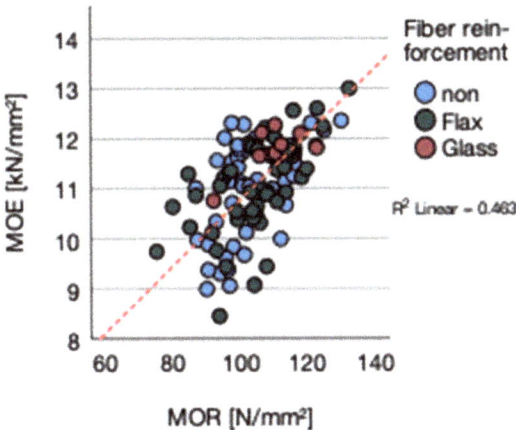

Figure 6. MOE vs. MOR.

MOE is influenced by density, as stated in the literature [25,27]. Flax fiber reinforcement improves MOE, depending on the type of adhesive. The position of the fiber reinforcement within the lay-up strongly influenced the improvements. Fiber reinforcement located closer to the outer layers or at the outside can better contribute to MOE performance due to higher tensile strength within the tension zone [8,14]. In contrast, MOR is strongly dependent on the strength of the surface layer [14].

3.6. Screw Withdrawal Resistance

The influence of fiber reinforcement (*p*-value 0.001) and the type of adhesive (*p*-value 0.000) as well as the interaction of both (*p*-value 0.000) are significant. The results for the screw withdrawal resistance display a divergent picture (Figure 7b). Fiber reinforcement improved the screw withdrawal resistance (SWR) median with the exception of PUR bonded flax reinforced plywood (Table 7).

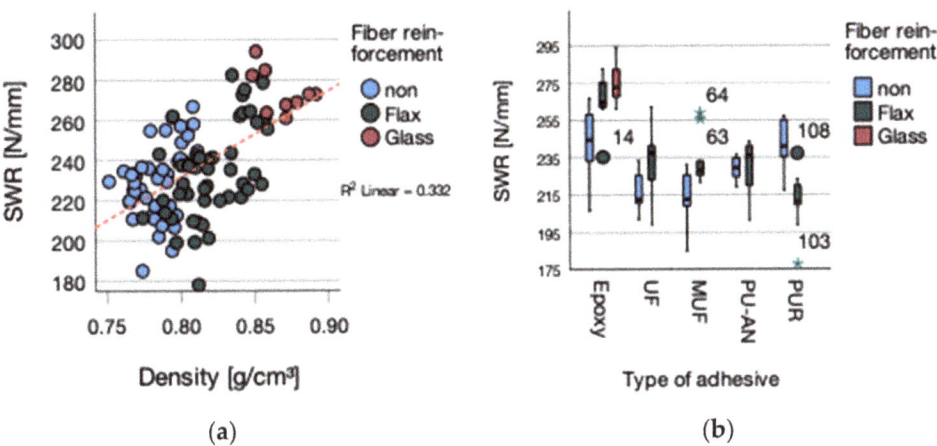

(a)   (b)

Figure 7. (**a**) SWR vs. density and (**b**) SWR grouped by type of adhesive.

Table 7. SWR and MSWF.

Adhesive	Reinforcement	N	SWR (N/mm)				MSWF (kN)			
			Min	Mean	Max	SD	Min	Mean	Max	SD
Epoxy	non	9	206.75	241.78	266.98	20.32	2.253	2.643	2.926	0.223
	flax	9	235.33	266.28	282.72	13.98	2.829	3.173	3.378	0.160
	glass	9	261.06	274.36	294.26	10.84	2.937	3.058	3.240	0.104
UF	non	9	202.07	217.53	233.21	11.21	2.186	2.322	2.488	0.113
	flax	9	199.32	232.79	262.20	17.72	2.268	2.627	2.894	0.180
MUF	non	9	184.91	213.04	231.10	15.41	1.993	2.292	2.507	0.181
	flax	9	221.70	233.37	258.95	14.02	2.498	2.665	2.952	0.158
PU-AN	non	9	219.24	229.14	236.55	6.59	2.331	2.444	1.522	0.067
	flax	9	201.40	228.36	243.71	15.63	2.348	2.638	2.807	0.175
PUR	non	9	217.40	241.56	257.15	14.80	2.337	2.605	2.769	0.162
	flax	9	178.00	211.56	237.22	16.44	2.102	2.488	2.802	0.189

SWR for the epoxy resin flax fiber reinforced mean of 266.28 (SD = 13.98) N/mm increased by 10.13%, respectively by 13.48% for glass fiber reinforced samples, compared to the reference mean of 241.78 (SD = 20.32) N/mm. SWR for the UF bonded flax fiber reinforced mean of 232.79 (SD = 11.21) N/mm improved by 7.02% in contrast to the reference mean of 217.53 (SD = 11.21) N/mm. The MUF glued flax fiber reinforced SWR mean of 233.37 (SD = 14.02) N/mm increased by 9.54% compared to the reference mean of 213.04 (SD = 15.41) N/mm. The isocyanate MDI based prepolymer adhesive (PU AN) bonded flax fiber reinforced SWR sample mean of 228.36 (SD = 15.63) N/mm decreased by 0.34% when comparing the median with an enhancement of 2.66%. The difference between mean and median could be explained by the differences in the SD values for the reference of 6.59 and 15.63 for the reinforced sample. The PUR bonded flax fiber reinforced SWR mean of 211.56 (SD = 14.80) N/mm declined by 12.42% compared to the reference mean of 241.56 (SD = 16.44) N/mm.

The maximum screw withdrawal force Fmax (MSWF) (Table 7) displayed for the epoxy resin bonded flax fiber reinforced plywood was an increase of 20.05% and respectively for glass fiber reinforcement of 15.70%. UF and MUF bonded flax fiber reinforced plywood enhanced by 13.14%, respectively 16.27%. PU AN increased the maximum screw withdrawal force by 7.94%, whereas PUR decreased by 4.49%.

The Pearson correlation (R2 0.332) between density and screw withdrawal resistance (Figure 7a) was significant with a p-value of 0.000. This is according to Wagenführ and Scholz (2008), who stated the relation between increasing density and enhanced screw withdrawal resistance [24].

In general, screw withdrawal strength is dependent on screw penetration length, screw diameter, angle between screw and wood fiber direction, wood species, wood moisture content, and temperature. This demonstrates that SWR perpendicular to the wood fiber orientation creates the highest values compared to fiber direction [35]. Furthermore, for laminar wood based composite structures, based on the research of Liu and Guan (2019), the location of fiber reinforcement influences the SWR, suggesting a combination of fiber reinforcement close to the plywood core plies and to the surface plies [36]. This is confirmed by comparing the results of maximum screw withdrawal force F_{max} to Bal et al. (2017). This demonstrates that MSWF improved by 13.65% respectively by 14.11%, if fiber reinforcement is located on the surface within the outer glue lines of the five layered (veneer thickness 2.7 mm) PF bonded poplar plywood reinforced with woven glass fiber fabric (areal weight 500 g/m^2) [7]. Similar effects have been reported by Kramár et al. (2020). The SWR for basalt fiber reinforcement located at the core layer of particleboards did not enhance the SWR. This was based on the assumption that the degree of compaction in particleboards is lower in the core layer than on the surface, thus affecting the SWR. In addition, this is due to the difference in density between the core (low density) and the surface layer (high density). Fiber reinforcement placed within the surface layer increased the SWR due to increased density caused by a higher degree of compaction within the surface layer [14].

This leads to the conclusion that increasing density is not the singular factor to influence the SWR. Research conducted by Maleki et al. (2017) highlighted that screw withdrawal perpendicular to grain displays a failure mode combination of splitting, caused by tension perpendicular to grain and rolling shear failure [37]. Further aspects such as glue line quality have to be taken into consideration. This is due to the fact that PUR reduced SWR and maximum screw pull-out force Fmax, indicating poor glue line bonding quality. In addition, research should focus on fiber and textile characteristics and screw pull-out behavior within the fiber adhesive matrix and the surrounding veneer plies for a deeper understanding of failure mode and interaction effects regarding SWR.

4. Conclusions

Comparing the percentage-based performance (Figure 8a; axis interval 5%) of the different adhesives with flax fiber reinforcement, it can be stated that epoxy resin is well suitable for improving MOE, MOR, TS, SS, and SWR. The UF, MUF, and isocyanate MDI based prepolymer adhesive (PU-AN) increased the performance of the mechanical properties of MOE, MOR, SS, and SWR, but lowered tensile strength compared to the singular references. PUR failed to suit flax fiber reinforcement.

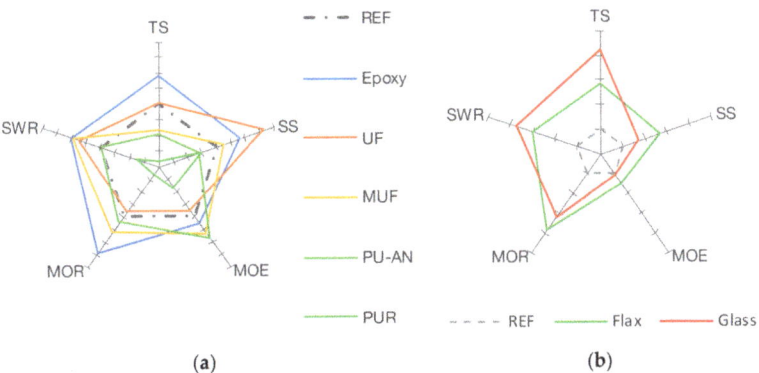

Figure 8. (**a**) Performance of flax fiber reinforcement (axis interval 5%). (**b**) Comparison of the flax and glass fiber reinforcement (axis interval 5%).

SWR was significantly influenced by the factor flax fiber reinforcement. For MOE, MOR, and SS, no significant influence of flax fiber reinforcement could be stated. TS was slightly influenced by the factor fiber reinforcement.

The results show the possibility of improving mechanical plywood properties by using reinforcing flax fiber fabrics bonded with different industrial standard adhesive systems (Figure 8b; axis interval 5%). Flax fiber reinforcement exceeded the glass fiber reinforced epoxy resin bonded plywood in terms of SS (+5.14%), MOE (+2.3%) and MOR (+3.54%). The SWR for flax reinforced epoxy resin bonded plywood was 3.35% lower and for TS, it was 6.96% lower.

Further research is mandatory to determine the influence of press parameters such as pressure, temperature, and time, in addition to factors like veneer and flax fiber fabric, moisture content or the influence of pre-treatment of the flax fiber fabric to improve bond ability. In addition, research on the fiber reinforcement location within plywood lay-up is necessary in order to optimize the mechanical properties of flax fiber reinforced plywood. Furthermore, research on the influence of formaldehyde emissions caused by UF [38] will have to be conducted.

Author Contributions: Conceptualization and methodology, J.J. and G.K.; Validation and formal analysis, J.J., G.K. and M.-C.B.; Investigation, J.J. and G.K.; Resources and funding acquisition, M.-C.B.; Writing—original draft preparation, review and editing, J.J., G.K., M.-C.B., A.P. and P.K.; Supervision and project administration, G.K. and M.-C.B. All authors have read and agreed to the published version of the manuscript.

Funding: This research received no external funding.

Acknowledgments: The authors are thankful for the support of Horatiu Noge, in conducting the fabrication of the test specimens and gratefully acknowledge the support of Thomas Wimmer for his contribution in conducting mechanical testing at Salzburg University of Applied Sciences at Campus Kuchl.

Conflicts of Interest: The authors declare no conflict of interest.

References

1. Mahut, J.; Reh, R. *Plywood and Decorative Veneers*; Technická Univerzita vo Zvolene: Zvolen, Slovakia, 2007.
2. Stark, N.M.; Cai, Z.; Carll, C. Chapter 11—Wood-Based Composite Materials Panel Products, Glued-Laminated Timber, Structural Materials. In *Wood Handbook—Wood as an Engineering Material*; General Technical Report FPL-GTR-282; U.S. Department of Agriculture, Forest Service, Forest Products Laboratory: Madison, WI, USA, 2010.
3. Paulitsch, M.; Barbu, M.C. *Holzwerkstoffe der Moderne*; DRW Verlag Weinbrenner: Leinfelden-Echterdingen, Germany, 2015.
4. Jorda, J.S.; Barbu, M.C.; Kral, P. Natural fiber reinforced veneer based products. *Pro Ligno* **2019**, *15*, 206–211.
5. Laufenberg, T.L.; Rowlands, R.E.; Krueger, G.P. Economic Feasibility of Synthetic Fiber Reinforced Laminated Veneer Lumber (Lvl). *For. Prod. J.* **1984**, *34*, 15–22.
6. Bal, B.C.; Bektaş, I.; Mengeloğlu, F.; Karakuş, K.; Demir, H. Ökkeş Some technological properties of poplar plywood panels reinforced with glass fiber fabric. *Constr. Build. Mater.* **2015**, *101*, 952–957. [CrossRef]
7. Bal, B.C. Propriedades de fixação de parafusos e pregos em painéis compensados de madeira reforçados com tecido de fibra de vidro. *Cerne* **2017**, *23*, 11–18. [CrossRef]
8. Liu, Y.; Guan, M.; Chen, X.; Zhang, Y.; Zhou, M. Flexural properties evaluation of carbon-fiber fabric reinforced poplar/eucalyptus composite plywood formwork. *Compos. Struct.* **2019**, *224*, 111073. [CrossRef]
9. Auriga, R.; Gumowska, A.; Szymanowski, K.; Wronka, A.; Robles, E.; Ocipka, P.; Kowaluk, G. Performance properties of plywood composites reinforced with carbon fibers. *Compos. Struct.* **2020**, *248*, 112533. [CrossRef]
10. Guan, M.; Liu, Y.; Zhang, Z. Evaluation of bending performance of carbon fiber-reinforced eucalyptus/poplar composite plywood by digital image correlation and FEA analysis. *J. Mater. Sci.* **2020**, *55*, 8388–8402. [CrossRef]
11. Baley, C.; Bourmaud, A.; Davies, P. Eighty years of composites reinforced by flax fibres: A historical review. *Compos. Part A Appl. Sci. Manuf.* **2021**, *144*, 106333. [CrossRef]
12. Speranzini, E.; Tralascia, S. Engineered lumber: LVL and solid wood reinforced with natural fibres. In Proceedings of the WCTE 2010, World Conference on Timber Engineering, Trentino, Italy, 20–24 June 2010; Volume 2, pp. 1685–1690.
13. Moezzipour, B.; Ahmadi, M. Physical and mechanical properties of reinforced ply wood with natural fibers. *J. Indian Acad. Wood Sci.* **2017**, *14*, 70–73. [CrossRef]
14. Kramár, S.; Mayer, A.K.; Schöpper, C.; Mai, C. Use of basalt scrim to enhance mechanical properties of particleboards. *Constr. Build. Mater.* **2020**, *238*, 117769. [CrossRef]
15. Jorda, J.; Kain, G.; Barbu, M.; Haupt, M.; Krišťák, L. Investigation of 3D-Moldability of Flax Fiber Reinforced Beech Plywood. *Polymers* **2020**, *12*, 2852. [CrossRef] [PubMed]
16. Valdes, M.; Giaccu, G.F.; Meloni, D.; Concu, G. Reinforcement of maritime pine cross-laminated timber panels by means of natural flax fibers. *Constr. Build. Mater.* **2020**, *233*, 117741. [CrossRef]
17. EN 323 *Wood-Based Panels—Determination of Density*; European Committee for Standardization: Brussels, Belgium, 2005.
18. EN 322 *Wood-Based Panels—Determination of Moisture Content*; European Committee for Standardization: Brussels, Belgium, 2005.
19. DIN 52377 *Prüfung von Sperrholz—Bestimmung des Zug-Elastizitätsmoduls und der Zugfestigkeit*; Deutsches Institut für Normung: Berlin, Germany, 2016.
20. EN 314-1 *Plywood—Bonding Quality—Test Methods*; European Committee for Standardization: Brussels, Belgium, 2005.
21. EN 314-2 *Plywood—Bonding Quality—Part 2—Requiements*; European Committee for Standardization: Brussels, Belgium, 2005.
22. EN 310 *Wood-Based Panels—Determination of Modulus of Elasticity in Bending and of Bending Strength*; European Committee for Standardization: Brussels, Belgium, 2005.
23. EN 320 *Particleboards and Fibreboards—Determination of Resistance to Axial Withdrawal of Screws*; European Committee for Standardization: Brussels, Belgium, 2011.
24. Wagenführ, A.; Scholz, F. *Taschenbuch der Holztechnik*; Carl Hanser Verlag: München, Germany, 2008.
25. Kollmann, F. *Anatomie und Pathologie, Chemie, Physik Elastizität und Festigkeit*; Springer: Berlin/Heidelberg, Germany, 1951.
26. Aydin, I.; Colakoglu, G.; Colak, S.; Demirkir, C. Effects of moisture content on formaldehyde emission and mechanical properties of plywood. *Build. Environ.* **2006**, *41*, 1311–1316. [CrossRef]

27. Niemz, P. *Physik des Holzes und der Holzwerkstoffe*; DRW Verlag Weinbrenner: Leinfelden-Echterdingen, Germany, 1993.
28. Bekhta, P.; Salca, E.-A.; Lunguleasa, A. Some properties of plywood panels manufactured from combinations of thermally densified and non-densified veneers of different thicknesses in one structure. *J. Build. Eng.* **2019**, *29*, 101116. [CrossRef]
29. Li, W.; Zhang, Z.; Zhou, G.; Leng, W.; Mei, C. Understanding the interaction between bonding strength and strain distribution of plywood. *Int. J. Adhes.* **2020**, *98*, 102506. [CrossRef]
30. Sbardella, F.; Lilli, M.; Seghini, M.; Bavasso, I.; Touchard, F.; Chocinski-Arnault, L.; Rivilla, I.; Tirillò, J.; Sarasini, F. Interface tailoring between flax yarns and epoxy matrix by ZnO nanorods. *Compos. Part A Appl. Sci. Manuf.* **2021**, *140*, 106156. [CrossRef]
31. Dodangeh, F.; Dorraji, M.S.; Rasoulifard, M.; Ashjari, H. Synthesis and characterization of alkoxy silane modified polyurethane wood adhesive based on epoxidized soybean oil polyester polyol. *Compos. Part B Eng.* **2020**, *187*, 107857. [CrossRef]
32. Somarathna, H.; Raman, S.; Mohotti, D.; Mutalib, A.; Badri, K. The use of polyurethane for structural and infrastructural engineering applications: A state-of-the-art review. *Constr. Build. Mater.* **2018**, *190*, 995–1014. [CrossRef]
33. Martínez, L.M.T.; Kharissova, O.V.; Kharisov, B.I. (Eds.) *Handbook of Ecomaterials*; Springer Nature: Cham, Switzerland, 2019.
34. Lavalette, A.; Cointe, A.; Pommier, R.; Danis, M.; Delisée, C.; Legrand, G. Experimental design to determine the manufacturing parameters of a green-glued plywood panel. *Eur. J. Wood Wood Prod.* **2016**, *74*, 543–551. [CrossRef]
35. Hübner, U.; Rasser, M.; Schickhofer, G. Withdrawal capacity of screws in European ash (*Fraxinus excelsior* L.). In Proceedings of the WCTE 2010, World Conference on Timber Engineering, Trentino, Italy, 20–24 June 2010; Volume 1, pp. 241–249.
36. Liu, Y.; Guan, M. Selected physical, mechanical, and insulation properties of carbon fiber fabric-reinforced composite plywood for carriage floors. *Eur. J. Wood Wood Prod.* **2019**, *77*, 995–1007. [CrossRef]
37. Maleki, S.; Najafi, S.K.; Ebrahimi, G.; Ghofrani, M. Withdrawal resistance of screws in structural composite lumber made of poplar (Populus deltoides). *Constr. Build. Mater.* **2017**, *142*, 499–505. [CrossRef]
38. Réh, R.; Krišťák, Ľ.; Sedliačik, J.; Bekhta, P.; Božiková, M.; Kunecová, D.; Vozárová, V.; Tudor, E.; Antov, P.; Savov, V. Utilization of Birch Bark as an Eco-Friendly Filler in Urea-Formaldehyde Adhesives for Plywood Manufacturing. *Polymers* **2021**, *13*, 511. [CrossRef] [PubMed]

Article

Experimental Study of the Influence of Selected Factors on the Particle Board Ignition by Radiant Heat Flux

Ivana Tureková [1], Martina Ivanovičová [1], Jozef Harangózo [1], Stanislava Gašpercová [2] and Iveta Marková [2,*]

[1] Department of Technology and Information Technologies, Faculty of Education, Constantine the Philosopher University in Nitra, Tr. A. Hlinku 1, 949 74 Nitra, Slovakia; ivana.turekova@ukf.sk (I.T.); mabumb@gmail.com (M.I.); jozef.harangozo@ukf.sk (J.H.)
[2] Department of Fire Engineering, Faculty of Security Engineering, University of Žilina, Univerzitná 1, 010 26 Zilina, Slovakia; stanislava.gaspercova@uniza.sk
* Correspondence: iveta.markova@uniza.sk; Tel.: +421-041-513-6799

Abstract: Particleboards are used in the manufacturing of furniture and are often part of the interior of buildings. In the event of a fire, particleboards are a substantial part of the fuel in many building fires. The aim of the article is to monitor the effect of radiant heat on the surface of particle board according to the modified procedure ISO 5657: 1997. The significance of the influence of heat flux density and particle board properties on its thermal resistance (time to ignition) was monitored. Experimental samples were used particle board without surface treatment, with thicknesses of 12, 15, and 18 mm. The samples were exposed to a heat flux from 40 to 50 kW·m^{-2}. The experimental results are the initiation characteristics such as of the ignition temperature and the weight loss. The determined factors influencing the time to ignition and weight loss were the thickness and density of the plate material, the density of the radiant heat flux and the distance of the particle board from the radiant source (20, 40, and 60 mm). The obtained results show a significant dependence of the time to ignition on the thickness of the sample and on the heat flux density. The weight loss is significantly dependent on the thickness of the particle board. Monitoring the influence of time to ignition from sample distance confirmed a statistically significant dependence. As the distance of the sample from the source increased, the time to ignition decreased linearly. As the distance of the sample from the source increased, the time to ignition increased.

Keywords: ignition; particleboard; radiant heat; thermal resistance; ANOVA

1. Introduction

Sheet board materials are among the most important wood products [1]. Their production encompasses utilization of wood of lower quality classes and obtaining suitable materials with improved physical and mechanical properties [2].

This product group contains wood-based boards for the use in building interiors, such as boards without surface treatment (raw) or with surface treatment (particleboards), plywood, fiberboard, and edge-glued wood panels [3,4].

Particleboard can be defined according to STN EN 309:2005 [4] as a molded wood material, produced by heat pressing of small wood particles (e.g., chips, shavings, sawdust, lamellas, etc.) or other lignocellulosic particles (e.g., flax shives, hemp shives, bagasse, etc.) with adhesives.

The processing wood of all woody plants occurring in Central Europe is used as a source of wood in the production of particleboards. These are less valuable forest assortments, industrial and residual waste, recycled wood, and other lignocellulosic materials [5].

Particleboards belong to a product group of board materials, but they are considered an input material in the furniture and construction industries [6]. In terms of quality assessment, particleboards have only few disadvantages, and flammability is among them [7–11].

The current state of technology and production techniques in particleboard production allow processing of practically all types of wood occurring in Central Europe using a suitable mixture [12–14].

Wood and sheet board materials represent a substantial part of the fuel in many building fires [15].

The assumption of a fire hazard requires an appropriate description of the fire ignition and fire development [16,17]. The initial process is ignition [18]. Flammability can be defined as the ability of materials to ignite when heated to elevated temperatures. It depends on many factors, mainly the critical heat flux and the thermal properties of materials. Currently, there are several methods for determining the flammability, fire-technical, and physical material properties, which are defined by relevant standards [19,20].

The aim of this article is to analyze the influence of heat flux density and particle board properties (thickness of 12 mm, 15 mm, 18 mm and board material density) on their thermal resistance (time to ignition) and ignition characteristics (ignition temperatures and weight loss). This dependence was also monitored when the distance of the sample from the radiant heat source changed, which represents an important safety factor in the ignition phase of real fires.

2. Materials and Methods

2.1. Samples

Particleboard research is part of improving their properties [21–24]. Separate attention is paid to the research of the physical and mechanical properties of the particleboards [25–28].

Particleboards with thicknesses of 12, 15, and 18 mm were used for experiments due to their practical applicability and popularity in practice (Figure 1). Selected materials are among the most widely used materials nowadays in the furniture and construction industry [29,30]. Particleboard samples were sourced from the company BUČINA DDD, Zvolen, Slovakia [31,32] under product name Particleboard raw unsanded (Table 1). Particleboards contain coniferous softwood chips, mainly spruce and urea-formaldehyde adhesive mixture.

Table 1. Physical and chemical properties and fire-technical characteristics of particleboards in thicknesses of 12–18 mm [31,32].

Parameters	Thickness of Particleboard Sample (mm)		
	12	15	18
Density ($kg \cdot m^{-3}$) (average)	690	713	644
Moisture (%)	5.05	5.25	5.45
Bending strength ($N \cdot mm^{-2}$)	13.2	12.5	12
Modulus of elasticity ($N \cdot mm^{-2}$)	2500	2450	2750
Swelling after 24 h (%)	3.5	3.5	3.5
Thermal conductivity ($W \cdot m^{-2} \cdot K^{-1}$)	0.10–0.14	0.10–0.14	0.10–0.14
Free formaldehyde content ($mg \cdot 100\ g^{-1}$) (Emission class E1)	6.5	6.5	6.5
Reaction to fire		D-s1, d0	

Large-size wood materials form the largest percentage of wood material in timber houses which means they can be directly exposed to fire [33].

Selected thicknesses of board materials are used in the construction and insulation of houses, in the construction of ceilings, soffits, partitions, etc.

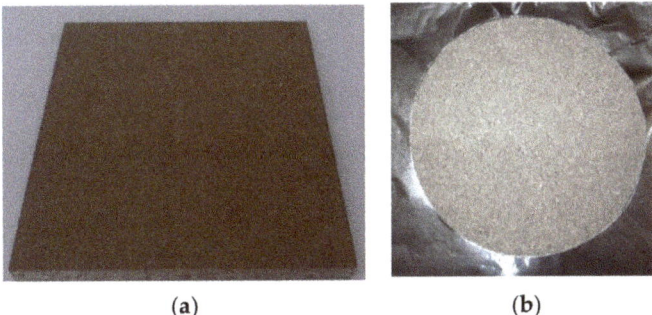

Figure 1. Example of experimental samples. (**a**) Particleboard; (**b**) sample prepared for measurements in accordance with ISO 5657 [34].

Particleboard samples were cut to specific dimensions (165 × 165) mm according to STN 5657: 1997 [34]. Selected board materials were kept at a specific temperature (23 °C ± 2 °C) and relative humidity (50 ± 5%).

2.2. Methodology

The density of the particleboards was determined according to STN EN 323: 1996 [35]. The time to ignition and weight loss depending on the selected level of heat flux density and thickness of board materials and the distance of selected board materials from the ignition source was determined according to the modified procedure ISO 5657: 1997 [34]. A detailed description of the modification and the course of the experiment is described in Tureková et al. [36].

The heating cone ensures heat flow in the range of 10 to 70 kW·m^{-2}. The heat acts in the center of the hole in the masking plate where the test sample is placed (Figure 2).

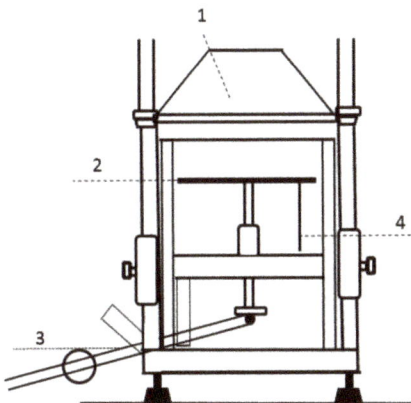

Figure 2. Scheme of the equipment for determination of flammability of materials at a heat flux of radiant heat of 10–50 kW·m^{-2} according to ISO 5657: 1997 [34]. Legend: 1-heating cone, 2-board for sample, 3-movable arm, 4-connection point for recording experimental data.

The heating cone temperatures were verified by a thermocouple that is in close and constant contact with the heating element tube, and the heat fluxes were determined on the basis of a calibration curve [36].

The samples were placed horizontally and exposed to a heat flux of 43 to 50 kW·m^{-2} by an electrically heated conical radiator. Orientation experiments determined the minimum heat flux required to maintain flame combustion.

The horizontally placed sample under the thermal cone is exposed to the selected heat flux and gradually thermally degrades. During the experiment, the course of degradation is monitored, which is manifested by weight loss. At the same time, time to ignition is monitored. Time to ignition was recorded while considering only the permanent ignition of the surface of the analyzed sample when exposed to a selected level of heat flux density.

Thermal inertia, which is closely related to the time to ignition, was calculated for each selected board material [37]. The higher the thermal inertia value, the slower the temperature rise on the surface of the board material and the later the ignition [38–40]. Thermal inertia was calculated according to Schieldge et al. [41]:

$$I = \lambda \cdot \rho \cdot c \cdot [J^2 \cdot m^{-4} \cdot s^{-1} \cdot K^{-2}] \tag{1}$$

where λ [W·m^{-1}·K^{-1}] is the thermal conductivity, ρ [kg·m^{-3}] is the board material density, and c [J·kg^{-1}·K^{-1}] is heat capacity.

The influence of the ignition source distance on the time to ignition of the board materials was monitored on particleboards with a thickness of 12 mm. The choice of thickness was made from a practical point of view. Particleboards with a thickness of 12 mm are the most commonly used materials in the construction industry in thermal insulation, timber houses, construction of ceilings and soffits [42].

The experiments were performed with the radiant heat fluxes of 44, 46, 48, and 50 kW·m^{-2}. The distance between the cone calorimeter and the particleboard was 20, 40, and 60 mm. The choice of distance was determined based on orientation experiments and changes in times to ignition were monitored even in case of minimal changes in distance from the ignition source. A preparation consisting of cement cubes measuring 20 × 20 × 20 mm was used to change the distance of the board material from the ignition source. The experiments were repeated five times.

Specific factors affecting time to ignition and weight loss are:
- Thickness and density of the board material;
- Radiant heat flux density;
- Distance of particleboards from the radiant heat source.

2.3. Mathematical and Statistical Processing of Data and Evaluation of Results

To evaluate the influence of the above-mentioned factors on the ignition temperature and weight loss, the obtained results were subjected to a statistical analysis. The obtained results of the ignition and weight loss temperatures were statistically evaluated by two-way analysis of variance (ANOVA) using the least significant difference (LSD) test (95%, 99% detectability level), (STATGRAPHICS software version 18/19 (Statgraphics Technologies, Inc., The Plains, VA, USA), with the following influence factors: board material thickness (12, 15 and 18 mm), radiant heat flux density (from 43 to 50 kW·m^{-2}), and distance of board materials from the ignition source (20, 40 a 60 mm).

3. Results and Discussion

The course of the experiment (Figure 3) according to ISO 5657: 1997 [34] confirmed the verified behavior of the material in terms of the classification "reaction to fire (D-s1, d0)" [43–45] (Figure 4). The priority of the experiment is to monitor the critical parameters of the ignition based on the change in board thickness (Figure 3 and Table 2).

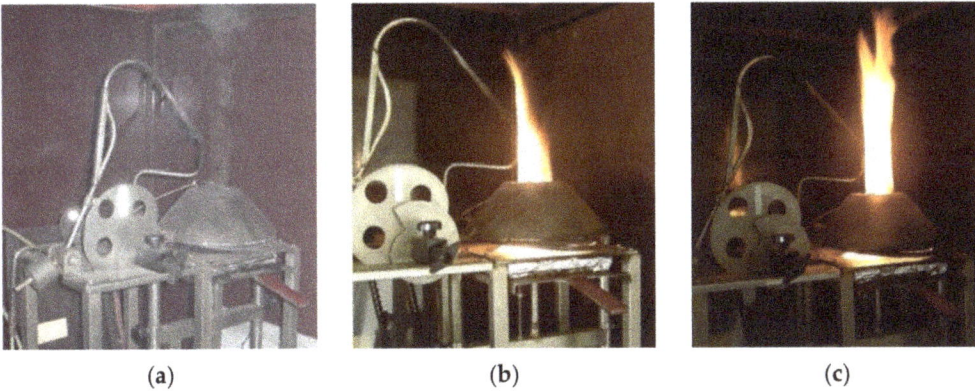

Figure 3. Course of measurement of time to ignition and weight loss for a particleboard sample with a thickness of 15 mm, heat flux intensity 45 kW·m^{-2}. Legend: (**a**) sample ignition (time to ignition 84 s); (**b**) burning of the sample in 100 s; and (**c**) burning of the sample in 120 s.

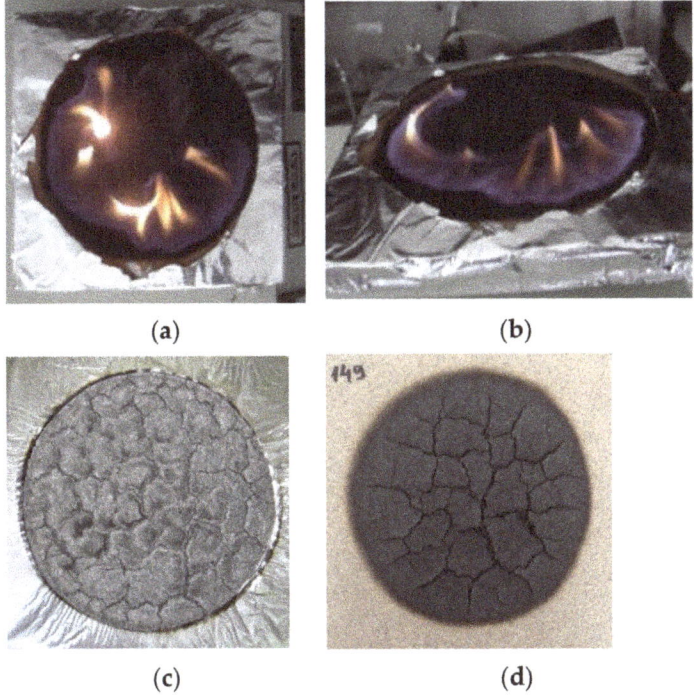

Figure 4. Combustion process of particleboards after their ignition by radiant heat (**a**) top view for sample with 15 mm thickness immediately after experiment; (**b**) side view for sample with 15 mm thickness immediately after ignition; (**c**) sample with 15 mm thickness taken out from the measuring device, placed at a distance of 20 mm after the end of the experiment; (**d**) cooled sample 10 min after the experiment, sample thickness of 18 mm.

Table 2. Time to ignition and mass loss in samples with different thickness using heat fluxes of 40 to 50 kW·m^{-2} at adistance of 20 mm.

Radiant Heat Flux (kW·m^{-2})	Thickness (mm)	Density (kg·m^{-3})	Thermal Inertia (kJ^2·m^{-4}·s^{-1}·K^{-2})	Time to Ignition (s)	Weight Loss (%)
40	12	689 ± 10	0.32 ± 0.002	130.6 ± 3.44	16.3 ± 0.3
	15	711 ± 10	0.31 ± 0.026	118.6 ± 3.83	13.5 ± 0.3
	18	644 ± 6	0.27 ± 0.080	139.0 ± 3.16	11.9 ± 0.2
41	12	689 ± 9	0.32 ± 0.002	114.0 ± 3.52	16.7 ± 0.29
	15	714 ± 11	0.33 ± 0.002	113.4 ± 4.03	13.8 ± 0.33
	18	644 ± 7	0.27 ± 0.082	131.2 ± 2.99	12.5 ± 0.21
42	12	688 ± 9	0.32 ± 0.002	95.2 ± 6.82	17.3 ± 0.39
	15	714 ± 10	0.33 ± 0.002	105.8 ± 3.06	14.3 ± 0.32
	18	645 ± 6	0.31 ± 0.001	122.4 ± 1.96	12.7 ± 0.18
43	12	691 ± 10	0.32 ± 0.002	89.0 ± 5.215	17.1 ± 0.52
	15	716 ± 11	0.33 ± 0.002	92.6 ± 3.441	14.6 ± 0.37
	18	642 ± 7	0.31 ± 0.008	117.0 ± 5.513	13.2 ± 0.17
44	12	691 ± 10	0.32 ± 0.002	80.0 ± 5.37	17.6 ± 0.41
	15	715 ± 10	0.33 ± 0.002	86.4 ± 4.88	15.4 ± 0.35
	18	643 ± 7	0.31 ± 0.002	102.8 ± 4.31	13.7 ± 0.24
45	12	691 ± 9	0.321 ± 0.002	78.2 ± 0.748	17.9 ± 0.30
	15	714 ± 11	0.327 ± 0.002	84.4 ± 2.057	15.3 ± 0.29
	18	645 ± 7	0.311 ± 0.002	92.2 ± 2.481	13.9 ± 0.29
46	12	690 ± 11	0.32 ± 0.002	71.6 ± 1.62	18.4 ± 0.52
	15	711 ± 9	0.33 ± 0.002	76.0 ± 2.28	15.7 ± 0.29
	18	644 ± 8	0.31 ± 0.001	89.0 ± 7.97	13.8 ± 0.56
47	12	690 ± 11	0.32 ± 0.002	66.4 ± 2.87	18.9 ± 0.29
	15	715 ± 10	0.33 ± 0.002	73.8 ± 0.80	16.2 ± 0.36
	18	645 ± 8	0.31 ± 0.002	75.6 ± 3.72	14.5 ± 0.34
48	12	689 ± 10	0.32 ± 0.002	64.4 ± 1.49	19.1 ± 0.34
	15	710 ± 8	0.33 ± 0.001	69.4 ± 1.96	16.3 ± 0.37
	18	644 ± 7	0.31 ± 0.001	75.0 ± 2.00	14.6 ± 0.22
49	12	692 ± 11	0.32 ± 0.002	60.6 ± 2.24	19.7 ± 0.44
	15	713 ± 10	0.33 ± 0.002	66.0 ± 2.28	16.6 ± 0.33
	18	644 ± 11	0.31 ± 0.002	67.2 ± 1.17	15.2 ± 0.13
50	12	689 ± 8	0.32 ± 0.001	59.8 ± 2.64	19.9 ± 0.41
	15	713 ± 11	0.33 ± 0.002	64.4 ± 2.50	16.5 ± 0.33
	18	643 ± 7	0.31 ± 0.001	66.8 ± 2.09	15.9 ± 0.94

3.1. Determination of Ignition Temperature and Weight Loss

The ability of the material surface to generate volatile gases when exposed to radiant heat as well as the ability of selected board materials to ignite when exposed to radiant heat fluxes caused by an ignition source were confirmed.

The density of samples ranged from 640 to 720 kg·m^{-3}. This range corresponds to the usual density of particleboards [46].

By comparing the calculated thermal inertia with the reported thermal inertia by Babrauskas [47,48], very similar results were confirmed. The thermal inertia values ranged from 0.31 to 0.33 kJ^2·m^{-4}·s^{-1}·K^{-2}. The difference was around 0.02 kJ^2·m^{-4}·s^{-1}·K^{-2} in specific particleboards.

In this case, it is not possible to look for the dependence of inertia on other parameters, as the ANOVA results show in Table 3.

Table 3. Analysis of variance for Col_4 Time to ignition-type III sums of squares.

Source	Sum of Squares	Df	Mean Square	F-Ratio	p-Value
Covariates					
Col_3 Thermal interaction	664.125	1	664.125	1.34	0.2481
Main Effects					
Col_2 Board thickness	4018.06	2	2009.03	4.06	0.0190
Residual	79,573.2	161	494.244		
Total	87,085.1	164			

All F-ratios are based on the residual mean square error.

Figure 5 shows a statistically significant dependence of the time to ignition on the sample thickness. This dependence was made for heat fluxes of 43–48 kW·m^{-2}.

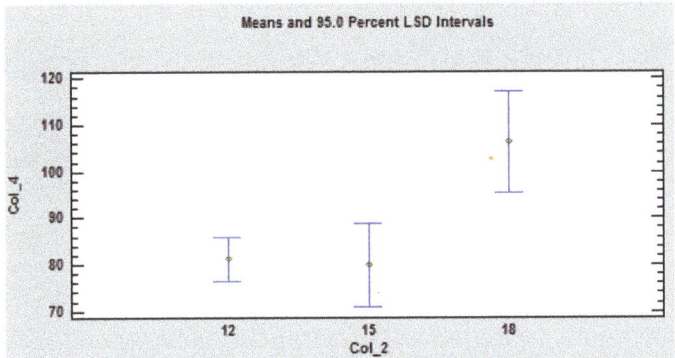

Figure 5. Graphical dependence of the ignition temperature on the thickness of the particleboard. Legend: axis "y"-Col_4 istime to ignition, axis "x"-Col_2 is thickness for heat flux interval <43,48> kW·m^{-2}. The values are statistically significant at $p \leq 0.05$ according to LSD ANOVA.

The ANOVA table (Table 3) decomposes the variability of Col_4 (Time to ignition) into contributions due to various factors. Since Type III sums of squares (the default) have been chosen, the contribution of each factor is measured having removed the effects of all other factors. The *p*-values test the statistical significance of each of the factors. Since one *p*-value is less than 0.05, this factor has a statistically significant effect on Col_4 at the 95.0% confidence level. Dependence of the decrease in time to ignition on the increase in heat flux and the increase in the particleboard thickness was confirmed (Figure 6).

These dependencies are statistically significant (Table 4). The *p*-values test the statistical significance of each of the factors. Since 2 *p*-values are less than 0.05, these factors have a statistically significant effect on Col_3 at the 95.0% confidence level (Figure 7).

Table 4. Analysis of variance for Col_3 time toignition-type III sums of squares.

Source	Sum of Squares	Df	Mean Square	F-Ratio	p-Value
Main Effects					
Col_1 Heat flux	15,376.7	10	1537.67	58.29	0.0000
Col_2 Board thickness	1405.32	2	702.658	26.64	0.0000
Residual	527.564	20	26.3782		
Total	17,309.6	32			

All F-ratios are based on the residual mean square error.

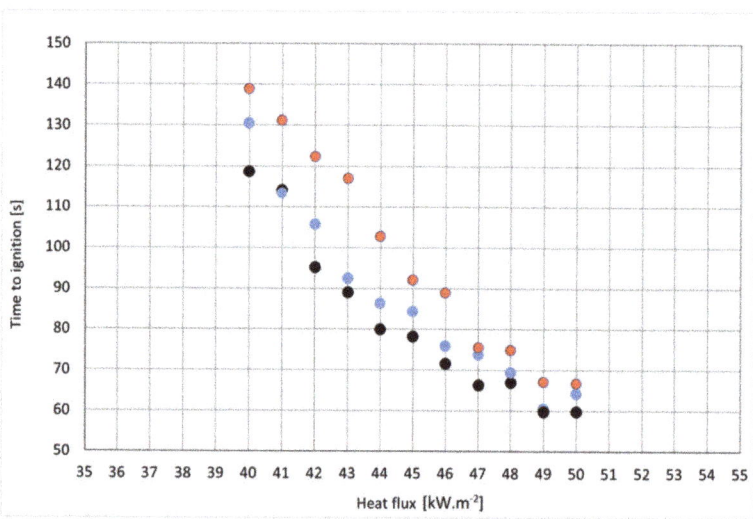

Figure 6. Graphical dependence of the time to ignition on the heat flux and thickness of the particleboard. Legend: black point-12 mm thickness; blue point-15 mm thickness; red point-18 mm thickness.

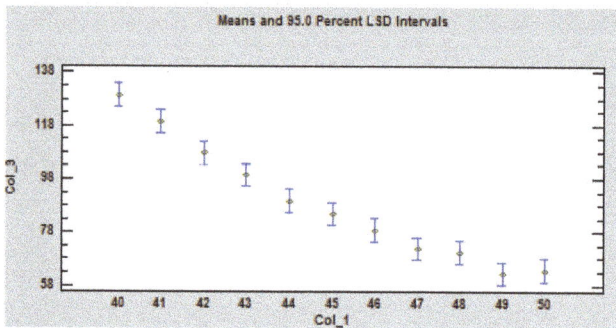

Figure 7. Graphical representation of the statistical evaluation-the influence of the sample thickness and heat flux on the time to ignition under the action of the radiant heat source on the particleboard. Legend: Col_1-heat flux; Col_2-thickness of particleboard samples as the variance of the values shown in blue; Col_3-Time to ignition. The values are statistically significant at $p \leq 0.05$ according to LSD ANOVA.

Figure 8 describes the weight loss in selected particleboard thicknesses when exposed to radiant heat flux (40–50 kW·m^{-2}). As the heat flux density increases, the value of the weight loss in the particleboard samples of the selected thicknesses increases on average by 0.4% (absolute % number) for a change of the heat flux of 1 kW·m^{-2}. The largest weight loss values were recorded in particleboards with a thickness of 12 mm.

The course of the increase in weight loss as a function of increasing radiant heat flux is statistically significant. This statement is based on a statistical analysis of the STATGRAPHICS Software Program version 18/19. The ANOVA method was used (Table 5, Figure 8), where the p-values test the statistical significance of each of the factors. Since 2 p-values are less than 0.05, these factors have a statistically significant effect on Col_3 at the 95.0% confidence level (ANOVA). confidence level (Figure 9).

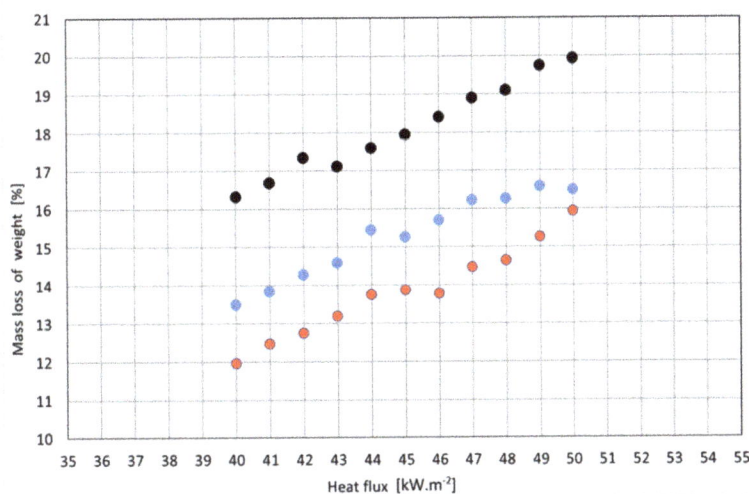

Figure 8. Graphical dependence of average values of weight loss on heat flux and particleboard thickness. Legend: black point-12 mm thickness; blue point-15 mm thickness; red point-18 mm thickness.

Table 5. Analysis of variance for Col_3 (mass loss)-type III sums of squares.

Source	Sum of Squares	Df	Mean Square	F-Ratio	p-Value
Main Effects					
Col_1 Heat flux	40.2861	10	4.028	76.04	0.0000
Col_2 Board thickness	103.654	2	51.826	978.28	0.0000
Residual	1.05955	20	0.0529		
Total	144.999	32			

All F-ratios are based on the residual mean square error.

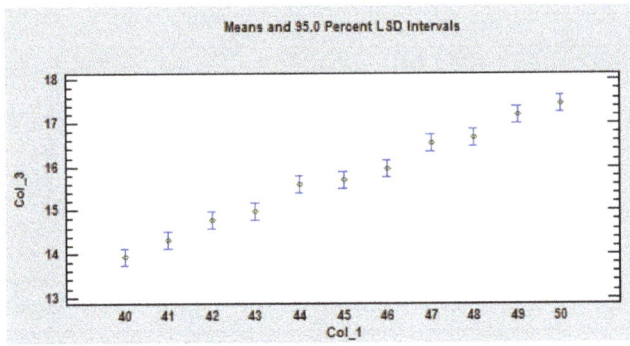

Figure 9. Graphical ANOVA for mass loss (Col_3). Legends: Col_1 is heat flux; Col_2 is thickness of particleboard samples as the variance of the values shown in blue. The values are statistically sig-nificant at $p \leq 0.05$ according to LSD ANOVA.

Valcheva and Savov [49] also presented scientific experiments covering characteristic features and the effect of different thicknesses of boards. The regression models describing the effect of thicknesses on main properties of medium-density particleboard are deduced and analyzed from the output data.

3.2. Monitoring the Effect of the Distance of Board Material from the Ignition Source

The distance of the particleboard from the radiant heat source (Figures 10 and 11) has influence on the time to ignition (Table 5). Particle boards were ignited at higher heat fluxes from 44 kW·m^{-2} at a distance of 40 mm (Figure 10a) particleboards were ignited only if the heat flux was at least 44 kW·m^{-2}; for 60 mm the lowest heat flux for ignition was 48 kW·m^{-2} (Figure 10d). The higher the heat flux, the shorter the time to ignition. Particleboards accumulated sufficient heat to allow the subsequent combustion without the action of an ignition source on the upper surface of the board material.

Figure 10. Measurements of time to ignition and weight loss of particleboards with thickness of 12 mm at (**a**–**c**) 40 mm from the ignition source; (**d**–**f**) 60 mm from the ignition source.

The obtained time to ignition has a decreasing character with a linear dependence. At the distance of 60 mm and heat fluxes of 44 and 46 kW·m^{-2}, ignition did not occur. However, the imaginary line through two points showing the ignition temperature values at the distance of 60 mm shows a different tendency. Ignition temperatures doubled. It can be assumed that with the increasing distance of the radiant heat source from the sample, the increase in time to ignition multiplies geometrically. Time to ignition is significantly dependent on the heat flux and sample thickness (Table 6, Figure 12), as confirmed by multifactor analysis (ANOVA).

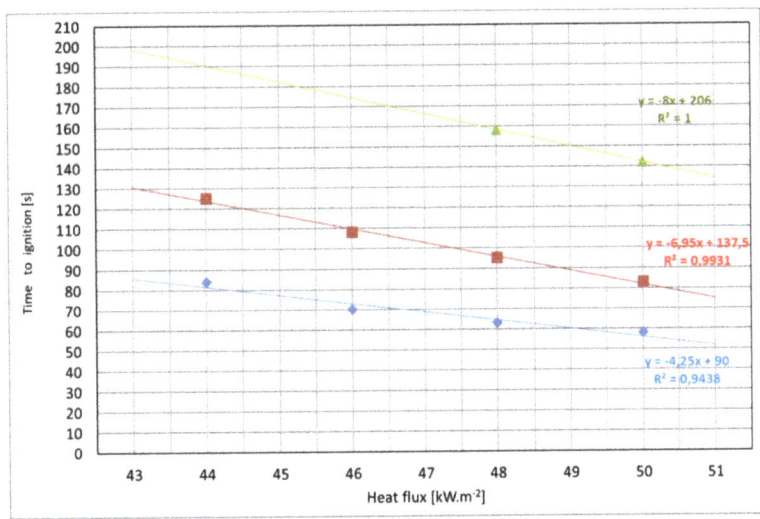

Figure 11. Graphical dependence of the time to ignition on the heat flux (44, 46, 48 and 50 kW·m^{-2}) and the distance of the particleboard with thickness of 12 mm from the ignition source. Legend: blue-20 mm; red-40 mm and green-60 mm.

Table 6. Analysis of variance for Col_6 (time to ignition)-type III sums of squares depending on sample thickness and distance from the source.

Source	Sum of Squares	Df	Mean Square	F-Ratio	p-Value
Main Effects					
Col_4 Distance from the source	47,980.0	2	23,990.0	534.73	0.0000
Col_5 Heat flux	6798.83	3	2266.28	50.51	0.0000
Residual	1974.02	44	44.8642		
Total	52,250.3	49			

All F-ratios are based on the residual mean square error.

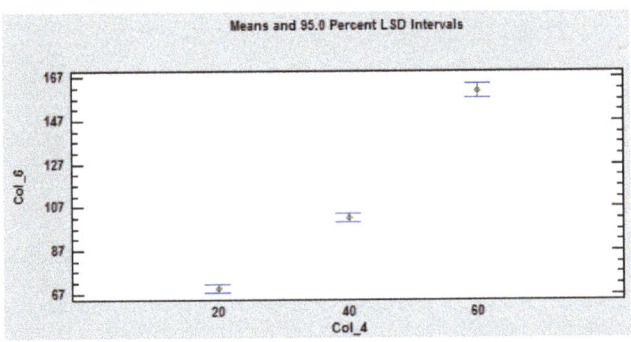

Figure 12. Graphical representation of the statistical evaluation-the influence of the sample position and heat flux on the time to ignition source on the particleboards. Legend: Col_4-position of the sample i.e., distance of the sample from the source; Col_6-time to ignition. The values are statistically significant at $p \leq 0.05$ according to LSD ANOVA.

Statistical analysis demonstrated a significant influence of factors such as distance of the heat-stressed sample (Figure 11) and heat flux (Figure 6) on the time to ignition under the action of radiant heat flux on the surface of particleboards.

Also, statistical analysis ANOVA showed effect of distance on time to ignition (Figure 12). The abbreviations Col_4–distance and Col_6–time to ignition, were used in the graphical representation of statistical results (Figure 12).

4. Conclusions

Based on the conducted experiments, the following results were obtained:

1. It was statistically confirmed that the time to ignition is significantly dependent on the thickness of the particleboard sample and the heat flux value. It was also possible to calculate the thermal inertia based on the measurements. The obtained results of the calculated inertia were very similar to the published values reported by Babrauskas [48];
2. It was confirmed that the weight loss was significantly dependent on the thickness of the particleboard. Selected thicknesses of particleboards which were exposed to radiant heat flux of 40–50 $kW·m^{-2}$ recorded on average by 0.4% (absolute % number) of weight loss with increasing heat flux density (for a change of the heat flux of 1 $kW·m^{-2}$). The largest weight loss values were recorded in particleboards with a thickness of 12 mm;
3. Statistically significant dependence was confirmed by monitoring the time to ignition and the distance of a sample with a thickness of 12 mm from the radiant heat source. At a distance of 60 mm and heat fluxes of 44 and 46 $kW·m^{-2}$, the particleboards with a thickness of 12 mm did not ignite.

Author Contributions: Conceptualization, I.T. and M.I.; methodology, I.T.; software, I.M.; validation, I.T., M.I. and J.H.; formal analysis, I.M.; investigation, I.T.; resources, I.T., J.H. and M.I.; data curation, M.I.; writing—original draft preparation, I.T.; writing—review and editing, I.M.; S.G.; project administration, S.G.; funding acquisition, I.T. All authors have read and agreed to the published version of the manuscript.

Funding: This article was supported by the Cultural and Educational Grant Agency of the Ministry of Education, Science, Research and Sport of the Slovak Republic on the basis of the project KEGA 0014UKF-4/2020 Innovative learning e-modules for safety in dual education and Institute Grant of University of Žilina No 12716.

Institutional Review Board Statement: Not applicable.

Informed Consent Statement: Not applicable.

Data Availability Statement: Not applicable.

Acknowledgments: This article was supported by the Project KEGA 0014UKF-4/2020 Innovative learning e-modules for safety in dual educationand Institute Grant of University of Žilina No 12716.

Conflicts of Interest: The authors declare no conflict of interest.

References

1. Ramage, M.H.; Burridge, H.; Busse-Wicher, M.; Fereday, G.; Reynolds, T.; Shaha, D.U.; Wu, G.; Yu, L.; Fleming, P.; Densley-Tingley, D.; et al. The wood from the trees: The use of timber in construction. *Renew. Sustain. Energy Rev.* **2017**, *68*, 333–359. [CrossRef]
2. Antov, P.; Kristak, L.; Reh, R.; Savov, V.; Papadopoulos, A.N. Eco-Friendly Fiberboard Panels from Recycled Fibers Bonded with Calcium Lignosulfonate. *Polymers* **2021**, *13*, 639. [CrossRef]
3. Langova, N.; Reh, R.; Igaz, R.; Kristak, L.; Hitka, M.; Joscak, P. Construction of Wood-Based Lamella for Increased Load on Seating Furniture. *Forests* **2019**, *10*, 525. [CrossRef]
4. *STN EN 309*; Particleboards. Definition and Classification. Slovak Office of Standards, Metrology and Testing: Bratislava, Slovakia, 2005.
5. Antov, P.; Savov, V.; Krišťák, L.; Réh, R.; Mantanis, G.I. Eco-Friendly, High-Density Fiberboards Bonded with Urea-Formaldehyde and Ammonium Lignosulfonate. *Polymers* **2021**, *13*, 220. [CrossRef] [PubMed]
6. Wang, M.G.; Tian, Y. Furniture of Environmental Protection Materials. *Appl. Mech. Mater.* **2015**, *727*, 197–200. [CrossRef]

7. Hoffman, A.; Muehlnikel, R. Experimental and numerical investigation of fire development in areal fire in a five-storey apartment building. *Fire Mater.* **2011**, *35*, 453–462. [CrossRef]
8. Krišťák, Ľ.; Igaz, R.; Brozman, D.; Réh, R.; Šiagiová, P.; Stebila, J.; Očkajová, A. Life Cycle Assessment of Timber Formwork: Case Study. *Adv. Mater. Res.* **2014**, *1001*, 155–161. [CrossRef]
9. Galla, Š. A analysis of a fire in a storehouse of fibreboards from the fire investigation point of view—Case study. *Košická Bezpeč. Rev.* **2015**, *1*, 26–31. (In Slovak)
10. Kubjatko, T.; Gortz, M.; Macurova, L.; Ballay, M. Synergy of Forensic and Security Engineering in Relation to the Model of Deformation Energies on Vehicles After Traffic Accidents. In Proceedings of the Transport Means—Proceedings of the International Conference, Trakai, Lithuania, 3–5 October 2018; pp. 1342–1348.
11. Iringová, A. Impact of fire protection on the design of energy-efficient and eco-friendly building envelopes in timber structures. In Proceedings of the Fire Protection, Safety and Security, Zvolen, Slovakia, 3–5 May 2017; pp. 79–85. [CrossRef]
12. Očkajová, A.; Kučerka, M. *Materials and Technologies 1. Wood Technology*, 1st ed.; UMB: Banská Bystrica, Slovakia, 2011; ISBN 978-80-557-0262-9. (In Slovak)
13. Warguła, Ł.; Dziechciarz, A.; Kaczmarzyk, P. The assessment of fire risk of non-road mobile wood chopping machines. *J. Res. Appl. Agric. Eng.* **2019**, *64*, 58–64.
14. Sydor, M.; Mirski, R.; Stuper-Szablewska, K.; Rogozinski, T. Efficiency of Machine Sanding of Wood. *Appl. Sci.* **2021**, *11*, 2860. [CrossRef]
15. Jin, C.D.; Li, J.; Zheng, R.X. Thermal and Combustion Characteristics of Binderless Fiberboard. *Adv. Mater. Res.* **2010**, *113*, 1063–1070. [CrossRef]
16. Marková, I.; Hroncova, E.; Tomaskin, J.; Turekova, I. Thermal analysis of granulometry selected wood dust particles. *BioResources* **2018**, *13*, 8041–8060. [CrossRef]
17. Szabová, Z.; Pastier, M.; Harangózo, J.; Chrebet, T. Determination of characteristics predicting the ignition of organic dusts. In *Occupational Safety and Hygiene II: 10th Annual Congress of the Portuguese Society of Occupational Safety and Hygiene on Occupational Safety an Hygiene (SPOSHO) Guimaraes, Portugal, 13–14 February 2014*; CRC Press: Boca Raton, FL, USA, 2014; pp. 143–145. ISBN 978-1-315-77352-0.
18. Turekova, I.; Markova, I. Ignition of Deposited Wood Dust Layer by Selected Sources. *Appl. Sci.* **2020**, *10*, 5779. [CrossRef]
19. Vandličková, M.; Markova, I.; Osvaldová, L.M.; Gašpercová, S.; Svetlík, J. Evaluation of African padauk (*Pterocarpus soyauxii*) explosion dust. *BioResources* **2020**, *15*, 401–414.
20. Lišková, Z.; Olbřímek, J. Comparison of Requirements in Slovak and Selected Foreign Legislation on the Issue of Safe Distance of the Wooden Building Structures from the Flue. *Appl. Mech. Mater.* **2016**, *820*, 396–401. [CrossRef]
21. Pedzik, M.; Auriga, R.; Kristak, L.; Antov, P.; Rogozinski, T. Physical and Mechanical Properties of Particleboard Produced with Addition of Walnut (*Juglans regia* L.) Wood Residues. *Materials* **2022**, *15*, 1280. [CrossRef]
22. Kristak, L.; Kubovsky, I.; Reh, R. New Challenges in Wood and Wood-Based Materials. *Polymers* **2021**, *13*, 2538. [CrossRef]
23. Bekhta, P.; Noshchenko, G.; Reh, R.; Kristak, L.; Sedliacik, J.; Antov, P.; Mirski, R.; Savov, V. Properties of Eco-Friendly Particleboards Bonded with Lignosulfonate-Urea-Formaldehyde Adhesives and pMDI as a Crosslinker. *Materials* **2021**, *14*, 4875. [CrossRef]
24. Krist'ak, L.; Reh, R. Application of Wood Composites. *Appl. Sci.* **2021**, *11*, 3479. [CrossRef]
25. Baskaran, M.; Azmi, N.A.C.H.; Hashim, R.; Sulaiman, O. Addition of Urea Formaldehyde Made from Oil Palm Trunk Waste. *J. Phys. Sci.* **2017**, *28*, 151–159. [CrossRef]
26. Lee, C.L.; Chin, K.L.; H'ng, P.S.; Chuah, A.L.; Khoo, P.S. Enhanced properties of single-layer particleboard made from oil palm empty fruit bunch fibre with additional water-soluble additives. *BioResources* **2021**, *16*, 6159–6173. [CrossRef]
27. Mirindi, D.; Onchiri, R.O.; Thuo, J. Physico-Mechanical Properties of Particleboards Produced from Macadamia Nutshell and Gum Arabic. *Appl. Sci.* **2021**, *11*, 11138. [CrossRef]
28. Bardak, S.; Nemli, G.; Bardak, T. The quality comparison of particleboards produced from heartwood and sapwood of European larch. *Maderas. Cienc. Y Tecnol.* **2019**, *21*, 511–520. [CrossRef]
29. Grigorieva, L.; Oleinik, P. Recycling Waste Wood of Construction. *Mater. Sci. Forum* **2016**, *871*, 126–131. [CrossRef]
30. Iringová, A.; Vandlíčková, D. Analysis of a Fire in an Apartment of Timber Building Depending on the Ventilation Parameter. *Civ. Environ. Eng.* **2021**, *17*, 549–558. [CrossRef]
31. *Safety Data Sheet Particleboard, Raw un-Sanded*; Bučina DDD: Zvolen, Slovakia, 2019.
32. List of technical data. In *Particleboard Pressed Flat, un-Sanded, without Surface Treatment*; Bučina DDD: Zvolen, Slovakia, 2019.
33. Osvald, A. Evaluation velocity of the facade fire-based wood. In Proceedings of the International Scientific Conference, Rajec, Slovakia, 10 November 2017; Volume 1, pp. 197–215. (In Slovak).
34. *ISO 5657*; Reaction to Fire Tests. Ignitability of Building Products Using Radiant Heat Source. ISO: Geneva, Switzerland, 1997.
35. *STN EN 323*; Wood-Based Panels. Determination of Density. Slovak Office of Standards, Metrology and Testing: Bratislava, Slovakia, 1996.
36. Tureková, I.; Marková, I.; Ivanovičová, M.; Harangozo, J. Experimental Study of Oriented Strand Board Ignition by Radiant Heat Fluxes. *Polymers* **2021**, *13*, 709. [CrossRef]
37. Morozov, M.; Strizhak, P.A. Researches of Advanced Thermal Insulating Materials for Improving the Building Energy Efficiency. *Key Eng. Mater.* **2016**, *683*, 617–625. [CrossRef]

38. Mitrenga, P.; Vandlíčková, M.; Dušková, M. Evaluation of the new fire retardants on wood by proposed testing method. In Proceedings of the International Conference on Engineering Science and Production Management (ESPM), Tatranská Štrba, Slovakia, 16–17 April 2015; pp. 481–485. [CrossRef]
39. MakovickáOsvaldová, L.; Janigová, I.; Rychlý, J. Non-Isothermal Thermogravimetry of Selected Tropical Woods and Their Degradation under Fire Using Cone Calorimetry. *Polymers* **2021**, *13*, 708. [CrossRef]
40. Liu, Z.H.; Zhao, Y.S.; Hu, Y.M. Research on Improved Thermal Inertia Model for Retrieving Soil Moisture. *Appl. Mech. Mater.* **2013**, *295*, 2075–2083. [CrossRef]
41. Schieldge, J.P.; Kahle, A.B.; Alley, R.E.; Gillespie, A.R. Use of thermal-inertia properties for material indetification. *Soc. Photo-Opt. Instrum. Eng.* **1980**, *238*, 350–357.
42. Shulga, G.; Neiberte, B.; Verovkins, A.; Jaunslavietis, J.; Shakels, V.; Vitolina, S.; Sedliačik, J. Eco-Friendly Constituents for Making Wood-Polymer Composites. *Key Eng. Mater.* **2016**, *688*, 122–130. [CrossRef]
43. Östman, B.A.L.; Mikkola, E. European Classes for the Reaction to Fire Performance of Wood Products. *Holz Roh Werkstoff* **2006**, *64*, 327–337. [CrossRef]
44. STN EN 13501-2; Fire Classification of Construction Products and Building Elements. Part 1: Classification Using Data from Reaction to fire tests. European Committe for Standartion: Brussels, Belgium, 2018.
45. STN EN ISO 11925-2; Reaction to Fire Tests—Ignitability of Building Products Subjected to Direct Impingement of Flame—Part 2: Single-Flame Source Test. Slovak Office of Standards, Metrology and Testing: Bratislava, Slovakia, 2021.
46. Valarmathi, T.N.; Palanikumar, K.; Sekar, S. Thrust Force Studies in Drilling of Medium Density Fiberboard Panels. *Adv. Mater. Res.* **2012**, *622*, 1285–1289. [CrossRef]
47. Babrauskas, V. *Ignition Handbook*, 1st ed.; Fire Science Publishers: Issaquah, WA, USA, 2003; ISBN 0-9728111-3-3.
48. Babrauskas, V. Ignition of wood: A Review of the State of the Art. *J. Fire Eng.* **2002**, *12*, 163–189. [CrossRef]
49. Valcheva, L.; Savov, V. The Effect of Thickness of Medium Density Fiberboard Produced of Hardwood Tree Species on their Selected Physical and Mechanical Properties. *Key Eng. Mater.* **2016**, *688*, 115–121. [CrossRef]

Article

Compressive Strength Properties Perpendicular to the Grain of Hollow Glue-Laminated Timber Elements

Nikola Perković *, Jure Barbalić, Vlatka Rajčić and Ivan Duvnjak

Structural Department, Faculty of Civil Engineering, University of Zagreb, 10000 Zagreb, Croatia
* Correspondence: nikola.perkovic@grad.unizg.hr

Abstract: Timber is one of the fundamental materials of human civilization, it is very useful and ecologically acceptable in its natural environment, and it fits very well with modern trends in green construction. The paper presents innovative hollow glued laminated (GL) timber elements intended for log-house construction. Due to the lack of data on the behavior of the hollow timber section in compression perpendicular to the grain, the paper presented involves testing the compression strength of elliptical hollow cross-section glue-laminated timber specimens made of softwood and hardwood, as well as full cross-section glue-laminated softwood timber specimens. The experimental research was carried out on a total of 120 specimens. With the maximal reduction of 26% compared to the full cross-section, regardless of the type of wood and direction of load, the compression strength perpendicular to the grain of hollow specimens decreases by about 55% compared to the full cross-section, with the coefficient $k_{c,90}$ equal to 1.0. For load actions at the edge and the middle of the element, $k_{c,90}$ factors were obtained with a value closer to those obtained for full cross-section, which indicates the same phenomenology, regardless of cross-sectional weakening. At the same time, the factors in the stronger axis are lower by about 10%, and in the weaker axis by about 30% compared to those prescribed by the Eurocode. Experimental research was confirmed by FEM analysis. Comparative finite element analysis was performed in order to provide recommendations for future research and, consequently, to determine the optimal cross-section form of the hollow GL timber element. By removing the holes in the central part of the cross-section, the stress is reduced. The distance of the holes from the edges defines the local cracking. Finally, if the holes are present only in the central part of the element, the behavior of the element is more favorable.

Keywords: timber; compression strength; perpendicular to grain; glulam; innovative; hollow; FEM

Citation: Perković, N.; Barbalić, J.; Rajčić, V.; Duvnjak, I. Compressive Strength Properties Perpendicular to the Grain of Hollow Glue-Laminated Timber Elements. *Polymers* **2022**, *14*, 3403. https://doi.org/10.3390/polym14163403

Academic Editors: Ľuboš Krišťák, Roman Réh and Ivan Kubovský

Received: 18 July 2022
Accepted: 17 August 2022
Published: 19 August 2022

Publisher's Note: MDPI stays neutral with regard to jurisdictional claims in published maps and institutional affiliations.

Copyright: © 2022 by the authors. Licensee MDPI, Basel, Switzerland. This article is an open access article distributed under the terms and conditions of the Creative Commons Attribution (CC BY) license (https://creativecommons.org/licenses/by/4.0/).

1. Introduction

Timber is a renewable, biodegradable, and environmentally friendly material that absorbs carbon dioxide from the atmosphere. During manufacturing, it requires little energy, it opens all kinds of new possibilities during operation in wooden structures, where the hitherto widespread use of concrete and bricks can be replaced. Based on all the above, it can be concluded that the construction industry is increasingly turning to timber as a construction material. Accordingly, there are numerous innovations and the production of factory wooden elements such as multi-layer cross-laminated timber (CLT—cross laminated timber), laminated veneer lumber (LVL—laminated veneer lumber), cross-glued veneer board (plywood), parallel glued " veneer noodles (PSL—parallel strand lumber), parallel glued wood "noodles" (LSL—laminated strand lumber), boards with oriented chipboard (OSB—oriented strand board), parallelly oriented chipboard (PLS—parallel strand board), chipboard boards (particleboard), fiber boards (—HB, MBH, MDF).

A typical log-house (or log-haus, Blockhaus, etc.) system represents a traditional construction system widely used in northern regions as well as in urban regions with a high seismic hazard such as the Mediterranean area [1]. The basic timber wall components vertically stacked one upon another are recognized as very efficient and reliable timber structures. The constructive principle of log-house is represented by the superposition of a series of timber logs, although several adapted construction systems were developed recently by Perkovic et al. [2].

In order to achieve an ecological approach, the idea of assembling a hollow glue-laminated timber wall element from slats, which were the waste product in process of carpentry production, was developed by Croatian company TERSA Ltd. Slats are arranged 20 mm thick and 120 mm in width lamellas. The lamellas are profiled before gluing into a single laminated beam, in such a way that combined creates an ellipse-shaped perforation between lamellas in the cross-section of 120 mm in width and 240 mm in height. This type of cross-section guarantees better behavior during the moistening and drying of wood, as well as better energy characteristics of load-bearing elements.

Although these elements have reliable properties [2], many structural aspects need further recognition. In the design of log-house structures with hollowed elements, one of the most important problems is the proper verification of the stress state at the places of concentrated force inputs, but also the bottom of the wall. Therefore, there is a need to evaluate the compression stresses and deformations perpendicular to the grain for different boundary conditions and cross-section axes orientations of hollow elements. Selected results from testing wall members are presented. The test results were compared with the current approach from Eurocode 5 [3].

1.1. State-of-the-Art

One point of discussion in the scientific community was whether standards should aim to maintain either well-defined basic material properties or reflect typical uses. Europe has opted for the former (the scientific) approach, on the assumption that it would then be possible to calculate the behavior in practical application situations, while US/Canada and Australia/New Zealand have chosen the latter (the technological) [4].

The compression perpendicular to the grain design approach presented in Eurocode 5 [3] is based on experiments by Madsen et al. [5]. Some modifications, as currently valid, were proposed by Blass and Görlacher [6]. According to this model, the load-bearing capacity of the element is obtained from effective contact area A_{ef}, characteristic compressive strength perpendicular to the grain $f_{c,90,k}$, and the factor $k_{c,90}$, which considers the load configuration, the possibility of splitting, and the degree of compressive deformation [3]. The effective contact area A_{ef} should be determined considering an effective contact length parallel to the grain, where the actual contact length, l, at each side is increased by 30 mm.

According to EC5 [3], the value of $k_{c,90}$ should be taken as 1.0, unless the conditions in the following paragraphs apply. In these cases, the higher value of $k_{c,90}$ specified may be taken, with a limiting value of 1.75. For members on continuous supports, provided that $l_1 \geq 2h$ (see Figure 1a), the value of $k_{c,90}$ should be taken as 1.25 for solid softwood timber and 1.5 for glued laminated softwood timber, where h is the depth of the member and l is the contact length. For members on discrete supports, provided that $l_1 \geq 2h$ (see Figure 1b) the value of $k_{c,90}$ should be assumed to be 1.5 for solid softwood timber and 1.75 for glued laminated softwood timber if $l \leq 400$ mm. Leijten et al. [4] pointed out the inconsistencies of the mentioned discontinuities and determined the coefficient $k_{c,90}$ based on empirical results. Finally, they proposed modified expressions for $k_{c,90}$, using the physical model of Van der Put [7].

The compressive strength of wood in the direction perpendicular to the grain, $f_{c,90,k}$, (CSPG) plays an important role and frequently governs the structural design. Obviously, CSPG depends on the type of wood and varies in radial and tangential directions [8–10]. Hoffmeyer et al. [11] concluded that the combined role of tensile and shear stresses perpendicular to the grain occurs in compression specimens of solid as well as glued laminated wood, where, at the design level, the 5% characteristic strength is not significantly different. Gehri [12] presented a study to verify the relationship between compressive strength and wood density, which is particularly evident when comparing healthy wood to rotten or insect-deteriorated wood [13]. Although wood is recognized as a building material due to highly technical-material characteristics in the direction parallel to the grains, it is necessary to note that elongated cells of the wood are stiffer and stronger when loaded along the axis of the cell rather than when loaded across it [14,15]. So, the modulus of elasticity in the direction perpendicular to the grain decreases by 30 times and strength by eight times for softwood and three times for hardwood [16].

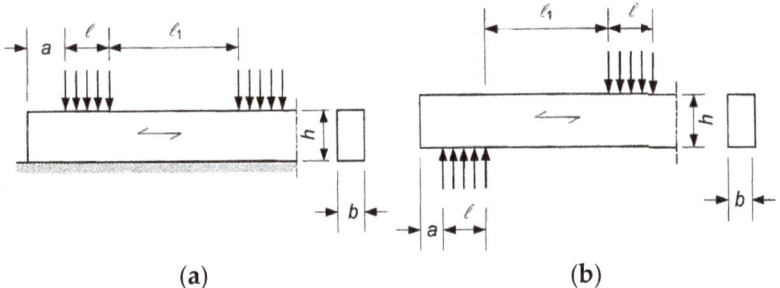

Figure 1. Member on: (**a**) Continuous supports; (**b**) discrete supports.

To determine the real $f_{c,90,k}$ value, the European (CEN) testing standard EN408 prescribes a method in which a block of timber is loaded in uniform compression over its entire surface. On the other hand, the American (ASTM) test standard D143 is based on the approach in which the test piece is a timber block, and the load is applied in the middle through a steel plate, where the test is primarily intended to simulate the behavior of a wood joint resting on a wall or foundation and does not intend to determine a physically correct perpendicular to grain strength [4]. In the absence of any physical model to modify the results and to account for situations deviating from the test set-up, modification factors were established and reported by Kunesh [17]. Madsen et al. [18] also took an interest in the relationship between deformation and compressive strength and recognized shortcomings of the ASTM method. Furthermore, Leijten [19] pointed out that in the Scandinavian countries, the standard characteristic bearing strength for a spruce wood element has double or even triple value than the stress at the proportional limit determined by tests, making values reported in the European standards questionable and very conservative. Further investigation is presented in [20]. Considering the above, the problem of a unified approach to determining the standard strength is obvious.

As a special issue, it should be highlighted the compression strength perpendicular to the grain in cross-laminated timber (CLT), where significant conclusions are given in [21–26].

A standard European test procedure for the determination of CSPG is defined by standard EN408 [27]. This procedure is based on former prescriptions of the fiber stress at the proportional limit, or the stress which causes a 1% deformation, first presented by Kolmann and Côté [28]. Using the test results, on the plot load/deformation (F-Δh) curve, a line (1) parallel to the linear part of the load-displacement and determined by values of 0.1 $F_{c,90, max}$ and 0.4 $F_{c,90, max}$ as intersections with the curve, needs to be defined. Finally, the ultimate load capacity, $F_{c,90, max}$ is defined as the intersection of curve and line (2), which is offset by 1% of the standardized specimen depth h and parallel to the line (1). The force corresponding to the upper limit of the linear segment of the load/displacement (F-Δh) curve is known as the proportional limit $F_{c,90, prop}$ [29].

1.2. Objectives

The main idea for this research came from the doubt about the sufficient bearing capacity of hollow elements to the compression perpendicular to the grain. Although the arrangement of the cavities is designed to ensure a regular force path to the support, the strength properties necessary for design could not be determined just on the wood class. In Figure 2, one can see the constitutive elements-lamella, and finally, the assembled cross-section of the innovative hollow glued laminated timber element.

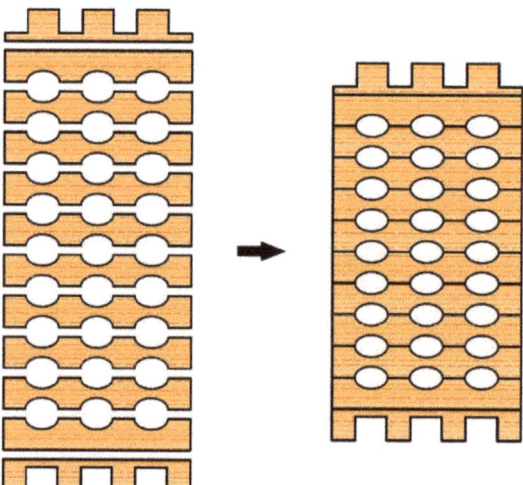

Figure 2. Individual lamellae of the element with elliptical holes and the cross-section of the assembled hollow glued laminated timber element.

The main objective of this research was to investigate the load-carrying capacity in the compression perpendicular to the grain of hollow glue-laminated softwood and hardwood timber element and to compare it with the requirements of the current European standard for full cross-section elements. According to the possible orientation of the elements, and thus of the cross-section orientation, it was necessary to test the specimens in both, strong-axis, and weak-axis directions. In order to compare the compressive strength and factor $k_{c,90}$, taking into account the load configuration, the possibility of splitting, and the degree of compressive deformation, depending on support and load type, several other test set-ups involving loading of the specimens' proportional rectangular surface only at the edge parts, as well as only at central parts, was investigated.

Variant combinations of lamellas allow for different percentages of perforation of the cross-section. In this paper, only the maximally perforated variant with elliptical holes was investigated and compared with normal (full) timber elements. Compared to the full GL elements, the cross-section area was reduced by 26%. Other variants were investigated by finite element modeling, with the goal of finding the optimal layout of the holes regarding stress distribution.

The elements are normally produced in two versions, made of softwood with the predominant use of European fir (*Abies alba*), and hardwood with the predominant use of European hornbeam (*Carpinus betulus*). Both types of hollow elements were tested, as well as full cross-section elements made of softwood, in a total of 120 specimens.

There is evidence that metal-to-wood compression inaccurately reflects typical wood-to-wood compression often present in structural applications [29,30]. However, a common method using metal-on-wood compression [10,31,32] was not applied. Instead, between the metal and the specimen, hardwood elements with prescribed contact surfaces were inserted. Although the digital image correlation (DIC) technique in the testing of structural elements has already been proved [33–35] for a better insight into the redistribution of stress, in this case, it would be useful only for certain types of samples. Therefore, it was not used in this test.

2. Materials and Methods

2.1. Test Setup

Preliminary research was done by Perkovic et al. [2], where different types of cavities were investigated, as well as the layout of the cavities themselves. In addition to elliptical cavities, the behavior of samples with circular holes was also investigated, and the conclusion was that the samples with circular cavities had significantly lower load-carrying capacity and less favorable failure modes. Consequently, the continuation of the research was carried out only with elliptical holes, as well as a modified cavities layout. First, this refers to the first and last lamella, which is shaped differently from the inner lamellas, with the aim of increasing resistance, considering that the highest normal stresses occur at the edges of the cross-section. Furthermore, it was concluded that the more favorable arrangement of the holes is such that they are set in columns, that is, that there is a "web" over which the load can be transferred from the top to the bottom. In addition, the type of adhesive was changed, considering that in the previous investigation [2], the fracture mode occurred in many samples due to the adhesive line, and thus the consistency of the results was disturbed. In this research, a PUR adhesive, *Kleberit 510*, intended for load-bearing timber structures, was used [36].

European standard EN 408 [27] was used for the CSPG evaluation. The production of the test specimens was designed so they match the actual shape of the element, and at the same time meet all the conditions prescribed by the standard. The loaded surfaces were carefully prepared to ensure that they are flat and parallel to each other and perpendicular to the axis of the test specimens. This preparation was performed after conditioning the timber. In the case of glued laminated elements, the test specimens provided for determining the base value of CSPG, are assigned in accordance with EN 408 [27]. In the case of glued laminated elements, height h of 200 mm, minimum width b_{min} of 100 mm, and the surface that is fully loaded $b \times l$ of 25,000 mm^2 is defined, to achieve a volume of 0.01 m^3 for the tested specimens. In addition to the specimens prescribed by the standard, additional specimens were defined and loaded on the edge and in the middle part of the element, in order to determine the distribution of force along the specimens.

The specimens were mounted vertically between the steel plates of the testing machine and the appropriate compression load. Due to the indentation of the end lamellas when the load is acting in a strong-axis direction, additional timber elements were made for this purpose, which on the one end corresponding to the indentation on the sample, and on the other end are flat and thus enable the introduction of loads over the entire surface. Here, the stronger axis represents the axis along which the lamellae are arranged. The length of the gauge, h_0 (approximately 0.6 h), is located centrally in the specimen height and no closer than b/3 of the loaded ends of the specimen, as shown in Figure 3.

The loading equipment used can measure the load to an accuracy of 1% of the load applied to the test specimen or, for loads less than 10% of the maximum load, to an accuracy of 0.1% of the maximum load. The universal testing machine Z600E with a capacity of 600 kN was used for testing. The test specimen has been loaded without eccentricity, which was achieved using spherically seated load heads. According to the standard [27], displacement control was used at different speeds from 3 to 6 mm/min, depending on the material and the position of the sample (loaded in strong-axis or weak-axis direction). The loading rate has been adjusted so that the maximum load $F_{c,90,max,est}$ or $F_{c,90,max}$ was reached within (300 ± 120) s. The test was stopped after reaching the compressive strength of the timber elements. This rate was determined from the results of preliminary tests.

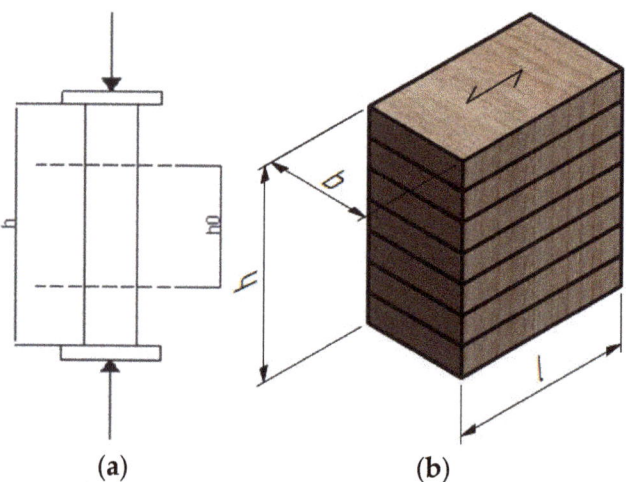

Figure 3. Test setup: (**a**) Load and gauge position; (**b**) specimen dimensions.

The compressive strength $f_{c,90}$ is determined from the equation:

$$f_{c,90} = F_{c,90,max}/bl, \qquad (1)$$

- $f_{c,90}$—compression strength (N/mm^2)
- $F_{c,90,max}$—maximal compression load parallel to the grain (N)
- b—width (mm)
- l—length (mm)

Compressive strength is calculated to an accuracy of 1%.

2.2. Type and Number of Samples

To fully consider the behavior of innovative hollow glue-laminated timber elements, all combinations of specimen positions and loads were investigated. This refers to the positioning and loading of specimens in both the strong- and weak-axis directions, and the position of the force with respect to the boundary conditions as well (loaded on the edge or in the middle). Comparative analysis was performed for full and hollow-timber cross sections made of softwood (fir). Furthermore, the analysis included hollow-timber cross-section elements made of hardwood (hornbeam). To obtain information on the base value of CSPG for full-timber cross sections made of hardwood, samples of solid hardwood were also analyzed, but only for the basic set-up, without examining the influence of the position of the load concerning the boundary conditions.

Before starting the experiment, the wood density and moisture content were measured on specially made cube specimens. A total of 24 samples were made, 12 of each type of wood (see Figure 4 and Table 1).

Table 1. Density of cube specimens.

	Width (mm)			Length (mm)			Height (mm)			Weight (g)			Density (kg/m^3)		
	Avg.	CoV. (%)	St. Dev.	Avg.	CoV. (%)	St. Dev.	Avg.	CoV. (%)	St. Dev.	Avg.	CoV (%)	St. Dev.	Avg.	CoV. (%)	St. Dev.
Softwood M1–M12	119.3	0.19	0.22	120.1	0.30	0.36	119.0	0.22	0.27	657.9	1.52	9.99	385.7	1.42	5.47
Hardwood T1–T12	119.3	0.21	0.25	119.9	0.35	0.42	118.5	0.89	1.06	1337.4	1.35	18.08	789.4	1.60	12.60

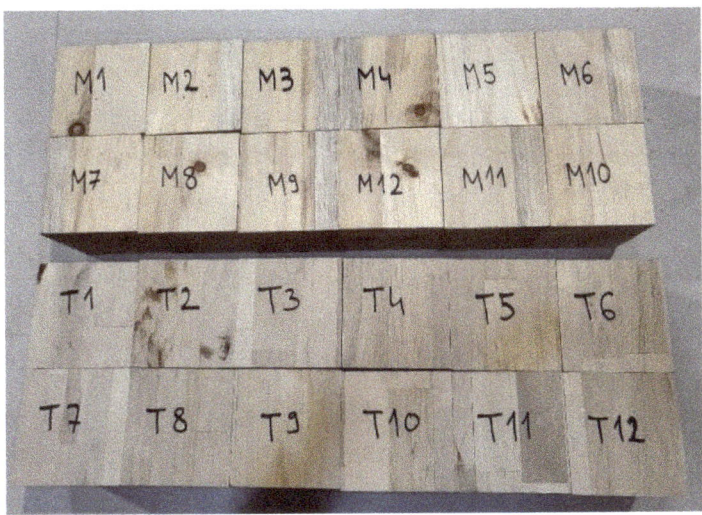

Figure 4. Cube specimens—density measurement.

Regarding the geometry (see Figure 5), the cross-sectional dimensions for all specimens were 120 × 240 mm, while the length was as follows: 105, 209, 400, 440, 520, and 640 mm.

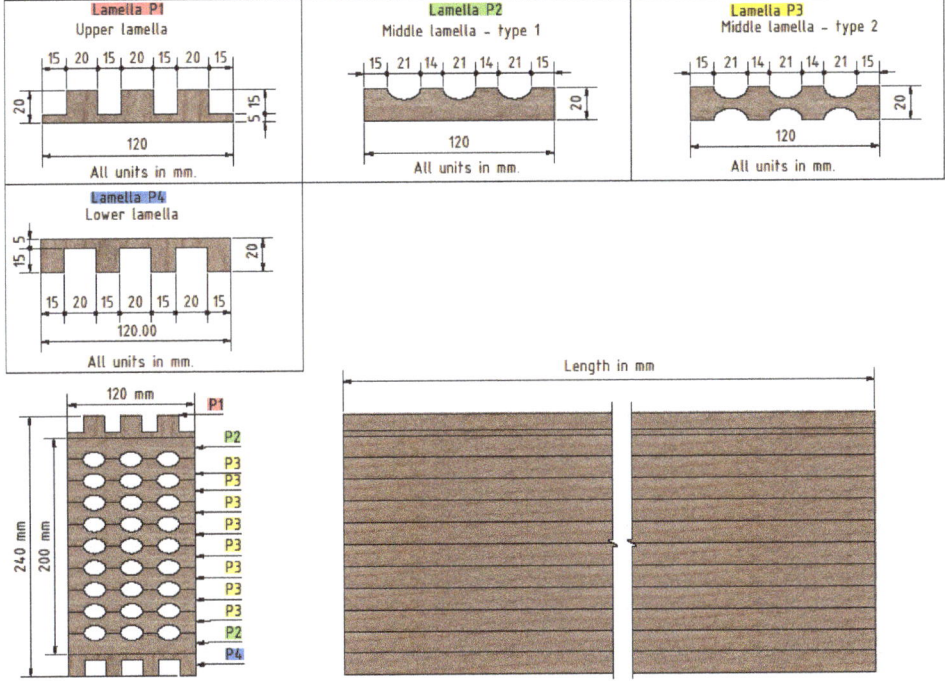

Figure 5. Lamellae and assembled specimen with elliptical cavities.

A total of 120 specimens were made, whereby 6 samples were made for each of the 20 groups. A particular group of specimens is characterized by the type of wood, the type of cross-section, and the length of the specimen, where the length of the specimen represents whether the specimen is loaded in the strong or weak axis direction. According to the schemes in Figure 6, softwood full and hollow timber cross-sections, as well as hardwood hollow timber cross-section specimens were tested (a total of 18 groups). Additionally, according to the schemes in Figure 6a,b, hardwood full-timber cross-section specimens were tested (a total of two groups).

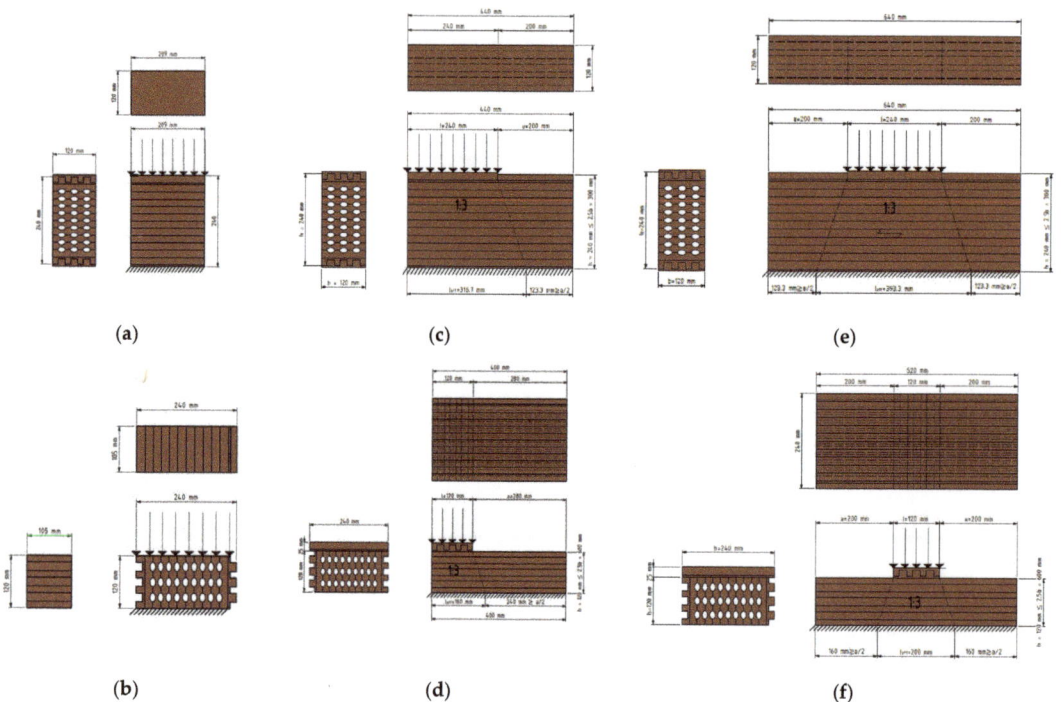

Figure 6. Positioning and loading of specimens, considering the axis direction and boundary conditions: (**a**) Specimen loaded over the entire surface in the strong-axis direction (four groups); (**b**) specimen loaded over the entire surface in the weak-axis direction (four groups); (**c**) specimen loaded on the edge in the strong-axis direction (three groups); (**d**) specimen loaded on the edge in the weak-axis direction (three groups); (**e**) specimen loaded in the middle area in the strong-axis direction (three groups); (**f**) specimen loaded in the middle area in the weak-axis direction (three groups).

While the hollow timber elements can be seen in the image above, the full normal ones are the same, but completely without the elliptical holes. As it can be seen in Figure 5, the first and last lamellae are serrated, so additional elements should be inserted to make the outer surfaces flat; the bottom one due to support, and the upper one due to force input. These additional elements are exactly in the same shape as lamellas P1 and P4 (Figure 5), but they are made of harder timber to avoid local embossing. Steel plates were placed on top of these additional elements, through which the load was applied (Figure 7).

Figure 7. Loading of specimens: (**a**) in strong-axis direction; (**b**) in weak-axis direction.

2.3. FEM Description

The numerical analysis was performed using the Dlubal RFEM [37] software package, more precisely, the RSECTION module. A parameterized input allowed entry of the cross-section dimensions and internal forces in such a way that they depend on certain variables [38]. The objective of the numerical analysis was to make a parametric analysis, the results of which would show where the highest stresses occur. Consequently, the optimal cross-section, which is between the two extremes, the full and hollow cross-section, is determined. For the sake of simplicity and easy comparison, all variant models are loaded with a pressure of 1 N/mm^2. The loading scheme and boundary conditions for FEM can be seen in Figure 8.

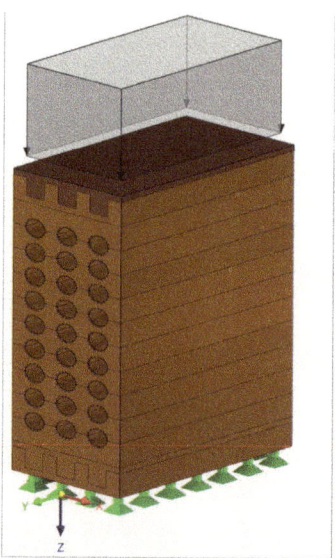

Figure 8. FEM—loading scheme.

Material properties used for modeling timber elements are shown in Table 2.

Table 2. FEM—material properties.

Moduli	Symbol	Value (N/mm^2)
Modulus of elasticity parallel	$E_{0, mean}$	11,500
Modulus of elasticity perpendicular	$E_{90, mean}$	300
Shear modulus	G_{mean}	650
Modulus of elasticity parallel	$E_{0,05}$	9600
Modulus of elasticity perpendicular	$E_{90,05}$	250
Shear modulus	G_{05}	540

3. Results

3.1. Experimental Work

Considering the large number of specimens and tests, it is not reasonable to accommodate all the graphs in this section. Therefore, only the characteristic results in form of load-displacement curves for each group of samples are presented (Figure 9), and the other values are presented in a tabular comparison. The final objective for all groups of specimens was to compare the load-carrying capacity and behavior of innovative hollow GL timber elements with normal GL timber elements. This was primarily referred to the specimens made of softwood, as a raw material more used in practice. Nevertheless, the important objective was to compare the characteristics of hollow softwood and hardwood specimens, as well as to determine the base CSPG value of the hardwood material.

Figure 9 shows the almost linear behavior of the specimens up to the yielding point, after the slope of the curve decreases, the displacement increases without an increase in force, and finally, failure of the timber occurs. Such behavior was common to all types of specimens, however, there are different failure modes for different types of specimens. In case of hollow timber specimens, failure has occurred at the weakest, or thinnest part of the cross-section, between the two elliptical cavities. In the case of normal GL timber specimens, the timber cracked when the compressive strength had been reached.

Furthermore, softwood specimens with elliptical cavities (ME) are expected to have the lowest stiffness, which can be seen from the slope of the curve. If we compare it with normal softwood specimens, without holes (MP), stiffness and strength reached by ME specimens are significantly lower. Moreover, due to the full cross-section, the horizontal part of the curve representing MP specimens indicates greater compression ductility. The physical manifestation of this is the imprinting of load cell into the timber element. Finally, hardwood specimens reached the highest stiffness and failure force, and the cause is higher timber density, i.e., compressive strength. However, an undesirable consequence of this is a brittle fracture. Although the principles leading to the failure of hardwood specimens with elliptical cavities (TE) or without holes (TP) are like those made of softwood, another difference could be seen. The hardwood specimens cracked at the finger joint due to the high strength of the timber, greater than glue. It was especially noticed on specimens loaded at the edge and in the middle.

In addition to the comparison of the load-displacement curves, the failure modes of the specimens were analyzed and compared. The failure modes can be divided into two characteristic groups, depending on whether it represents a hollow- or full-timber cross-section. The main failure mode of the innovative hollow GL timber elements was the timber failure of the area between the holes, in the direction of the applied load Figure 10a,b). Because elliptical cavities are arranged in columns, the load transmission is simple, along the ridges of solid wood. The cracks are mainly a straight line connecting the tops of the arcs of the ellipses. This indicates a proper path of load transmission and that the failure occurred during crushing in the cavity area. In the case of normal GL timber elements, failure occurred when the compressive strength perpendicular to the grains is reached, and fracture followed the stress trajectory (Figure 10d,e). Test of compressive strength perpendicular to the grain for specimens loaded in the direction of the stronger axis (Figure 10a,b,d,e) indicated that the behavior of these specimens have been similar to specimens loaded in the direction of the weaker axis (Figure 10c,f). Again, an almost linear behavior was observed, which turned into a curve, that indicates the yielding of the material. In this case, too, it could be observed that the cracks in hollow timber specimens were predominantly vertical, in the direction of the force, connecting the cavities.

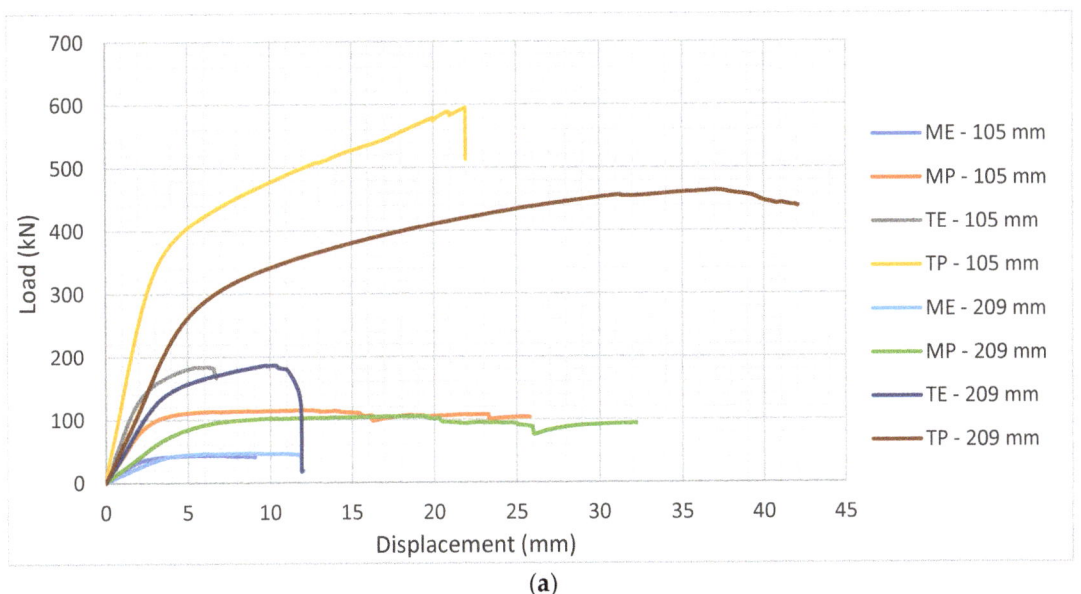

(a)

Figure 9. *Cont.*

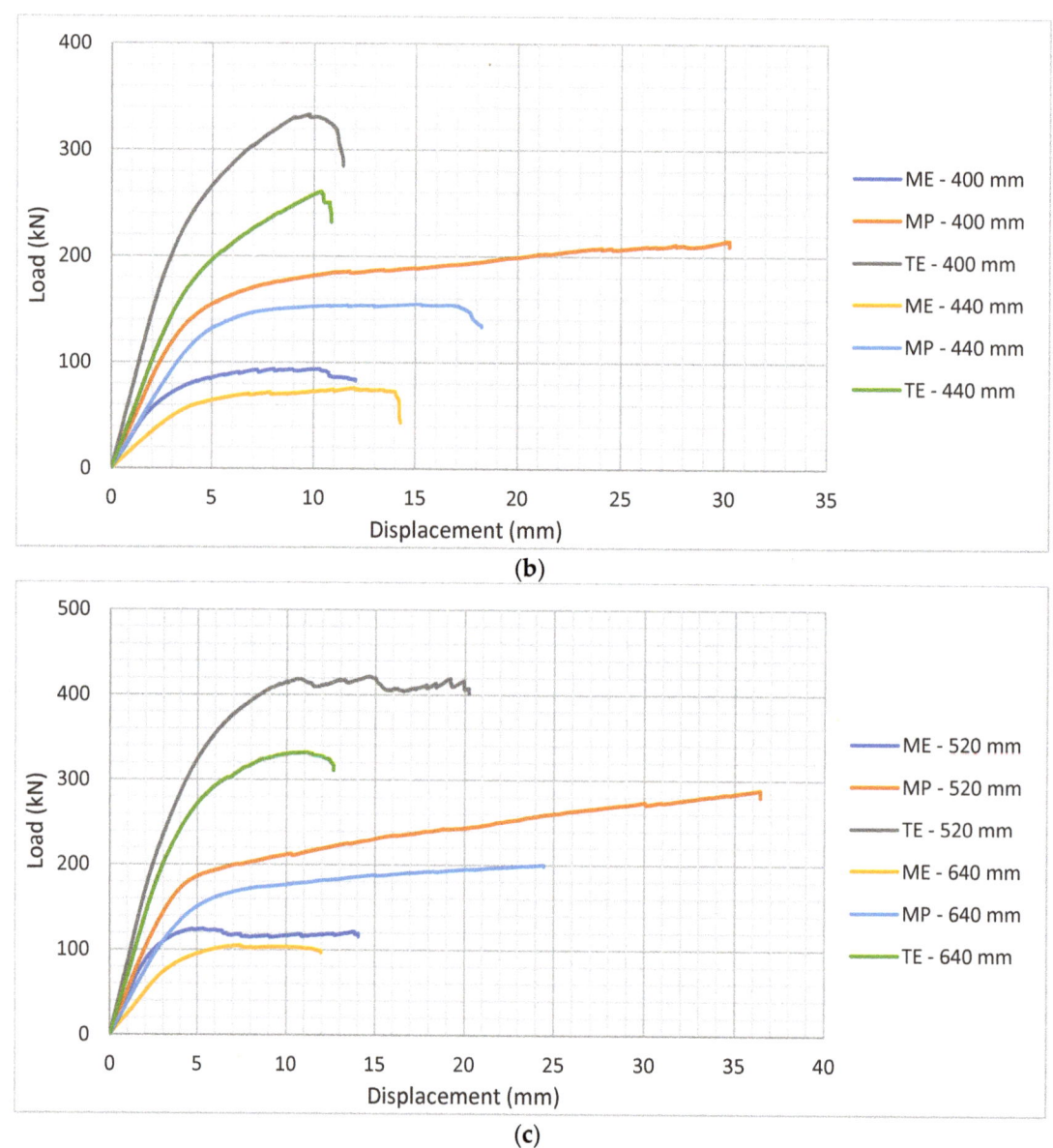

Figure 9. Characteristic load-displacement curves for each group of specimens: (**a**) Standardized specimens; (**b**) specimens loaded on edge of the element; (**c**) specimens loaded in the middle of the element.

Figure 10. Failure modes characteristic for group of specimens: (**a**) ME-209 mm; (**b**) TE-209 mm; (**c**) ME-400 mm; (**d**) MP-209 mm; (**e**) TP-209 mm; (**f**) MP-400 mm.

3.2. FEM Analysis

At the very beginning, it was important to verify the FEM analysis, in order to be able to carry out a parametric analysis for future research. For this purpose, experimentally investigated specimens were analyzed, and the result is shown in Figure 11.

As it can be seen in Table 3, $f_{c,90,k}$ for MP-209 mm was determined to be 4.07 MPa (in the upper corner), and for ME-209 mm, $f_{c,90,k}$ = 1.83 MPa. Those stresses initially appeared in the upper corner and on the perimeter of the holes in the case of hollow GL specimens, which was also confirmed in the FE mode (Figure 11). By evaluating the results of the FEM analysis, the initial σ_z stress for MP-209 mm was 4.067 MPa (Figure 11a) for the value of failure load 106.7 kN (Table 4) and 1.869 MPa (Figure 11b) for MP-209 mm specimen and the value of failure load 47.7 kN (Table 4). Furthermore, in Figure 10a,d, the failure mode and primary cracks are shown. This was also confirmed by the FEM model (check the stress trajectories in Figure 11).

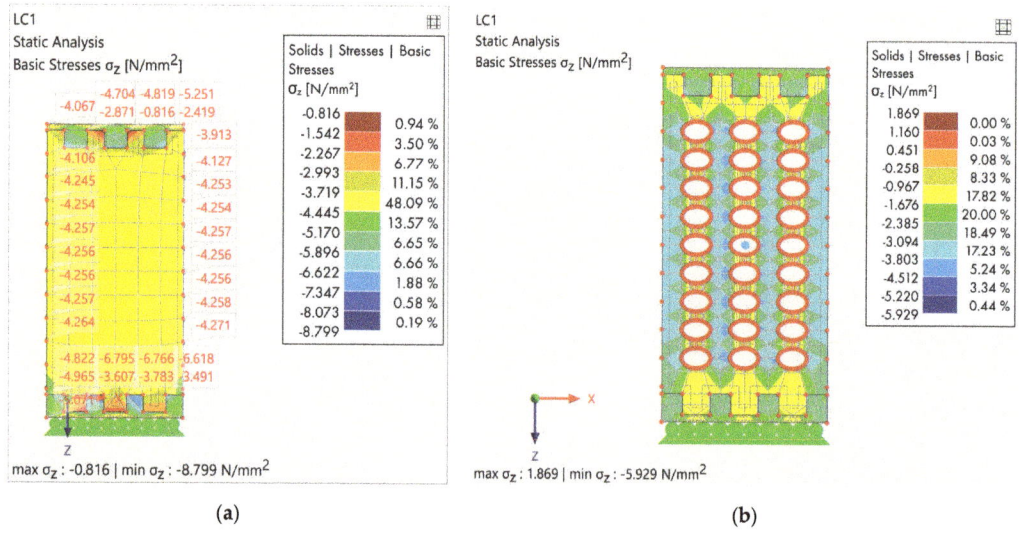

Figure 11. FE verification: (**a**) MP-209 mm; (**b**) ME-209 mm.

Table 3. List of CSPGs for different load modes and boundary conditions with associated $k_{c,90}$ factors.

Type of Cross-Section	MP		ME		TE		TP	
Specimen Length (mm)	$f_{c,90,k}$ (MPa)	$k_{c,90}$	$f_{c,90,k}$ (MPa)	$k_{c,90}$	$f_{c,90,k}$ (MPa)	$k_{c,90}$	$f_{c,90,k}$ (MPa)	$k_{c,90}$
105	4.17	1.00	1.90	1.00	6.75	1.00	15.08	1.00
209	4.07	1.00	1.83	1.00	6.58	1.00	12.96	1.00
400	5.90	1.42	2.67	1.40	9.59	1.42	/	/
440	5.03	1.24	5.90	1.45	8.16	1.24	/	/
520	6.08	1.46	2.95	1.55	10.42	1.54	/	/
640	5.90	1.45	2.78	1.51	9.72	1.48	/	/

ME—softwood hollow, MP—softwood normal, TE—hardwood hollow, TP—hardwood normal.

In the next step, a parametric analysis was made. All results of the parametric analysis were evaluated and visualized in an appealing graphical form (Figures 12–15). As can be seen in the figures, the analysis was carried out step by step, from the model with the highest percentage of cavities to the model without cavities.

Table 4. Failure force—comparison.

Specimen Length (mm)	Type of Cross-Section	Average Failure Force (kN)	CoV. (%)	St. Dev.	F_{max}-Ratio in Relation to ME	F_{max}-Ratio in Relation to TE
105	ME	45.2	9.07	4.1	1.00	0.54
	MP	118.4	7.09	8.4	2.62	1.41
	TE	185.2	8.26	15.3	4.10	2.21
	TP	622.19	6.22	38.7	13.77	7.43
209	ME	47.7	5.03	2.4	1.00	0.54
	MP	106.7	9.09	9.7	2.24	1.22
	TE	184.1	1.90	3.5	3.86	2.10
	TP	453.80	2.29	10.4	9.51	5.17

Table 4. Cont.

Specimen Length (mm)	Type of Cross-Section	Average Failure Force (kN)	CoV. (%)	St. Dev.	F$_{max}$-Ratio in Relation to ME	F$_{max}$-Ratio in Relation to TE
400	ME	87.3	9.51	8.3	1.00	0.31
	MP	219.3	8.76	19.2	2.51	0.78
	TE	320.2	6.09	19.5	3.67	1.15
	TP	/	/	/	/	/
440	ME	75.3	1.73	1.3	1.00	0.39
	MP	157.4	8.64	13.6	2.09	0.82
	TE	255.3	10.77	27.5	3.39	1.33
	TP	/	/	/	/	/
520	ME	126.5	5.06	6.4	1.00	0.24
	MP	282.4	5.24	14.8	2.23	0.54
	TE	413.7	3.50	14.5	3.27	0.79
	TP	/	/	/	/	/
640	ME	106.9	8.70	9.3	1.00	0.31
	MP	196.0	4.23	8.3	1.83	0.57
	TE	323.8	6.61	21.4	3.03	0.94
	TP	/	/	/	/	/

ME—softwood hollow, MP—softwood normal, TE—hardwood hollow, TP—hardwood normal.

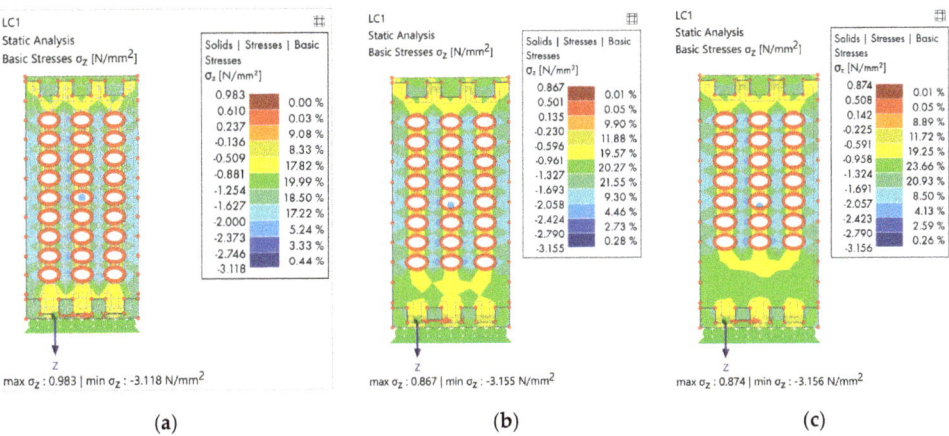

Figure 12. Results of FEM analysis—stress: (**a**) Completely perforated timber element; (**b**) the first lamella without cavities; (**c**) the second lamella without cavities.

The analysis was made with several groups of models, and all models refer to softwood. The goal is to make a comparative analysis of the stress due to the geometrical distribution of the cavities. The first group of models refers to models where cavities were gradually removed in rows, starting from the bottom lamella (Figure 12).

The next group of models (Figure 13) is reflected in the variability of the holes in alternating rows.

The third group of models (Figure 14) is shown in the variation of the columns of cavities.

Finally, the last group of models (Figure 15) refers to specimens that are the opposite of the previous group, i.e., the holes only present in the central area, while the final model is a normal GL timber specimen, without holes.

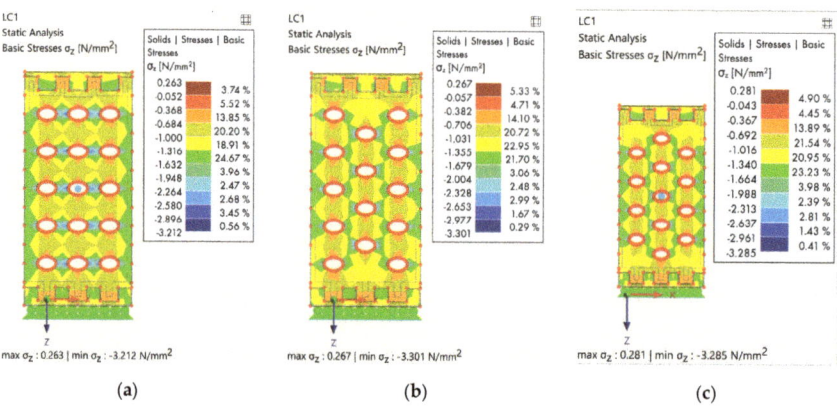

Figure 13. Results of FEM analysis—stress: (**a**) Each subsequent lamella without cavities; (**b**) alternating arrangement of holes, type 1; (**c**) alternating arrangement of holes, type 2.

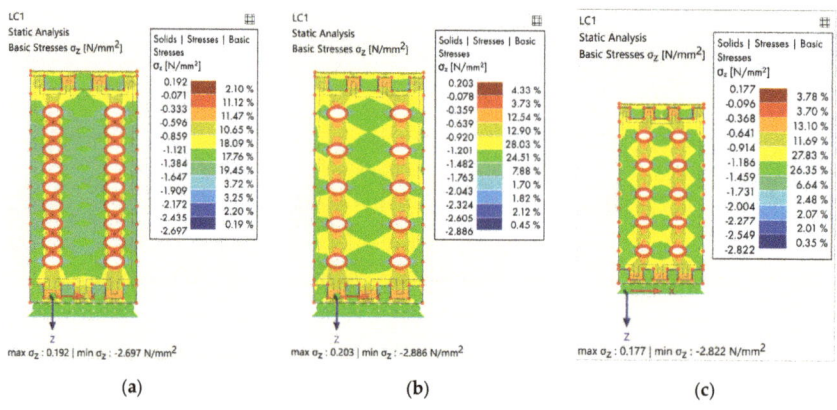

Figure 14. Results of FEM analysis—stress: (**a**) The central part without cavities; (**b**) the central part and each subsequent lamella without cavities; (**c**) the outer part and each subsequent lamella without holes.

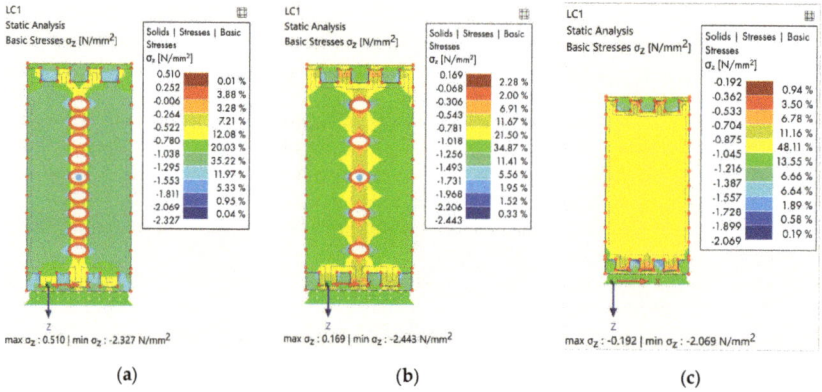

Figure 15. Results of FEM analysis—stress: (**a**) The central part with cavities; (**b**) the central part and each subsequent lamella with cavities; (**c**) normal—without cavities.

4. Discussion

4.1. Experimental Work

The ratio of failure forces of all specimens is shown in Table 4. As expected, the groups of hardwood hollow timber specimens showed the greatest ultimate force followed by groups of softwood specimens without holes, and finally groups of softwood hollow timber specimens. The hollowed hardwood timber specimens showed a higher load capacity even than the softwood specimens without holes due to approximately three times higher CSPG. The reason that groups of softwood hollow timber specimens had the lowest load capacity lies in the small distance between the cavities, that is, the small thickness of the solid wood between the cavities, which would transmit the load from the top to the bottom of the sample.

Furthermore, compressive strengths perpendicular to the grain for each group of specimens, with associated $k_{c,90}$ factors are given in Table 3.

4.2. FEM Analysis

The first model (Figure 12a) in the first group shows the stress concentration where the maximum stress occurs in the area between the holes and is 3.118 N/mm². When the first row of cavities had been removed (Figure 12b), it minimally affected the stress distribution; however, although at the bottom of the sample the stress was lower, the maximum stress was similar to the first one (3.155 N/mm²). By removing the holes on the next lamella (Figure 12c), the stress distribution was more favorable both locally and globally, especially at the bottom part of the specimen.

The second group of FE models shows the variability of the holes in alternating rows. Figure 13a shows the stress distribution when the cavities on each successive lamella were removed. The stress was less on the vertical timber areas between the cavities, but that is why the stress was slightly higher on the timber horizontal areas between the cavities compared to the first group of models. The second (Figure 13b) and third models in this group were very similar, although the third model (Figure 13c) in this group was slightly better due to the absence of cavities on the outer parts. The maximum stress for the first model in this group was 3.212 N/mm², while the stress in the second model was 3.301 N/mm².

The next group is shown in Figure 14a, where the cavities are left out in the middle cross-section area, and this was reflected in the stress distribution. The maximum stress was lower compared to the previous models (2.697 N/mm²) and, the stress distribution is more favorable because there are no stress concentrations in the central part. When the holes in each subsequent lamella are omitted (Figure 14b,c), the global stress distribution was more favorable, but due to the smaller number of holes, slightly higher stress occurred at the edges of the ellipse, caused by the flow of the principal stresses.

For the last group, it can be observed that the most favorable specimen in terms of stress was the just-mentioned normal specimen (Figure 15c), while the second specimen (Figure 15b) in this group showed better behavior compared to the first (Figure 15a), and the reason for this was the lower perforation of the specimen and, accordingly, less stress concentration.

5. Conclusions

From the presented study, it can be concluded that the CSPG of softwood, for a full laminated cross-section loaded in the direction of the stronger axis, is equal to 4.07 MPa and the CSPG of hardwood is equal to 12.96 MPa, with the coefficient $k_{c,90}$ equal to 1.0. For load action at the edge of the element, the factor $k_{c,90}$ = 1.24 was obtained, as lower by 20% than the value prescribed in Eurocode 5 [3] of 1.55. For the load action at the middle of the element, the factor $k_{c,90}$ = 1.45 was obtained, which is lower by 12% than the value prescribed in [3] of 1.66. The CSPG of softwood, for a hollowed laminated cross-section loaded in the direction of the stronger axis, decreases by about 55% compared to the full cross-section, with a value of 1.83 MPa, and for hardwood, it decreases by about 50%, to a value of 6.58 MPa, with the coefficient $k_{c,90}$ equal to 1.0. For load actions at the edge and the middle of the element, $k_{c,90}$ factors were obtained with a value closer to those obtained for full cross-section, which indicates the same phenomenology, regardless of cross-sectional weakening.

In addition, it can be concluded that the CSPG of softwood, for a full laminated cross-section loaded in the direction of the weaker axis, is equal to 4.17 MPa and the CSPG of hardwood is equal to 15.08 MPa, with a coefficient of $k_{c,90}$ equal to 1.0. For load action at the edge of the element, the factor $k_{c,90}$ = 1.42 was obtained, as lower for 30% than the value prescribed in [3] of 2.07. For load action at the middle of the element, the factor $k_{c,90}$ = 1.46 was obtained, which is lower by 35% than the value prescribed in [3] of 2.21. The CSPG of softwood, for a hollowed laminated cross-section loaded in the direction of the weaker axis, decreases by about 55% compared to the full cross-section, with a value of 1.90 MPa, and for hardwood, it decreases by about 55%, to a value of 6.75 MPa, with the coefficient $k_{c,90}$ equal to 1.0. It can be concluded that the properties are similar to the situation when the cross-section is loaded in the direction of the stronger axis.

Moreover, it can be concluded that the degree of hollowness is proportional to the CSPG regardless of the type of wood. Moreover, the weakening does not affect the transfer of force with respect to the boundary conditions and position of the load, i.e., the $k_{c,90}$ factors are approximately similar for hollowed and full cross-sections. However, in order to better understand it, it is necessary to study the stress distribution and force path in more detail using the DIC measurement method. As mentioned in the introduction, the factor $k_{c,90}$ is difficult to determine unequivocally for different boundary conditions. This research presented that the values given in European standards [3] still cannot be applied uniformly. So, further research is necessary for the correction of factors regarding the type of wood, type of section, etc.

Finally, FE analysis confirmed the experimental work. The results of the comparative numerical analysis indicated how the arrangement and layout of the cavities affect the stress distribution. It has been proven that by removing certain rows or columns of holes, we can favorably influence stress distribution. If the first lamellae are full, without cavities, this has a positive effect on the overall behavior of the element, that is, it will crack at a higher force. By avoiding cavities in every subsequent lamella, stress concentration is reduced and the area between the two cavities is increased, which ultimately results in a higher load capacity of the element. If the central part of the cross-section is without holes, the stress is reduced, but special attention should be paid to the distance of the holes from the edges so that local cracking does not occur. In the end, if the cavities are present only in the central part of the element, the behavior of the element is more favorable, but the question arises as to how meaningful it is to make such patterns and how many advantages there are compared to the specimen without cavities, considering that the perforation of this kind of specimen is much lower compared to the previously studied samples. In the continuation of the research, it is planned to conduct an experimental investigation of variant solutions for innovative hollow glued laminated timber elements.

6. Patents

The producer of the timber elements, a company (Tersa Ltd from Croatia), is in the application process for an intellectual property patent so that this product and system are protected.

Author Contributions: Conceptualization, N.P., J.B., I.D., and V.R., methodology, N.P., J.B., and V.R.; software, N.P.; validation, N.P.; formal analysis, N.P.; investigation, N.P. and J.B.; resources, V.R.; data curation, N.P. and J.B.; writing—original draft preparation, N.P. and J.B.; writing—review and editing, N.P., J.B., I.D., and V.R.; visualization, N.P. and J.B.; supervision, V.R. and I.D.; project administration, J.B.; funding acquisition, V.R. All authors have read and agreed to the published version of the manuscript.

Funding: This research was funded by European structural and investment funds, Operational programme competitiveness, and cohesion, The Call "Increasing the Development of New Products and Services Arising from Research and Development Activities-Phase II" with The Call Reference Number "KK.01.2.1.02" from December 2019, that is, through The Project "Development of a modular house using innovative timber elements" with Grant Number "KK.01.2.1.02.0060".

Institutional Review Board Statement: Not applicable.

Informed Consent Statement: Not applicable.

Data Availability Statement: Data available on request due to restrictions, e.g., privacy or ethics. The data presented in this study are available on request from the corresponding author.

Acknowledgments: All tested specimens were produced by TERSA Ltd. from Črnkovci, Croatia, directed by Vlado Šakić, an author of the idea for innovative hollow elements. All tests were done with the help of the staff of the Structural Testing Laboratory, Faculty of Civil Engineering, University of Zagreb, especially Marin Bodulušić. The authors would like to thank the mentioned parties for their fruitful cooperation.

Conflicts of Interest: The authors declare no conflict of interest. The funders had no role in the design of the study; in the collection, analyses, or interpretation of data; in the writing of the manuscript, or in the decision to publish the results.

References

1. Sciomenta, M.; Bedon, C.; Fragiacomo, M.; Luongo, A. Shear Performance Assessment of Timber Log-House Walls under In-Plane Lateral Loads via Numerical and Analytical Modelling. *Buildings* **2018**, *8*, 99. [CrossRef]
2. Perković, N.; Rajčić, V.; Pranjić, M. Behavioral Assessment and Evaluation of Innovative Hollow Glue-Laminated Timber Elements. *Materials* **2021**, *14*, 6911. [CrossRef] [PubMed]
3. EN 1995-1-1:2004/A2:2014; Eurocode 5: Design of Timber Structures—Part 1-1: General—Common Rules and Rules for Buildings. CEN: Brussels, Belgium, 2004.
4. Leijten, A.J.M.; Larsen, H.J.; van der Put, T.A.C.M. Structural Design for Compression Strength Perpendicular to the Grain of Timber Beams. *Constr. Build. Mater.* **2010**, *24*, 252–257. [CrossRef]
5. Madsen, B.; Leijten, A.J.M.; Gehri, E.; Mischler, A.; Jorissen, A.J.M. *Behaviour of Timber Connections*; Vancouver Ltd.: Vancouver, BC, Canada, 2000.
6. Blass, H.J.; Görlacher, I.R. Compression Perpendicular to the Grain. In Proceedings of the 8th World Conference of Timber Engineering, Lahti, Finland, 14–17 June 2004; Finnish Association of Civil Engineers RIL: Lathi, Finland, 2004; pp. 435–440.
7. Roh Werkst, H.; van der Put, T.A.C.M. Derivation of the Bearing Strength Perpendicular to the Grain of Locally Loaded Timber Blocks. *Holz Roh-Werkst.* **2008**, *66*, 409–417.
8. Pozgaj, A.; Chovanec, D.; Kurjatko, S.; Babiak, M. *Struktúra a Vlastnosti Dreva*; Priroda: Bratislava, Slovakia, 1993.
9. Tabarsa, T.; Chui, Y.H. Characterizing Mircroscopic Behaviour of Wood in Radial Compression. Part 2: Effect of Species and Loading Direction. *Wood Fiber Sci.* **2001**, *33*, 223–232.
10. Kretschmann, D.E. *Influence of Juvenile Wood Content on Shear Parallel, Compression, and Tension Transverse to Grain Strength and Mode I Fracture Toughness for Loblolly Pine*; Research Paper FPL-RP-647; U.S. Department of Agriculture, Forest Service, Forest Products Laboratory: Madison, WI, USA, 2008; Volume 647, p. 27.
11. Hoffmeyer, P.; Damkilde, L.; Pedersen, T.N. Structural Timber and Glulam in Compression Perpendicular to Grain. *Holz Roh-Werkst.* **2000**, *58*, 73–80. [CrossRef]
12. Gehri, E. Timber in Compression Perpendicular to the Grain. In Proceedings of the International Conference of IUFRO S5.02 Timber Engineering, Copenhagen, Denmark, 18–20 June 1997; BKM: Copenhagen, Denmark, 1997; pp. 16–17.
13. Verbist, M.; Branco, J.M.; Nunes, L. Characterization of the Mechanical Performance in Compression Perpendicular to the Grain of Insect-Deteriorated Timber. *Buildings* **2020**, *10*, 14. [CrossRef]
14. Bodig, J.; Jayne, B.A. *Mechanics of Wood and Wood Composites*; Van Nostrand Reinhold Company Inc.: New York, NY, USA, 1982.
15. Dinwoodie, J.M. *Timber: Its Nature and Behavior*; E & FN Spon: New York, NY, USA, 2000.
16. EN 384:2016; Structural Timber-Determination of Characteristic Values of Mechanical Properties and Density. CEN: Brussels, Belgium, 2016.
17. Kunesh, R.H. Strength and Elastic Properties of Wood in Transverse Compression. *For. Prod. J.* **1967**, *18*, 65–72.
18. Madsen, B.; Hooley, R.F.; Hall, C.P. A Design Method for Bearing Stresses in Wood. *Can. J. Civ. Eng.* **2011**, *9*, 338–349. [CrossRef]
19. Leijten, A.J.M. The Bearing Strength Capacity Prediction by Eurocode 5 and Other Potential Design Code Models. In Proceedings of the World Conference on Timber Engineering (WCTE 2016), Vienna, Austria, 22–25 August 2016; Vienna University of Technology: Vienna, Austria, 2016; pp. 1–8.
20. Leijten, A.J.M.; Jorissen, A.J.M.; de Leijer, B.J.C. The Local Bearing Capacity Perpendicular to Grain of Structural Timber Elements. *Constr. Build. Mater.* **2012**, *27*, 54–59. [CrossRef]

21. Serrano, E.; Enquist, B. Compression Strength Perpendicular to Grain in Cross-Laminated Timber (CLT). In Proceedings of the World Conference on Timber Engineering (WCTE 2010), Trentino, Italy, 20–24 June 2010; Trees and Timber Institute, National Research Council: Trentino, Italy, 2010; pp. 1–8.
22. Brandner, R. Cross Laminated Timber (CLT) in Compression Perpendicular to Plane: Testing, Properties, Design and Recommendations for Harmonizing Design Provisions for Structural Timber Products. *Eng. Struct.* **2018**, *171*, 944–960. [CrossRef]
23. Gasparri, E.; Lam, F.; Liu, Y. Compression Perpendicular to Grain Behavior for the Design of a Prefabricated CLT Façade Horizontal Joint. In Proceedings of the WCTE 2016—World Conference on Timber Engineering, Vienna, Austria, 22–25 August 2016; Vienna University of Technology: Vienna, Austria, 2016.
24. Brandner, R.; Schickhofer, G. Properties of Cross Laminated Timber (CLT) in Compression Perpendicular to Grain. In Proceedings of the International Network on Timber Engineering Research (INTER), Bath, UK, 1–4 September 2014; KIT: Karlsruhe, Germany, 2014; pp. 1–13.
25. Franzoni, L.; Lebée, A.; Lyon, F.; Forêt, G. Elastic Behavior of Cross Laminated Timber and Timber Panels with Regular Gaps: Thick-Plate Modeling and Experimental Validation. *Eng. Struct.* **2017**, *141*, 402–416. [CrossRef]
26. Tian, Z.; Gong, Y.; Xu, J.; Li, M.; Wang, Z.; Ren, H. Predicting the Average Compression Strength of CLT by Using the Average Density or Compressive Strength of Lamina. *Forests* **2022**, *13*, 591. [CrossRef]
27. *EN 408:2012*; Timber Structures—Structural Timber and Glued Laminated Timber—Determination of Some Physical and Mechanical Properties. CEN: Brussels, Belgium, 2012. Available online: https://standards.iteh.ai/catalog/standards/cen/6ffae6c9-5eaf-4c84-8bf3-5132cbfc563c/en-408-2010a1-2012 (accessed on 2 July 2022).
28. Kollmann, F.F.P.; Côté, W.A. *Principles of Wood Science and Technology*; Springer: Berlin/Heidelberg, Germany, 1968.
29. Brol, J.; Kubica, J.; Weglorz, M. The Problem of Compressive Strength in Direction Perpendicular to the Grains on Example of Tests of the Load-Bearing Capacity of the Continuously Supported Timber-Frame Sill Plates. *Materials* **2020**, *13*, 1160. [CrossRef] [PubMed]
30. Basta, C.T.; Gupta, R.; Leichti, R.J.; Sinha, A. Characterizing Perpendicular-to-Grain Compression (C) Behavior in Wood Construction. *Holzforschung* **2011**, *65*, 845–853. [CrossRef]
31. Kathem, A.; Tajdar, H.; Arman, K. *Compression Perpendicular to Grain in Timber—Bearing Strength for a Sill Plate*; Linnaeus University: Växjö, Sweden, 2014.
32. Leijten, A.J.M.; Jorissen, A.J.M. Global Test Standards and Code Design Rules for Compressive Strength Perpendicular to Grain. In Proceedings of the World Conference of Timber Engineering (WCTE 2010), Trento, Italy, 20–24 June 2010; CNR-IVALSA: Trento, Italy, 2010.
33. Stoilov, G.; Pashkouleva, D.; Kavardzhikov, V. Digital Image Correlation for Monitoring of Timber Walls. *Int. J. NDT Days* **2019**, *II*, 417–422.
34. Speranzini, E.; Marsili, R.; Moretti, M.; Rossi, G. Image Analysis Technique for Material Behavior Evaluation in Civil Structures. *Materials* **2017**, *10*, 770. [CrossRef] [PubMed]
35. Hu, M.; Johansson, M.; Olsson, A.; Oscarsson, J.; Enquist, B. Local Variation of Modulus of Elasticity in Timber Determined on the Basis of Non-Contact Deformation Measurement and Scanned Fibre Orientation. *Eur. J. Wood Wood Prod.* **2014**, *73*, 17–27. [CrossRef]
36. Kleiberit. Deutsches Institut fur Bautechnik. Available online: https://www.kleiberit.com/fileadmin/Content/Documents/EN/Info_Sheets/510_Tragender_Holzbau_GB_US.pdf (accessed on 13 July 2022).
37. Structural Analysis Software. Dlubal. Available online: https://www.dlubal.com/en (accessed on 13 July 2022).
38. RSECTION: Section Properties and Stress Analysis. Dlubal Software. Available online: https://www.dlubal.com/en/products/cross-section-properties-software/rsection (accessed on 13 July 2022).

Article

Withdrawal Performance of Nails and Screws in Cross-Laminated Timber (CLT) Made of Poplar (*Populus alba*) and Fir (*Abies alba*)

Farshid Abdoli [1,*], Maria Rashidi [2,*], Akbar Rostampour-Haftkhani [3,*], Mohammad Layeghi [4] and Ghanbar Ebrahimi [4]

[1] Department of Wood and Paper Science, Faculty of Natural Resources, University of Tarbiat Modares, Tehran 14117-13116, Iran
[2] Centre for Infrastructure Engineering, Western Sydney University, Sydney 2000, Australia
[3] Wood Science and Technology, Department of Natural Resources, Faculty of Agriculture and Natural Resources, University of Mohaghegh Ardabili, Ardabil 56199-11367, Iran
[4] Department of Wood and Paper Science and Technology, College of Natural Resources, University of Tehran, Karaj 14179-35840, Iran; mlayeghi@ut.ac.ir (M.L.); gh.ebrahimi@ut.ac.ir (G.E.)
* Correspondence: abdolifarshid@gmail.com (F.A.); m.rashidi@westernsydney.edu.au (M.R.); arostampour@uma.ac.ir (A.R.-H.)

Abstract: Cross-laminated timber (CLT) can be used as an element in various parts of timber structures, such as bridges. Fast-growing hardwood species, like poplar, are useful in regions where there is a lack of wood resources. In this study, the withdrawal resistance of nine types of conventional fasteners (stainless-steel nails, concrete nails and screws, drywall screws, three types of partially and fully threaded wood screws, and two types of lag screws), with three loading directions (parallel to the grain, perpendicular to the surface, and tangential), and two layer arrangements (0-90-0° and 0-45-0°) in 3-ply CLTs made of poplar as a fast-growing species and fir as a common species in manufacturing of CLT was investigated. Lag screws (10 mm) displayed the highest withdrawal resistance (145.77 N), whereas steel nails had the lowest (13.13 N), according to the main effect analysis. Furthermore, fasteners loaded perpendicular to the grain (perpendicular to the surface and tangential) had higher withdrawal resistance than those loaded parallel to the grain (edge). In terms of the layer arrangement, fasteners in CLTs manufactured from poplar wood (0-45-0°) had the greatest withdrawal resistance, followed by CLTs manufactured from poplar wood in the (0-90-0°) arrangement, and finally, those made from fir wood in the (0-90-0°) arrangement. The fastener type had the most significant impact on the withdrawal resistance, so changing the fastener type from nails to screws increased it by about 5–11 times, which is consistent with other studies. The results showed that poplar, a fast-growth species, is a proper wood for manufacturing CLTs in terms of fastener withdrawal performance.

Keywords: cross-laminated timber; withdrawal resistance; nails; screws; loading direction; layer arrangement; bridges

Citation: Abdoli, F.; Rashidi, M.; Rostampour-Haftkhani, A.; Layeghi, M.; Ebrahimi, G. Withdrawal Performance of Nails and Screws in Cross-Laminated Timber (CLT) Made of Poplar (*Populus alba*) and Fir (*Abies alba*). *Polymers* **2022**, *14*, 3129. https://doi.org/10.3390/polym14153129

Academic Editor: Ľuboš Krišťák

Received: 30 June 2022
Accepted: 28 July 2022
Published: 31 July 2022

Publisher's Note: MDPI stays neutral with regard to jurisdictional claims in published maps and institutional affiliations.

Copyright: © 2022 by the authors. Licensee MDPI, Basel, Switzerland. This article is an open access article distributed under the terms and conditions of the Creative Commons Attribution (CC BY) license (https://creativecommons.org/licenses/by/4.0/).

1. Introduction

Sustainable building approaches using renewable resources such as wood and wood-based products have grown in popularity in recent decades [1–3]. Cross-laminated timber (CLT) is an engineered wood product (EWP) formed from sized lumbers and orthogonally laminated. This relatively new form of EWP is reported to be extensively employed in many projects, including mid-rise and even high-rise buildings, due to its unique cross-wise layups that can tolerate large loads and stresses either in-plane or out-of-plane [4–6]. CLT might be regarded as a stand-alone element because of its high degree of prefabrication and load-carrying capabilities. It is commonly utilized as a wall, floor diaphragm, roof, and other structural components. Various connectors, such as angle brackets and hold-downs with fasteners (nails and screws), might be used to join these parts. Connections, like a fuse in an electrical circuit, are frequently the source of ductility and energy dissipation in the structure in case of overloading because of the brittle failure behavior of wood when stressed in tension or shear [7]. The connections between structural parts must reinforce the structures to withstand gravity or vertical loads induced by self-weight, live loads, wind, and seismic loads, and then transmit these forces to the foundation [8–10].

Most failures begin at the connections, usually the weakest sections of timber structures [11]. Hence, it is vital to investigate how connections and fasteners behave under different loads. Nails or screws in CLT connectors are exposed to lateral and withdrawal loads, or a combination of both. When exposed to withdrawal loads, fasteners may pull out (withdraw) of wooden members or rupture in tension. Head pull-through is uncommon in CLT connectors.

The diameter, thread geometry, penetration depth of a fastener, and load-to-grain angle are the main factors that influence the withdrawal resistance of a fastener [12–14]. The latter is owed to the orthotropic nature of wood and wood-based products like CLT. Commonly, withdrawal loading is characterized as parallel or perpendicular to the grain, where the latter includes both radial and tangential loading, as shown in Figure 1.

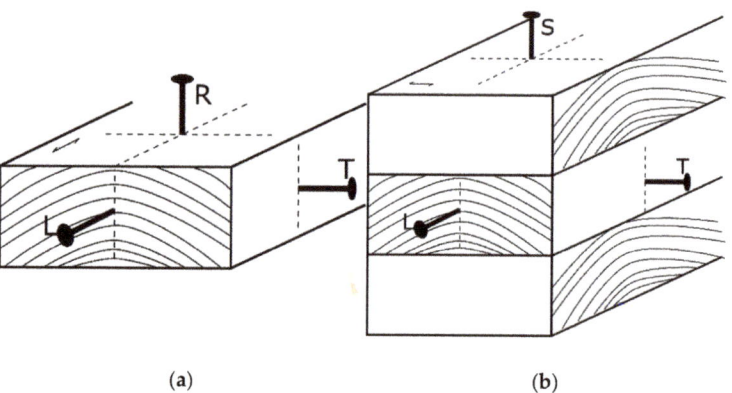

(a) (b)

Figure 1. (a) Withdrawal directions in sawn timber: (L) parallel to the grain or longitudinal, (R) perpendicular to the grain in radial direction, (T) perpendicular to the grain in tangential direction (side face). (b) Loading direction in CLT: (L) longitudinal edge loading (parallel to grain), (T) tangential edge loading (perpendicular to grain), (S) perpendicular to the surface loading.

In CLT, the loading directions are characterized as edge loading and surface loading, where edge loading can occur parallel to the grain (L) or in a tangential direction (T). The type of loading is mainly governed by the application of the CLT panel (wall-to-wall, wall-to-floor, etc.).

Uibel and Blaß [15–17] first studied the lateral and withdrawal resistance of screws, nails, and dowels in CLT made of European spruce (*Picea abies*). Since these first studies in the early 2000s, several more studies have examined the impacts of different parameters on the withdrawal performance of fasteners in CLT. Design guidance for CLT connections is given in several CLT Handbooks [18–20] and by CLT and fastener manufacturers. Yet, due to the many possible parameter combinations, most studies only considered a limited combination of grain orientations; loading directions; fastener types and fastener features (including thread geometry, diameter, and penetration depth); and wood species. The selected literature related to the present study is summarized here to contextualize the experimental program presented in this paper. Li et al. [21] investigated the withdrawal resistance of bamboo scrimber specimens modified by embedment length, screw diameter, and screw angle. They revealed that unlike wood, the tensile strength and stiffness of the bamboo scrimber in the radial and tangential directions are comparable. By increasing the density of CLT, the withdrawal capacity and strength of self-tapping screws enhanced from 19.7 to 57.3% [22].

Several studies examined the influence of the load-to-grain angle on the withdrawal capacity in sawn timber. One of the most comprehensive works on fully threaded self-tapping screws was published by Blaß and Bejtka [23] considering embedment, penetration depth, load-to-grain angle, lateral loading, and minimum fastener spacing requirements, all in sawn spruce (*Picea abies*).

Teng et al. [24] investigated the effect of the load-to-grain angle on the withdrawal resistance of screws and nails in sawn larch (*Larix gemlimii Rupr.*) and spruce (*Picea glauca*). They concluded that the insertion angle had a significant effect; however, there was no significant difference between radial and tangential directions. Furthermore, they found a positive correlation between density and withdrawal strength. Ringhofer et al. [25] studied the effect of combined axial and lateral loading in sawn Norway spruce (*Picea Abies*), highlighting the influence of the penetration depth and potential splitting for screws inserted parallel to grain (end grain).

Similar observations were made for CLT, where the withdrawal strength of self-tapping screws in the CLT side face (surface) is much greater than that of screws in the CLT narrow face (edge) [14]. This is also reflected by the failure modes observed in these different loading directions. Li et al. [5] investigated the withdrawal resistance of a single self-tapping screw in CLT manufactured from radiata pine (*Pinus radiata*). According to their results, column-like pull out of timber components was observed when self-tapping screws were axially withdrawn parallel to the grain direction, whilst the tearing failure of adjacent fibers occurred when they were withdrawn perpendicular to the grain direction.

However, the loading situation in CLT is often more complex since fasteners may penetrate one or more layers leading to more complex stress states, especially when fasteners are loaded in a combination of axial withdrawal and lateral shear loading. Hossain et al. [26] studied combined axial and lateral loading for screws in spruce-pine-fir CLT. Brown et al. [27] studied the effect of different angles and penetration lengths of inclined screws in CLT made of New Zealand radiata pine (*Pinus radiata*) and Douglas fir (*Pseudotsuga menziesi*), giving recommendations on the maximum penetration length to avoid screw rupture, and the ratio of different load-to-grain angles to optimize ductility [28].

Yet, even more, simple parameters, such as timber density, require more careful consideration in CLT. It is generally accepted that there is a positive correlation between fastener withdrawal strength and timber density [29]. However, in CLT, different layers may exhibit different densities. Mahdavifar et al. [12] discovered that the density of the face layer had a substantial impact on the fastener withdrawal resistance of wood screws and ring shank nails in hybrid CLTs produced from Douglas fir and lodgepole pine.

In CLTs made of Japanese larch (Larix kaempferi (Lamb.) Carr.), the withdrawal strength of self-tapping screws increased with increasing the timber density and effective length, and the withdrawal failure mode was a mix of shear cracks parallel to the grain due to fiber bending and tension perpendicular to the grain [30].

Furthermore, the surface conditions of the fastener have a significant impact on withdrawal capacity. Izzi et al. observed that a threaded shank enhanced the withdrawal capability of nails in CLTs manufactured from spruce when compared to smooth shank nails [31]. The friction between the threaded shank and the surrounding wood in CLT then determines the load-bearing mechanism of the annular-ringed shank nailing joints loaded in withdrawal [32]. According to Ceylan and Girginannular [33], threads make fasteners easy to insert and hard to pull-out. Rezvani et al. [34] showed that fully threaded screws in angle bracket connections outperformed partially threaded screws in terms of uplift, in-plane shear, and out-of-plane stress. D'Arenzo et al. found that adding inclined fully threaded screws in the bottom corner of traditional angle brackets significantly increased the performance for tensile loading, avoiding the usual low withdrawal resistance of the annular-ringed nails in the bottom plate [35].

In addition to fastener diameter and surface conditions, fastener penetration depth needs to be considered. Li et al. [36] found that increasing penetration depth by 50 mm (75 mm vs. 125 mm) increased the withdrawal resistance of radiata pine CLTs in the narrow face with self-tapping screws. In wood-plastic composite panels, screw withdrawal resistance rose when screw diameter, loading rate, and penetration depth increased [37].

Finally, the wood species used for CLT manufacturing significantly impacts their properties. Some hardwood species, such as poplar, grow fast, which is advantageous in countries with a shortage of wood supplies, such as Iran [38]. The shortage of wood supplies is a global problem. Although several research studies have investigated the withdrawal performance of CLTs made of softwood species such as pine, fir, and spruce, there are few findings on CLTs made of poplar wood. Most studies so far focused on the structural properties of poplar CLT, including flexural behavior and rolling shear [39–44].

The material properties of wood and wood-based products, such as density, manufacturing technique, and environmental conditions, have been shown to affect the withdrawal behavior of fasteners [45]. Hübner et al. [46] confirmed that the withdrawal resistance is affected by differences in wood species (or material properties). Moreover, Taj et al. [47] reported that anatomical diversity in wood has a substantial influence on fastener withdrawal resistance. According to Ringhofer et al. [13,48], the density of hardwood boards, geometric qualities and screw penetration depth, the number of layers that the screw penetrated, the angle, and gap insertion are significant factors to consider in the withdrawal performance of CLTs. Furthermore, methods for determining the withdrawal strength of self-tapping screws in solid and EWP were proposed.

Apart from the aforementioned characteristics, Yermán et al. [49] evaluated the withdrawal resistance of nails in modified and unmodified pine wood in terms of cyclic moisture changes. The results indicated that under moisture fluctuations, nail withdrawal ability seems to be lost owing to a combination of corrosion and the mechano-sorptive process of the nails backing out with alternating wetting and drying. When self-tapping screws were inserted in CLTs made from spruce, the withdrawal resistance was reduced by increasing the moisture [50].

In summary, most studies are on the withdrawal resistance of self-tapping screws and nails in softwood CLTs made of pine, fir, spruce, or larch. No research has been conducted on connections in poplar CLT to the authors' knowledge. As a result, the aim of this study is to examine the withdrawal performance of 3-ply CLTs made from poplar as a fast-growing species with various fasteners (seven types of screws and two types of nails) in three withdrawal loading directions (parallel to grain and perpendicular to the grain, both radial and tangential) in two-layer arrangements of 0-90-0° and 0-45-0°. Furthermore, the findings are compared to data obtained from CLTs made of fir, which is a common softwood species used in CLT manufacturing with a similar density to poplar.

2. Material and Methods

2.1. Wood and Manufacturing of CLT

Poplar *(Populus alba)* and fir *(Abies alba)* wood with oven-dried densities of 381 and 390 kg/m^3, respectively, were used in this research. For poplar, the modulus of elasticity, modulus of rupture, and shear strength parallel to the grains were 7380 MPa, 59 MPa, and 4.96 MPa, respectively. Similarly, they were 6658 MPa, 59.6 MPa, and 6.52 MPa for fir, respectively.

The poplar logs were cut into plain sawn boards with dimensions of 2000 mm × 110 mm × 25 mm (Length × Width × Thickness). Plain sawn fir boards with the same dimension were also prepared. Afterward, boards were air-dried at a temperature of 20 °C and a relative humidity of 65% until a constant weight was achieved. After drying, the boards were sawn and planed to the final cross-section size of 90 mm × 16 mm (Width × Thickness).

Poplar and fir boards without any wood defects or significant knots were selected and layered in 0/90/0° and 0-45-0° arrangements for poplar CLTs, and 0-90-0° for fir CLTs. The boards were both surface and edge-glued with one-component polyurethane adhesive (at a spread rate of 300 g/m^2) and cold-pressed for 150 min at a pressure of 1 MPa using hydraulic equipment. The CLT panels were stored for several weeks at a relative humidity of 65% and a temperature of 25 °C.

2.2. Fasteners

Seven types of screws and two types of nails as fasteners were employed in this investigation. The characteristics of fasteners are given in Table 1. Note that in addition to typical timber fasteners, concrete nails and screws, as well as drywall screws, were employed to study the effect of different surface conditions. The different withdrawal directions with respect to the CLT samples are displayed in Figure 2.

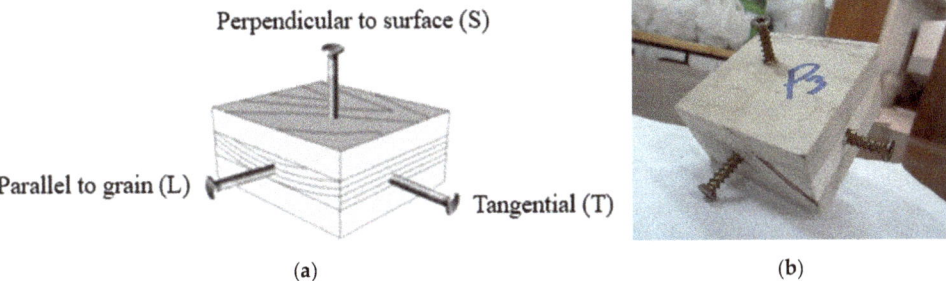

Figure 2. Loading directions (**a**) and placements of fasteners (**b**) in the CLT samples for the withdrawal tests.

Smaller blocks with a dimension of 75 mm × 75 mm × 48 mm were cut from the CLT panels, and a fastener was inserted in the three directions as shown in Figure 2a: parallel to the grain in the middle layer (L), the tangential direction of the middle layer (T), and perpendicular to the surface of the CLT panel (S). Fasteners were inserted at a 32 mm penetration depth into the CLT samples, as depicted in Figure 2b. The fasteners were placed to eliminate any gaps and to meet the requirements of boundary conditions, and end and edge distances stipulated in previous studies [14,51]. In other words, it means that the distance between the screws in different directions was such that they did not affect each other.

Table 1. Characteristics of fasteners used for withdrawal tests in CLT members.

	Steel Nail	Concrete Nail	Drywall Screw	Wood Screw	Galvanized SPAX Screw	Self-Tapping Screw	Concrete Screw	Lag Screw (8 mm)	Lag Screw (10 mm)
Length (mm)	63.5	66	61.8	62.5	59.16	68.7	60	65.5	69
Major diameter (mm)	3.75	4.16	4.2	4.7	4.46	6.44	7.8	7.75	9.43
Minor diameter (mm)	-	-	2.7	3.9	2.9	4.7	4.7	5.6	6.6
Pre-drilled hole diameter (mm)	3	3	2	3.5	2	3.5	3.5	5	6.5

2.3. Experimental Setup

In order to measure the withdrawal resistance of the fasteners, CLT samples were installed in an Instron testing machine model 4486 (Norwood, MA, USA), (Figure 3). A constant displacement rate of 6 mm/min was applied according to ASTM D 1761 [52]. Finally, the withdrawal resistance of the fasteners was calculated according to Equation (1):

$$W = P_{max}/L \quad (1)$$

where W is the withdrawal resistance of the fastener (N/mm), P_{max} is the maximum load (N), and L is the penetration depth of the fastener in CLT sample.

Figure 3. Instron machine for testing withdrawal resistance of the fasteners.

2.4. Statistical Analysis

Based on a completely randomized full factorial design, data from the experiments were statistically analyzed to determine the main and interaction effects between the nine fasteners, three loading directions, and three types of CLT panels, including poplar CLTs with two-layer arrangements (0-90-0° and 0-45-0°), and fir CLTs with only one arrangement of layers (0-90-0°). The full factorial design allows studying each independent variable with respect to the response variable (in this case, withdrawal resistance W), as well as the interaction between variables. Each variable may take on different values, and combinations of these values are called "treatment". In the present study, 81 treatments were analyzed, each with six repetitions, for a total of 486 tests (162 specimens for each direction). Duncan's multiple range test was performed to show the statistical differences between the treatment means at a 95% confidence level. SPSS software version 25 was used to conduct the statistical data analysis.

3. Results and Discussion

Table 2 gives the mean values of the withdrawal resistance of fasteners parallel to the grain (L) and perpendicular to the grain (T and S) in two arrangements for poplar CLTs and one arrangement for fir CLT. Screws displayed 7.5 to 11 times higher withdrawal resistance than nails. Concrete nails displayed a higher withdrawal capacity than steel nails. Furthermore, wood screws had the lowest withdrawal resistance among screws, whereas lag screws (10 mm) had the greatest. Fasteners in poplar CLTs with a 0-45-0° arrangement had the maximum withdrawal resistance, whereas the lowest withdrawal resistance was seen in fir CLTs. In terms of loading direction, fasteners perpendicular to the grain directions (S and T) demonstrated a higher withdrawal resistance than those parallel to the grain (L) direction. The highest withdrawal resistance was observed for fasteners inserted in the S direction.

Table 2. Withdrawal resistance of fasteners in all arrangements and loading directions.

Fastener	Loading Direction	Withdrawal Resistance (N)		
		Poplar (0-90-0°)	Poplar (0-45-0°)	Fir (0-90-0°)
Steel nail	L	10 (0.6)	14 (2.7)	9 (1.5)
	T	14 (4)	15 (1.6)	13 (3.3)
	S	13 (4.3)	19 (3.5)	11 (2.4)
Concrete nail	L	18 (2.3)	17 (3.3)	21 (7.2)
	T	14 (2.9)	23 (8.9)	16 (4.7)
	S	23 (3.7)	23 (10.2)	19 (7.9)
Wood screw	L	47 (9.2)	77 (11.4)	47 (12.6)
	T	77 (5.9)	79 (10.4)	73 (23)
	S	95 (10.2)	93 (17.3)	75 (14.3)
Drywall screw	L	92 (15.9)	82 (19.7)	86 (16.5)
	T	98 (16)	79 (17.8)	94 (8.6)
	S	99 (7.2)	104 (7.3)	91 (5.5)
SPAX galvanized screw	L	83 (10.6)	98 (18.7)	71 (6)
	T	127 (18.5)	108 (21.3)	106 (15.6)
	S	129 (10.4)	116.1 (10.3)	115 (10)
Self-tapping screw	L	97 (14)	123 (16.1)	91 (13.6)
	T	140 (5.8)	123 (20.8)	119 (22.9)
	S	147 (19.9)	153 (13.3)	119 (18.2)
Concretescrew	L	89 (42)	146 (15.5)	76 (6)
	T	152 (25.7)	118 (18.7)	127 (8)
	S	131 (11.7)	138 (14.4)	126 (6)
Lag screw (8 mm)	L	126 (9.4)	126 (14.4)	137 (8)
	T	78 (10.5)	134 (20)	98 (26)
	S	147 (6.2)	150 (20)	124 (9.2)
Lag screw (10 mm)	L	159 (10.1)	182 (20.8)	111 (19.3)
	T	89 (32.5)	161 (23.3)	119 (37.7)
	S	172 (24.7)	172 (29.2)	141 (18.5)

The numbers in parenthesis show standard deviation.

Table 3 shows the analysis of variance of the main and interaction effects of CLT types, fastener types, and loading directions on the withdrawal resistance of CLTs. The findings revealed that CLT types, fastener types, and loading directions all had a significant influence on the withdrawal performance of the CLTs. Moreover, the interaction effects of CLT types * * fastener type, CLT types * loading direction, fastener type * loading direction, and CLT types * type of fastener * loading direction on withdrawal performance were significant.

Table 3. Analysis of variance's main and interaction effects of the layer arrangement, fastener type, loading direction on the withdrawal resistance of CLTs.

Source	Type III Sum of Squares	df	Mean Square	F	Sig.
CLT types	21,815.903	2	10,907.951	45.480	0.000 **
Fastener types	965,774.335	8	120,721.792	503.337	0.000 **
Loading directions	30,160.979	2	15,080.490	62.877	0.000 **
CLT types * Fastener types	22,264.690	16	1391.543	5.802	0.000 **
CLT types * Loading directions	5502.278	4	1375.569	5.735	0.000 **
Fastener types * Loading directions	41,220.657	16	2576.291	10.742	0.000 **
CLT types * Fastener types * Loading directions	43,430.328	32	1357.198	5.659	0.000 **

** significant at 99% confident level; * significant at 95% confident level; ns: not significant.

3.1. Effects of CLT Types, Fastener Types, and Loading Directions on Withdrawal Resistance

The main effects of the CLT types, types of fasteners, and loading directions on the withdrawal performance of CLTs are shown in Figure 4A–C. According to the results, there was a significant difference in the withdrawal performance between CLT types, despite poplar and fir CLTS having similar specific gravity. CLTs manufactured with poplar wood showed higher withdrawal resistance in both arrangements than those made from fir wood (Figure 4A).

According to Figure 4A, poplar CLTs with the 0-45-0° arrangement had the highest withdrawal resistance (99.2 N), while those made from fir wood had the lowest (82.8 N). In other words, the withdrawal resistance of the CLTs made of poplar in the arrangement of 0-45-0° was 8.6 percent more than those in the arrangement of 0-90-0°. It might be due to the more substantial involvement of fasteners with timber fibers in 0-45-0°. Furthermore, the withdrawal resistance of poplar CLTs in the 0-90-0° configuration was 10.2 percent higher than that of fir CLTs. According to Brandner et al. [53,54], the pull-out strength increased linearly when the load grain angle rose from 0 to 30. Screw pull-out strengths varied between 19 and 24 percent for load-grain angles between 0 and 90.

Duncan's test showed that there was a significant difference among the means. Variations in recorded values of fastener withdrawal resistance for two species with close specific gravity may be related to the anatomical structure of these two species. The presence of parenchyma rays in hardwoods like poplar may explain this difference [47], and it is hypothesized that these rays resist the withdrawal of fasteners. In other words, parenchyma rays are transverse elements in the wood that cause more involvement with the fasteners when they are exposed to the withdrawal force.

Figure 4B depicts the main effect of the loading direction on the withdrawal resistance of the CLTs. The findings showed a significant difference in withdrawal resistance of CLTs in all loading directions. More particularly, the withdrawal resistance was greatest when the loading was in the S direction (101.7 N). Loading in the L direction, on the other hand, resulted in the lowest withdrawal resistance (82.8 N), and the difference was 22.8 percent. Duncan's test revealed a significant difference between the means of withdrawal in three directions. More details are discussed in the failure mode section. Since wood and wood-based products are orthotropic materials, anatomical variations in each direction can lead to significantly different properties such as the withdrawal performance in directions [47]. Li et al. [36] investigated the withdrawal resistance of self-tapping screws inserted in CLTs' narrow faces (T and L). Their results indicated that the direction of CLT (T and L) affects the withdrawal resistance of self-tapping screws.

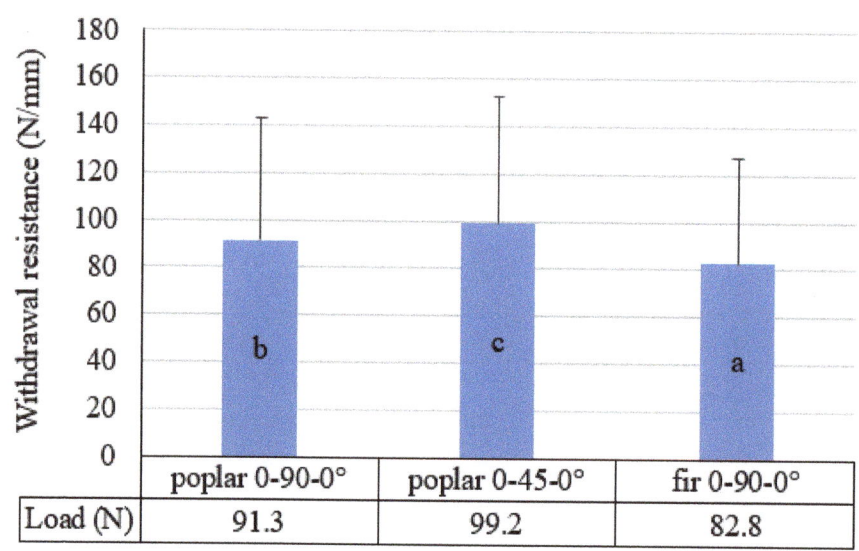

(A) CLT types

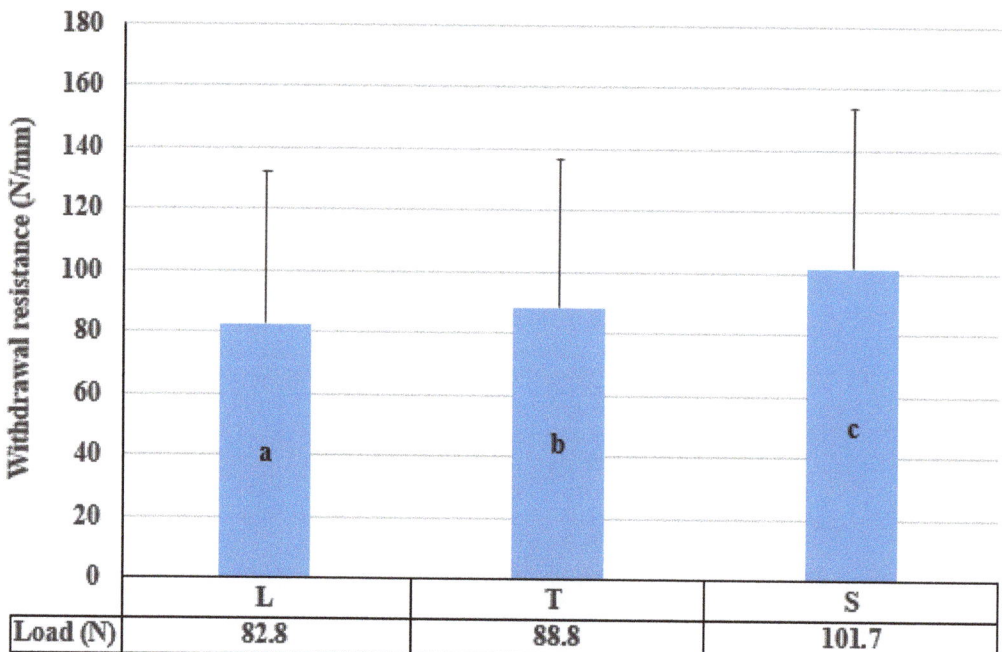

(B) Loading directions

Figure 4. *Cont.*

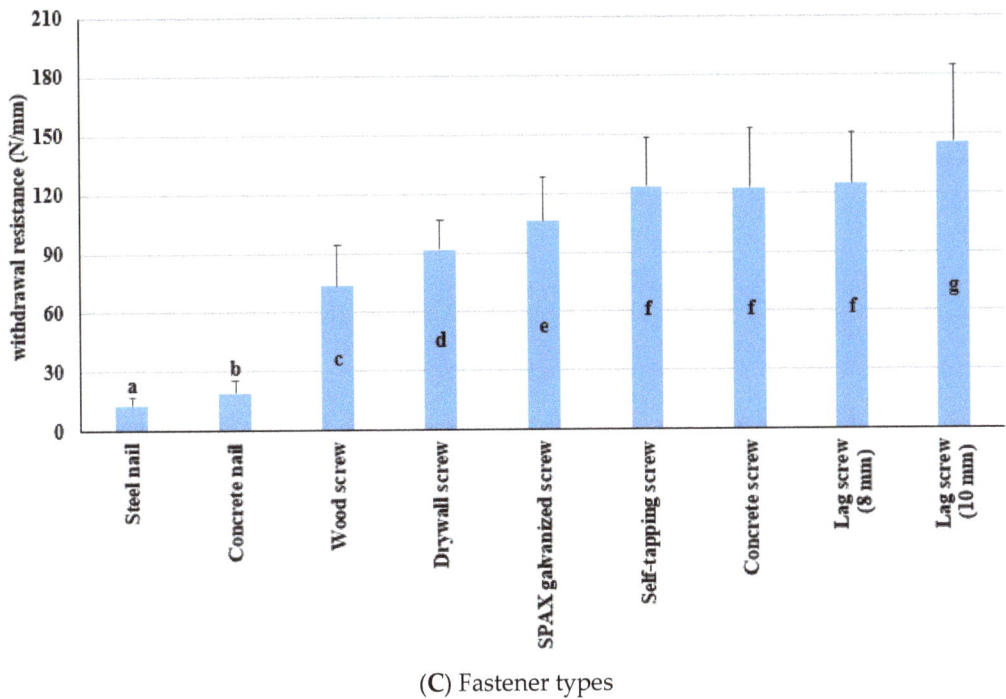

(**C**) Fastener types

Figure 4. Effects of layer arrangements (**A**), loading directions (**B**), and types of fasteners (**C**) on withdrawal resistance.

The withdrawal resistance of different fasteners in CLTs is shown in Figure 4C. The most effective variable in the withdrawal performance of CLTs was the fastener type. According to the results, steel nails had the lowest withdrawal resistance (13.13 N), while lag screws (10 mm) had the maximum withdrawal resistance (145.77), implying an 11 times difference. This confirms previous findings regarding the surface condition and diameter of different fasteners.

Duncan's test indicated that there was no statistically significant difference between self-tapping screws (d = 6.4 mm), concrete screws (d = 7.8 mm), and lag screws (d = 7.8 mm), despite small differences in outer screw diameter.

On the other hand, the difference between wood, drywall, and SPAX galvanized screws was significant, despite smaller differences in the outer diameter (ranging between 4.2 and 4.7 mm). This may be owed to the different surface properties, including the difference between inner and outer diameter, thread gauge, and fastener material.

Concrete nails outperformed steel nails in terms of withdrawal resistance, with a statistically significant difference (47.2%). This result is consistent with the findings of Izzi et al. [31]. Concrete nails feature a low-profile thread on their shanks, providing a higher withdrawal resistance than steel nails, which do not have any threads. Furthermore, the diameter of the concrete nail is greater than that of the steel nail.

Overall, increasing the diameter of the fasteners enhanced their withdrawal resistance. Lag screws (10 mm) had the highest withdrawal resistance. Previous research found similar findings [12,21,33]. According to Gehloff [55], the diameter factor might significantly affect the fasteners' withdrawal capacity. The withdrawal capacity of glulam rose by 70%, according to Abukari et al. [56,57], when the diameter of the screws was raised from 6 mm to 12 mm (100%).

Lag screws have a wide screw head diameter, which increases their resistance against head pull-through [58]. Self-tapping screws vary from primarily laterally loaded screws, such as hexagon head coach screws, in that they are optimized for loading in the axial direction [48].

Despite having a similar diameter, wood screws proved to be less resistant to withdrawal force than drywall screws due to their tapers, cut shank, and threads. According to Hoelz et al. [59], in addition to diameter, additional geometric characteristics such as flank distance and thread height affect fastener pull-out resistance. In this aspect, drywall screws had a greater flank distance and thread height than wood screws, resulting in a stronger resistance under the withdrawal load for drywall screws.

3.2. Interaction Effects of CLT Types, Fastener Types, and Loading Directions

Figure 5A displays the interaction effect of CLT types and fastener types under withdrawal force in CLTs. Fasteners in CLTs manufactured of poplar wood with the arrangement of 0-45-0° exhibited a stronger withdrawal resistance than other arrangements. Fasteners in 0-90-0° fir CLTs had the lowest withdrawal resistance. In terms of interaction between fastener type and CLT type, lag screws (10 mm) in the 0-45-0° arrangement had the highest withdrawal resistance (173.4 N), whereas steel nails in the 0-90-0° arrangement of CLTs made of fir wood had the lowest (10.7 N).

In general, fasteners exhibited greater withdrawal resistance as their diameter increased, and this trend was most pronounced for 0-45-0° poplar CLTs and 0-90-0° fir CLTs.

Altering the arrangement from 0-90-0° to 0-45-0° in CLTs made of poplar wood, the withdrawal resistance of steel nails and concrete nails changed 27% and 16%. Moreover, the withdrawal resistance of wood screws, lag screws (8 mm), and lag screws (10 mm) changed about 14%, 17%, and 24%, respectively. Furthermore, no significant difference was observed for other fasteners.

In changing the CLT types from poplar to fir, no significant difference was observed in the withdrawal resistance of concrete nails, drywall screws, and lag screws (8 mm); however, other fasteners showed 11–15% more withdrawal resistance in CLTs made from poplar wood in the arrangement of 0-90-0°.

Figure 5B depicts the interaction effect of CLT types and loading directions. The findings indicated that fasteners loaded perpendicular to the grains (S and T) had greater withdrawal resistance than fasteners loaded parallel to the grains (L). More specifically, the lowest withdrawal resistance (72.1 N) was reported in fir CLTs when the withdrawal was parallel to the grain (L). However, the greatest withdrawal load was observed in poplar CLTs with an arrangement of 0-45-0° (107.6 N) when the withdrawal loading direction was perpendicular to the surface (S).

CLT samples perpendicular to surface loading (S) had the highest withdrawal resistance, followed by samples with tangential loading direction (T), while the final one was parallel to the grain (L). The angle of fibers in the middle layer of poplar CLTs with the 0-45-0°-layer arrangement in contact with the fasteners was the same (45°) under both loading directions (L and T); hence, there was no significant difference between the mentioned loading directions. Figure 6d shows the related failure mode (inclined shear). In the CLTs made of poplar wood (0-90-0° and 0-45-0°), the biggest difference was observed in the L direction since altering the arrangement of the middle layer to 45° resulted in an increase of the withdrawal resistance (20%). In addition, between poplar and fir CLTs (both with 0-90-0° arrangement), the biggest difference was obtained from the S direction, while the lowest one was obtained from the L direction. In the poplar CLTs (0-90-0°), the difference between the S direction and L and T directions was about 33% and 21%, respectively. For the arrangement of 0-45-0° (poplar CLTs), these differences were about 12% and 15%, respectively. However, in the fir CLTs, the difference was about 26% and 1%, which means in softwoods such as fir, the difference in withdrawal resistance between the directions of S and T is insignificant. Conversely, for hardwoods such as poplar, the significant difference between S and T directions may be related to the placement of the parenchyma rays in the radial direction of the wood and perpendicular to the annual rings.

Figure 5C depicts the interaction impact of fastener types and loading directions. As expected, screws exhibited a higher withdrawal resistance than nails. Concrete nails had a higher withdrawal resistance than steel nails, although the difference was statistically insignificant in all loading directions and CLT types. Wood screws had the lowest withdrawal resistance among screws, whereas lag screws (10 mm) had the greatest, and this difference was statistically significant. Furthermore, the withdrawal resistance of fasteners perpendicular to the grain direction (T and S) was more than that of fasteners parallel to the grain direction (L). Except for lag screws (8 and 10 mm), the highest withdrawal resistance of the fasteners was recorded in the S direction loading perpendicular to the surfaces of the CLTs. By changing the fastener type, the withdrawal resistance changed more in the L loading direction compared to the others (S and T). The smallest change in the withdrawal resistance of the fasteners occurred in the T direction.

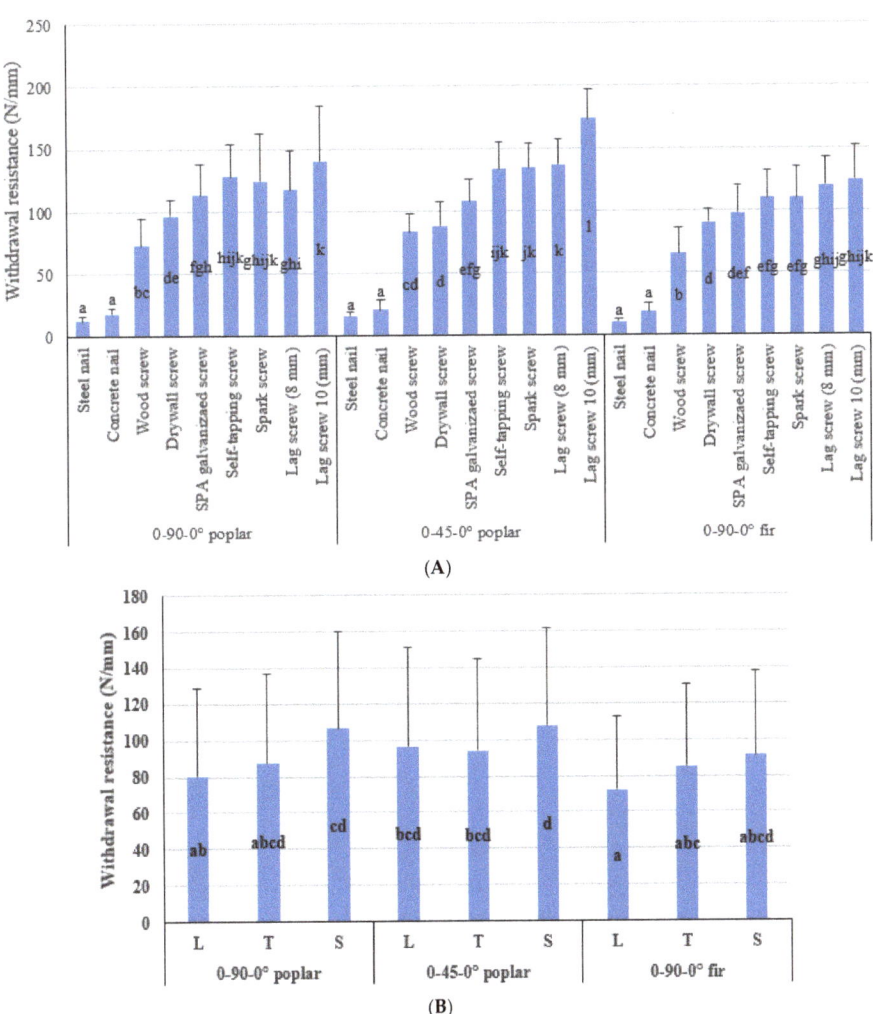

Figure 5. *Cont.*

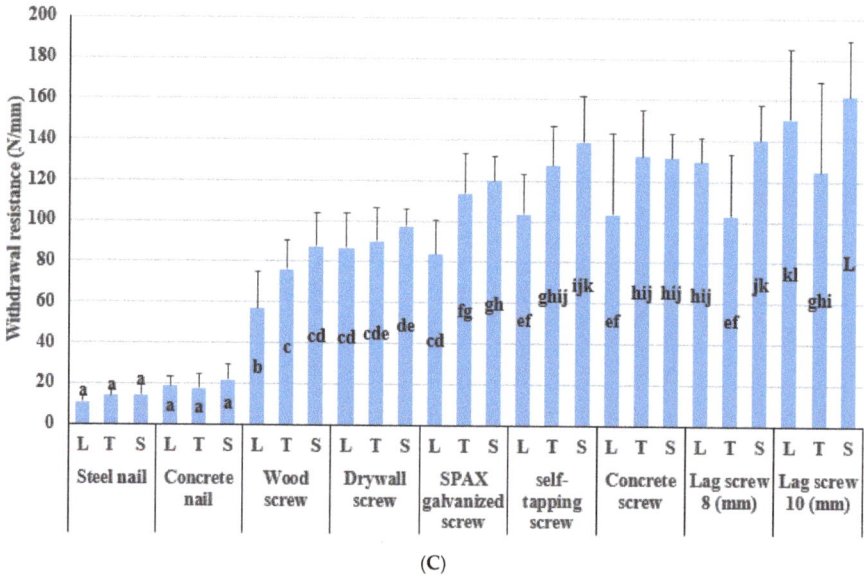

Figure 5. Interaction effects layer arrangements, loading directions, and types of fasteners on the withdrawal resistance of CLTs. (**A**) CLT types * Fastener types. (**B**) CLT types * Loading directions. (**C**) Fastener types * Loading directions.

Figure 6. Typical failure modes of CLT members under the withdrawal load parallel to the grains (L direction). (**a**–**c**) Column-like pull-out of wood fibers, (**d**) inclined shear failure along with the grains.

3.3. Failure Modes

Typical failure modes resulting from the withdrawal force parallel to the grain (L) are shown in Figure 6a–d; they are the column-like pull-out of wood fibers. Fasteners in this direction showed the lowest withdrawal resistance. It is worth differentiating between nails that simply spread fibers and screws that cut fibers. Li et al. [36] stated that shear stress formed around the threads under the withdrawal load parallel to the grain in the local wood regions, and nearby timber fibers were easily pulled out after stress limits were surpassed. The same failure mechanism was also reported by Ringhofer et al. [60].

The withdrawal resistance of drywall screws parallel to the grain (L) was found to be greater than that of the wood screws despite their outer diameters being similar (4.2 and 4.7 mm, respectively). Compared to drywall screws, wood screws have tapered shanks and cut threads. It seemed that this thread geometry significantly influenced shear failure and grain failure mode patterns as well as resulting in withdrawal resistance. The failure modes of the drywall and wood screws parallel to the grain (L) are shown in Figure 6a,b: drywall screws withdrew more wood fibers since their flank distance and thread height were greater than those of the wood screws, which is in agreement with previous studies [13,59]. The shape of threads also affected the shear failure along with the grain's mode pattern. Concrete screws threads, for instance, operate as saw teeth, as can be seen in Figure 6c.

Hoelz et al. [59] stated that the failure mechanisms in the thread contact depend on the orientation of the wood fiber to the screw thread. In the present study, fasteners in the layer arrangement of 0-45-0° produced inclined shear failure along with the grain's mode pattern (Figure 6d). In this case, the wood fibers emerged from one side of the fasteners due to the 45° angle of the wood fibers. None of the specimens exhibited fastener failure, e.g., head shear off or tensile rupture of the shank. In other words, all test failures included the withdrawal failures of the fasteners on wooden parts rather than the tensile failures of the fasteners themselves.

Typical failure modes resulting from the withdrawal force in the tangential (T) direction are shown in Figure 7a–d; they involve the tearing of adjacent fibers around the fasteners. All specimens in this direction showed the splitting failure of the wood layer's fibers, and the failure only happened around the fasteners. Pang et al. [61] reported the same results. Fasteners in this loading direction (T) demonstrated a higher withdrawal resistance than in the L direction. According to Figure 7, greater damage occurred in CLT specimens in this direction than parallel to the grain (Figure 6). This is also reflected in the withdrawal resistance: the higher the withdrawal resistance of the fasteners, the more damage that occurred in the timber. For example, drywall screws exhibited a greater withdrawal resistance than wood screws in the T direction, and lifted more surrounding wood fibers than wood screws (Figure 7b–d). According to Li et al. [36], the difference in failure modes might be explained by the various stress levels at the interface between the screw threads and the wood components, leading to a combination of shear and tensile fiber failure when the fasteners were withdrawn perpendicular to the grain.

Typical withdrawal failure types caused by the force perpendicular to the CLT surface (S) are illustrated in Figure 8a–f, representing withdrawal failure accompanied by substantial fiber deformation. In other words, in this direction (S), the areas of the failures (Figure 7a–d) were larger than those in the L direction (Figure 6a–d). Fasteners showed the greatest withdrawal resistance in this loading direction (S). The existence of cross-section parenchyma rays on the tangential surfaces of CLT samples, particularly in poplar wood, may explain this [47]. As a result, more interactions occurred between wooden tissue and fasteners, resulting in an increased withdrawal resistance in the fasteners. As seen in Figure 8, greater damages occurred in the CLT specimens in this direction than in the other orientations (Figures 6 and 7), especially for fasteners with high diameters. It means that fasteners with higher diameters cause more damage to CLT under the withdrawal load than fasteners with low diameters. According to Figure 8a,b, lag screws (10 mm) caused more damage than other fasteners, such as wood screws (Figure 8c), lag screws with an 8 mm diameter (Figure 8d,e), and drywall screws (Figure 8f) under the withdrawal load. Following that, lag screws with an 8 mm diameter (Figure 8d,e) caused more damage than wood screws (Figure 8c) and drywall screws (Figure 8f). It is worth noting that the position of fasteners in CLT members is important since they will be subjected to varied loads such as withdrawal. Consequently, when fasteners are positioned on the tangential surface, greater space between them is preferable, resulting in more resistance during withdrawal loading.

Figure 7. Typical failure modes of CLT members under the withdrawal load parallel to the grains (T direction). (**a–d**) Tearing of adjacent fibers around the fasteners.

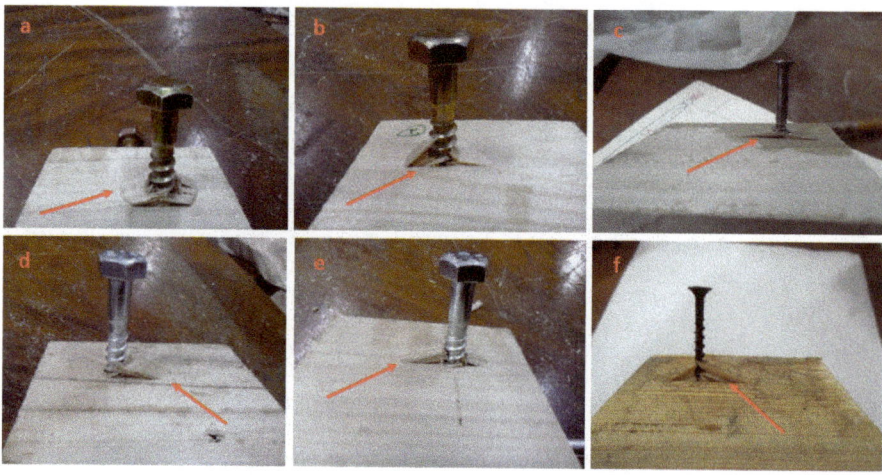

Figure 8. Typical failure modes of CLT members under the withdrawal load parallel to the grains (S direction). (**a–d**) Withdrawal failure accompanied by substantial fiber deformation.

4. Conclusions

The following are the key conclusions that can be derived from the experimental results:

- In terms of the main effect, fastener type was the most effective factor in the withdrawal performance of CLT. Following that were loading direction and CLT type, respectively.
- In terms of the interaction effect, CLT type * fastener type was the most effective factor in the withdrawal performance of CLT. Following that were fastener type * loading direction and CLT type * loading direction.
- In addition to fir wood, fasteners inserted in poplar (a fast-growing species) showed satisfactory withdrawal resistance.
- Diameter had a great influence on the withdrawal resistance of the fasteners.
- In terms of the loading direction, which is crucial in CLT connection design (wall to wall, wall to floor, etc.), fasteners in the S direction had the highest withdrawal resistance, followed by fasteners in the T direction. In this regard, obtaining data about various fasteners in all loading directions could be valuable for finding optimal fasteners for each direction.
- Different layer arrangements were examined to improve the low withdrawal resistance in the L direction of the CLT. The results showed that the difference between withdrawal resistance in L and T directions was reduced by changing the arrangement, thereby improving them in the L direction.
- Failure modes in different CLT directions and different fastener types should be considered for achieving a better withdrawal resistance. Further research is recommended to characterize the impact of thread height and gauge (flank distance).
- The higher the diameter of fasteners, the higher the damaged area for each fastener, which directly correlates with withdrawal capacity. The damaged area in the S direction was higher than in the T and L directions. Therefore, applying these findings to the design of angle brackets connected with nails or screws or a combination of them is recommended.
- The results revealed that screws with larger diameters showed high withdrawal resistance. However, it is important to consider where to install these fasteners on the CLT because of how much more damage they might do. The design of the angle brackets could benefit from these insights. Therefore, using fasteners with smaller diameters in the low end and edge distances of the angle brackets and fasteners with larger diameters in the higher end and edge distances is recommended.

Author Contributions: Conceptualization, F.A. and A.R.-H.; methodology, A.R.-H.; software, F.A.; validation, F.A., M.R. and A.R.-H.; formal analysis, F.A.; investigation, M.R.; resources, A.R.-H.; data curation, A.R.-H.; writing—original draft preparation, F.A.; writing—review and editing, M.R.; supervision, G.E.; project administration, M.L.; funding acquisition, M.R. All authors have read and agreed to the published version of the manuscript.

Funding: This research received no external funding.

Institutional Review Board Statement: Not applicable.

Informed Consent Statement: Not applicable.

Data Availability Statement: The data presented in this study are available on request from the corresponding author.

Conflicts of Interest: The authors declare no conflict of interest.

References

1. Lyu, C.; Gilbert, B.; Guan, H.; Karampour, H.; Gunalan, S. Finite element modelling of the progressive collapse of post-and-beam mass timber building substructures under edge and corner column removal scenarios. *J. Build. Eng.* **2022**, *49*, 104012. [CrossRef]
2. Navaratnam, S.; Thamboo, J.; Ponnampalam, T.; Venkatesan, S.; Chong, K.B. Mechanical performance of glued-in rod glulam beam to column moment connection: An experimental study. *J. Build. Eng.* **2022**, *50*, 104131. [CrossRef]
3. Wang, Y.; Lian, W.; Benjeddou, O. Experimental and numerical investigation on withdrawal connectors usage for lateral resistance of timber shear walls structure. *J. Build. Eng.* **2021**, *44*, 103266.
4. Karacabeyli, E.; Gagnon, S. *Canadian CLT Handbook*; Digital; FPInnovations: Pointe-Claire, QC, Canada, 2019.

5. Li, X.; Ashraf, M.; Subhani, M.; Kafle, B.; Kremer, P. Resistance of Cross Laminated Timber Members under Axial Loading—A Review of Current Design Rules. In *CIGOS 2019, Innovation for Sustainable Infrastructure*; Springer: Berlin/Heidelberg, Germany, 2020; pp. 179–184.
6. Li, X.; Subhani, M.; Ashraf, M.; Kafle, B.; Kremer, P. A current-state-of-the-art on design rules vs test resistance of Cross Laminated Timber members subjected to transverse loading. In *CIGOS 2019, Innovation for Sustainable Infrastructure*; Springer: Berlin/Heidelberg, Germany, 2020; pp. 185–190.
7. Ottenhaus, L.-M.; Jockwer, R.; van Drimmelen, D.; Crews, K. Designing timber connections for ductility—A review and discussion. *Constr. Build. Mater.* **2021**, *304*, 124621. [CrossRef]
8. Leng, Y.; Xu, Q.; Harries, K.A.; Chen, L.; Liu, K.; Chen, X. Experimental study on mechanical properties of laminated bamboo beam-to-column connections. *Eng. Struct.* **2020**, *210*, 110305. [CrossRef]
9. Wang, T.-H.; Chung, Y.-L.; Wang, S.-Y.; Chang, W.-S. Glue-laminated bamboo for dowel-type moment-resisting connections. *Compos. Struct.* **2021**, *267*, 113848. [CrossRef]
10. Rashidi, M.; Hoshyar, A.N.; Smith, L.; Samali, B.; Siddique, R. A comprehensive taxonomy for structure and material deficiencies, preventions and remedies of timber bridges. *J. Build. Eng.* **2021**, *34*, 101624. [CrossRef]
11. Wang, J.; He, J.-X.; Yang, Q.-S.; Yang, N. Study on mechanical behaviors of column foot joint in traditional timber structure. *Struct. Eng. Mech. Int. J.* **2018**, *66*, 1–14.
12. Mahdavifar, V.; Sinha, A.; Barbosa, A.R.; Muszynski, L.; Gupta, R. Lateral and withdrawal capacity of fasteners on hybrid cross-laminated timber panels. *J. Mater. Civ. Eng.* **2018**, *30*, 04018226. [CrossRef]
13. Ringhofer, A.; Brandner, R.; Schickhofer, G. *A Universal Approach for Withdrawal Properties of Self-Tapping Screws in Solid Timber and Laminated Timber Products*; Institute of Timber Engineering and Wood Technology, Graz University of Technology: Graz, Austria, 2015.
14. Ringhofer, A.; Brandner, R.; Blaß, H.J. Cross laminated timber (CLT): Design approaches for dowel-type fasteners and connections. *Eng. Struct.* **2018**, *171*, 849–861. [CrossRef]
15. Uibel, T.; Blaß, H.J. Edge joints with dowel type fasteners in cross laminated timber. In Proceedings of the CIB-W18 Meeting, Bled, Slovenia, 28–31 August 2007.
16. Uibel, T.; Blaß, H.J. Joints with dowel type fasteners in CLT structures. In Proceedings of the Focus Solid Timber Solutions-European Conference on Cross Laminated Timber (CLT), Bath, UK, 21–22 May 2013; pp. 119–136.
17. Uibel, T.; Blaß, H.J. Load carrying capacity of joints with dowel type fasteners in solid wood panels. In Proceedings of the CIB-W18 Meeting, Florence, Italy, 28–31 August 2006.
18. Gagnon, S.; Pirvu, C. *CLT Handbook: Cross-Laminated Timber*; FPInnovations: Pointe-Claire, QC, USA, 2011.
19. Karacabeyli, E.; Gagnon, S.; Pirvu, C. *Canadian CLT Handbook: Cross-Laminated Timber*; FPInnovations: Pointe-Claire, QC, USA, 2019.
20. Bogensperger, T.; Moosbrugger, T.; Schickhofer, G. *BSPhandbuch, Holz-Massivbauweise in Brettsperrholz*; Verlag der Technischen Universität Graz: Graz, Austria, 2010.
21. Li, H.; Qiu, H.; Wang, Z.; Lu, Y. Withdrawal resistance of the self-tapping screws in engineered bamboo scrimber. *Constr. Build. Mater.* **2021**, *311*, 125315. [CrossRef]
22. Khai, T.D.; Young, J.G. Withdrawal capacity and strength of self-tapping screws on cross-laminated timber. *Structures* **2022**, *37*, 772–786. [CrossRef]
23. Blaß, H.J.; Bejtka, I.; Uibel, T. *Tragfähigkeit von Verbindungen mit Selbstbohrenden Holzschrauben mit Vollgewinde*; KIT Scientific Publishing: Karlsruhe, Germany, 2006; Volume 4.
24. Teng, Q.; Que, Z.; Li, Z.; Zhang, X. Effect of installed angle on the withdrawal capacity of self-tapping screws and nails. In Proceedings of the World Conference of Timber Engineering, Seoul, Korea, 20–23 August 2018.
25. Ringhofer, A.; Burtscher, M.; Gstettner, M.; Sieder, R. Self-Tapping Timber Screws Subjected to Combined Axial and Lateral Loading. In *INTER-International Network on Timber Engineering Reserach*; Timber Scientific Publishing: Karlsruhe, Germany, 2021; pp. 95–112.
26. Hossain, A.; Popovski, M.; Tannert, T. Cross-laminated timber connections assembled with a combination of screws in withdrawal and screws in shear. *Eng. Struct.* **2018**, *168*, 1–11. [CrossRef]
27. Brown, J.; Li, M.; Karalus, B.; Stanton, S. *Withdrawal Behaviour of Self-Tapping Screws in New Zealand Cross-Laminated Timber*; University of Canterbury: Christchurch, New Zealand, 2020.
28. Brown, J.R.; Li, M.; Tannert, T.; Moroder, D. Experimental study on orthogonal joints in cross-laminated timber with self-tapping screws installed with mixed angles. *Eng. Struct.* **2021**, *228*, 111560. [CrossRef]
29. Ross, R.J. *Wood Handbook: Wood as an Engineering Material*; General Technical Report FPL-GTR-190; Forest Products Laboratory, USDA Forest Service: Washington, DC, USA, 2010; p. 509.
30. Xu, J.; Zhang, S.; Wu, G.; Gong, Y.; Ren, H. Withdrawal Properties of Self-Tapping Screws in Japanese larch (*Larix kaempferi* (Lamb.) Carr.) Cross Laminated Timber. *Forests* **2021**, *12*, 524. [CrossRef]
31. Izzi, M.; Polastri, A.; Fragiacomo, M. Modelling the mechanical behaviour of typical wall-to-floor connection systems for cross-laminated timber structures. *Eng. Struct.* **2018**, *162*, 270–282. [CrossRef]
32. Izzi, M.; Flatscher, G.; Fragiacomo, M.; Schickhofer, G. Experimental investigations and design provisions of steel-to-timber joints with annular-ringed shank nails for Cross-Laminated Timber structures. *Constr. Build. Mater.* **2016**, *122*, 446–457. [CrossRef]
33. Ceylan, A.; Girgin, Z.C. Comparisons on withdrawal resistance of resin and phosphate coated annular ring nails in CLT specimens. *Constr. Build. Mater.* **2020**, *238*, 117742. [CrossRef]

34. Rezvani, S.; Zhou, L.; Ni, C. Experimental evaluation of angle bracket connections in CLT structures under in-and out-of-plane lateral loading. *Eng. Struct.* **2021**, *244*, 112787. [CrossRef]
35. D'Arenzo, G.; Rinaldin, G.; Fossetti, M.; Fragiacomo, M. An innovative shear-tension angle bracket for cross-laminated timber structures: Experimental tests and numerical modelling. *Eng. Struct.* **2019**, *197*, 109434. [CrossRef]
36. Li, X.; Ashraf, M.; Subhani, M.; Ghabraie, K.; Li, H.; Kremer, P. Withdrawal resistance of self-tapping screws inserted on the narrow face of cross laminated timber made from Radiata Pine. *Structures* **2021**, *31*, 1130–1140. [CrossRef]
37. Haftkhani, A.R.; Ebrahimi, G.; Tajvidi, M.; Layeghi, M. Investigation on withdrawal resistance of various screws in face and edge of wood–plastic composite panel. *Mater. Des.* **2011**, *32*, 4100–4106. [CrossRef]
38. Hematabadi, H.; Behrooz, R.; Shakibi, A.; Arabi, M. The reduction of indoor air formaldehyde from wood based composites using urea treatment for building materials. *Constr. Build. Mater.* **2012**, *28*, 743–746. [CrossRef]
39. Hematabadi, H.; Madhoushi, M.; Khazaeyan, A.; Ebrahimi, G.; Hindman, D.; Loferski, J. Bending and shear properties of cross-laminated timber panels made of poplar (*Populus alba*). *Constr. Build. Mater.* **2020**, *265*, 120326. [CrossRef]
40. Hematabadi, H.; Madhoushi, M.; Khazaeian, A.; Ebrahimi, G. Structural performance of hybrid Poplar-Beech cross-laminated-timber (CLT). *J. Build. Eng.* **2021**, *44*, 102959. [CrossRef]
41. Kramer, A.; Barbosa, A.R.; Sinha, A. Viability of hybrid poplar in ANSI approved cross-laminated timber applications. *J. Mater. Civ. Eng.* **2014**, *26*, 06014009. [CrossRef]
42. Wang, Z.; Fu, H.; Chui, Y.-H.; Gong, M. Feasibility of using poplar as cross layer to fabricate cross-laminated timber. In Proceedings of the World Conference on Timber Engineering, Quebec City, QC, Canada, 10–14 August 2014.
43. da Rosa Azambuja, R.; DeVallance, D.B.; McNeel, J. Evaluation of Low-Grade Yellow-Poplar (*Liriodendron tulipifera*) as Raw Material for Cross-Laminated Timber Panel Production. *For. Prod. J.* **2022**, *72*, 1–10. [CrossRef]
44. Wang, Z.; Dong, W.; Wang, Z.; Zhou, J.; Gong, M. Effect of macro characteristics on rolling shear properties of fast-growing poplar wood laminations. *Wood Res.* **2018**, *63*, 227–238.
45. Mohammad, M.; Blass, H.; Salenikovich, A.; Ringhofer, A.; Line, P.; Rammer, D.; Smith, T.; Li, M. Design approaches for CLT connections. *Wood Fiber Sci.* **2018**, *50*, 27–47. [CrossRef]
46. Hübner, U.; Rasser, M.; Schickhofer, G. Withdrawal capacity of screws in European ash (*Fraxinus excelsior* L.). In Proceedings of the 11th World Conference on Timber Engineering, Trentino, Italy, 20–24 June 2010; pp. 241–250.
47. Taj, M.A.; Kazemi Najafi, S.; Ebrahimi, G. Withdrawal and lateral resistance of wood screw in beech, hornbeam and poplar. *Eur. J. Wood Wood Prod.* **2009**, *67*, 135–140. [CrossRef]
48. Ringhofer, A.; Brandner, R.; Schickhofer, G. Withdrawal resistance of self-tapping screws in unidirectional and orthogonal layered timber products. *Mater. Struct.* **2015**, *48*, 1435–1447. [CrossRef]
49. Yermán, L.; Ottenhaus, L.-M.; Montoya, C.; Morrell, J.J. Effect of repeated wetting and drying on withdrawal capacity and corrosion of nails in treated and untreated timber. *Constr. Build. Mater.* **2021**, *284*, 122878. [CrossRef]
50. Silva, C.; Branco, J.M.; Ringhofer, A.; Lourenço, P.B.; Schickhofer, G. The influences of moisture content variation, number and width of gaps on the withdrawal resistance of self tapping screws inserted in cross laminated timber. *Constr. Build. Mater.* **2016**, *125*, 1205–1215. [CrossRef]
51. Brandner, R.; Flatscher, G.; Ringhofer, A.; Schickhofer, G.; Thiel, A. Cross laminated timber (CLT): Overview and development. *Eur. J. Wood Wood Prod.* **2016**, *74*, 331–351. [CrossRef]
52. *ASTM-D1761-20*; Standard Test Methods for Mechanical Fasteners in Wood and Wood-Based Materials. Document Center Inc.: Belmont, CA, USA, 2020.
53. Brandner, R. Properties of axially loaded self-tapping screws with focus on application in hardwood. *Wood Mater. Sci. Eng.* **2019**, *14*, 254–268. [CrossRef]
54. Brandner, R.; Ringhofer, A.; Reichinger, T. Performance of axially-loaded self-tapping screws in hardwood: Properties and design. *Eng. Struct.* **2019**, *188*, 677–699. [CrossRef]
55. Gehloff, M. *Pull-Out Resistance of Self-Tapping Wood Screws with Continuous Thread*; University of British Columbia: Vancouver, BC, Canada, 2011.
56. Abukari, M.H. *The Performance of Structural Screws in Canadian Glulam*; McGill University: Montréal, QC, Canada, 2012.
57. Abukari, M.H.; Coté, M.; Rogers, C.; Salenikovich, A. Withdrawal resistance of structural screws in Canadian glued laminated timber. In Proceedings of the World Conference on Timber Engineering 2012 (WCTE 2012), Auckland, New Zealand, 15–19 July 2012.
58. Chybiński, M.; Polus, Ł. Withdrawal strength of hexagon head wood screws in laminated veneer lumber. *Eur. J. Wood Wood Prod.* **2022**, *80*, 541–553. [CrossRef]
59. Hoelz, K.; Dörner, P.-T.; Hohlweg, J.; Matthiesen, S. Influence of thread parameters on the withdrawal capacity of wood screws to optimize the thread geometry. *Eur. J. Wood Wood Prod.* **2022**, *80*, 529–540. [CrossRef]
60. Ringhofer, A.; Augustin, M.; Schickhofer, G. Basic steel properties of self-tapping timber screws exposed to cyclic axial loading. *Constr. Build. Mater.* **2019**, *211*, 207–216. [CrossRef]
61. Pang, S.-J.; Ahn, K.-S.; Kang, S.G.; Oh, J.-K. Prediction of withdrawal resistance for a screw in hybrid cross-laminated timber. *J. Wood Sci.* **2020**, *66*, 79. [CrossRef]

Article

The Effect of Openings' Size and Location on Selected Dynamical Properties of Typical Wood Frame Walls

Marcin Szczepanski [1,2], Ahmed Manguri [1,3], Najmadeen Saeed [3,4] and Daniel Chuchala [2,5,*]

1. Faculty of Civil and Environmental Engineering, Gdansk University of Technology, 80-233 Gdansk, Poland; marcin.szczepanski@pg.edu.pl (M.S.); ahmed.manguri@pg.edu.pl (A.M.)
2. EkoTech Center, Gdansk University of Technology, 80-233 Gdansk, Poland
3. Civil Engineering Department, University of Raparin, Ranya 46012, Kurdistan Region, Iraq; najmadeen_qasre@uor.edu.krd
4. Civil Engineering Department, Tishk International University, Erbil 44001, Kurdistan Region, Iraq
5. Faculty of Mechanical Engineering and Ship Technology, Gdansk University of Technology, 80-233 Gdansk, Poland
* Correspondence: daniel.chuchala@pg.edu.pl

Abstract: The wooden frame constructions are now popular in many developed countries of the world. Many of these locations where such buildings are constructed are exposed to seismic and other shocks which are generated by human activities. This paper discusses the effect of the size and location of openings in the wooden frame walls under dynamic loadings. Natural frequencies of such frames with and without openings have been determined. Three 14 m high walls with different widths, including 3, 6, and 12 m, have been considered. Dynamic analysis has been made using finite element method structural analysis software Dlubal RFEM 5.17. The results show that the effect of the size and location of the openings on the natural frequency is significant. Numerically speaking, the relative change of the natural frequencies of a wall without and with an opening in a specific place could be up to 30%. In addition, the change of the natural frequency for the location of the openings is more sensitive than that to the sizes. Furthermore, the appropriate sizes and locations of openings of the wooden frame walls have been suggested. The appropriate size and place were found to be small openings in the top of the walls.

Keywords: wood-frame; OSB boards; composite materials in buildings; dynamic loads; natural frequency; openings

Citation: Szczepanski, M.; Manguri, A.; Saeed, N.; Chuchala, D. The Effect of Openings' Size and Location on Selected Dynamical Properties of Typical Wood Frame Walls. *Polymers* **2022**, *14*, 497. https://doi.org/10.3390/polym14030497

Academic Editor: Ľuboš Krišťák

Received: 5 January 2022
Accepted: 18 January 2022
Published: 26 January 2022

Publisher's Note: MDPI stays neutral with regard to jurisdictional claims in published maps and institutional affiliations.

Copyright: © 2022 by the authors. Licensee MDPI, Basel, Switzerland. This article is an open access article distributed under the terms and conditions of the Creative Commons Attribution (CC BY) license (https://creativecommons.org/licenses/by/4.0/).

1. Introduction

The wooden frame constructions are now popular in developed countries such as USA, Canada, Australia, and many European countries. This is due to the very good mechanical properties of wood and wood-based materials, which are increasingly used in the building industry [1]. The environmental aspect is also important, as timber-frame houses have a significant impact on reducing CO_2 emissions compared to concrete buildings [2]. Additionally, wooden structures perform better in certain climatic conditions [3]. Wood-frame wall structures consist of a frame made of sawn timber and an oriented standard boards (OSB) casing, with the center filled with thermal insulation material (Figure 1). The frames for walls are usually made of pine or spruce timber, which is properly sorted to meet the requirements for mechanical properties [4]. The boards surrounding the frame are most often OSB boards, which have good physical and mechanical properties [5,6]. However, other solutions are often proposed based on the use of wood waste or other biomasses [7–9]. The inside of the frame is filled with an insulating material that achieves various functions: thermal insulation [10], acoustic insulation [11,12], and structural reinforcement [13]. Insulation materials are often mineral wool [10], polyurethane foams (PU foams) [13–15], or other solutions based on waste from various biomasses [15–17].

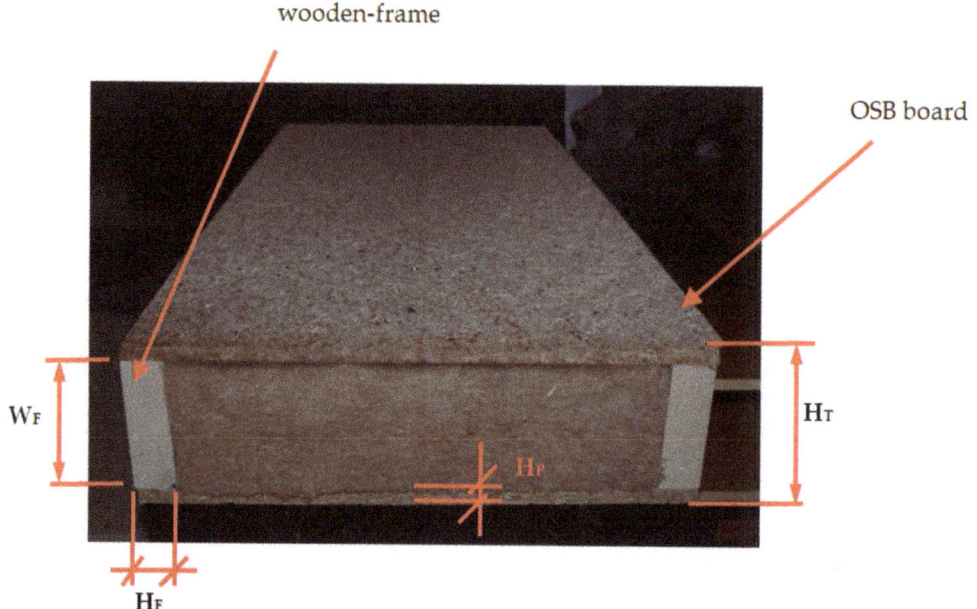

Figure 1. Example wall module for standard wooden-frame construction, where W_F is width of frame beam, H_F is height of frame beam, H_P is thickness of plate, H_T is thickness of wall.

In general, shear walls are the main elements of the timber structures that resist the lateral loads and earthquake excitation loads [18]. Different techniques have been implemented to evaluate and enhance the earthquake resistance of timber structures. X-Lam buildings' ductile behavior has been measured under dynamic loads, and the buildings were modeled to perform non-linear time-history analyses for design purposes [19]. Furthermore, CLT (Cross Laminate Timber) wall panels have been tested to understand their behavior under a quasi-static lateral load [20]. Some of these studies have been conducted through numerical models.

Numerical modeling is considered an economical and easy to perform analysis rather than experimental works. Numerical calculations use finite element methods that give relatively accurate results [14]. Nowadays, the focus of the researchers on these models due to their efficiency, time, and cost-saving has been increased. Generally speaking, there are two types of wooden frame structures. The first one is prefabricated wooden frame construction, in which the structure is built by connecting colossal, prefabricated elements [21]. The second is a light frame construction system using many small and closely spaced members that can be assembled by nailing [22]. The latter will be considered in this study.

In order to improve the quality of occupants' lives, engineers are trying to modernize the structures. One of such ways is adding additional openings or widening the current ones. The openings have a substantial impact on the structures' dynamic behavior [23]. Since openings can significantly affect the structural properties of buildings, researchers studied this effect on overall stiffness of CLT structures [23] and shear properties of wooden walls [24,25]. The effect of size and shape of openings on the stiffness and shear behaviors of CLT walls has also been estimated [26]. The aim of this work is the analysis of natural frequencies of vibrations based on different sizes and locations of openings in wooden frame walls. Furthermore, the appropriate size and location of the openings in the wooden walls will be suggested. Since, in the modern life, the openings are performed in the walls for different purposes, they have negative impact on the structural performance if they are carried out in arbitrary places.

2. Methodology

In this section, the properties and dimensions of the walls and the openings are introduced.

2.1. Wall Geometries and Material Properties

Three walls with different width, 3 (W1), 6 (W2), and 12 m (W3), each have 14 m height and are used as numerical models in this research; these are typical dimensions of the prefabricated walls. The walls were modeled as connected panels with dimensions of 0.60 m length, a height of 1.20 m, and a thickness of 0.15 m. The panels were constructed as four connected wooden boards made of the coniferous timber, Scots Pine (*Pinus sylvestris* L.) of C18 strength classes [27], which create a frame covered on both sides with oriented standard boards (OSB3). The properties of the frame [28] and the sheathing [13,29], according to Eurocode 5 [30], have been tabulated in Table 1. To make modeling of the walls as real as possible, the support reactions were assumed to be hinged (i.e., the transitions were inhibited while the rotations were permitted).

Table 1. Material properties of different elements used in the wooden frame wall [4,5].

Element	Material	Density (kg/m^3)	Elasticity Modulus (GPa)
Frame	Pine Wood of class C18	430	9.0-along fibers 0.3-across fibers
Sheathing	OSB 3	713.8	4.93-along fibers 1.98-across fibers

2.2. Numerical Models of Wooden Walls

For each wall, several scenarios of openings in terms of location and size have been evaluated. Firstly, the three walls with 3, 6, and 12 m wide walls without openings have been tested (see Figure 2). Secondly, the openings assumed to be 0.9 × 2 m (Op1) in different locations on W1, W2, and W3. For W2, the illustration is given in Figure 3. Same locations allocated for 1.8 × 2 m (Op2) for W1, W2, and W3.

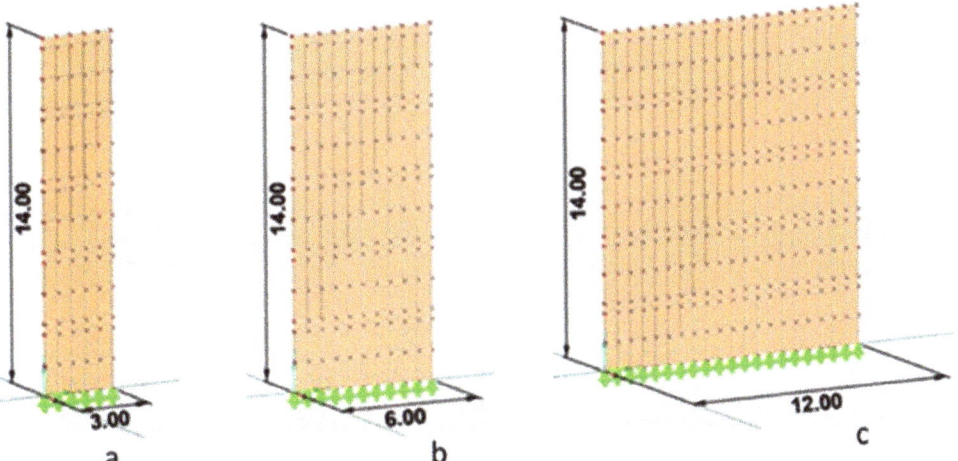

Figure 2. Dimensions of analyzed walls: (**a**) W1, (**b**) W2, (**c**) W3.

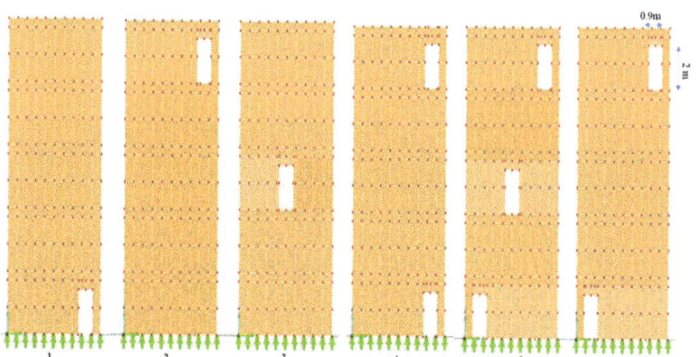

Figure 3. Location of openings on W2. 1–6 are the different positions of the openings with dimensions 0.9 × 2 m.

Another type of system opening is the horizontal arrangement in each story. It means that the holes were placed next to each other at the same distance. The distance between two adjacent openings is set to be different for the three various width walls. For W1, the distance between two Op1s is equal to 0.3 m and for Op2s is 0.6 m, whereas for W2, the two adjacent Op1s have the distance equal to 0.48 m and for two adjacent Op2s, their distance is 0.8 m. Furthermore, the distance between two Op1s and Op2s were set to be 0.39 and 0.5 m, respectively, for W3. Figure 4 shows the openings in top and bottom level of W1, W2, and W3.

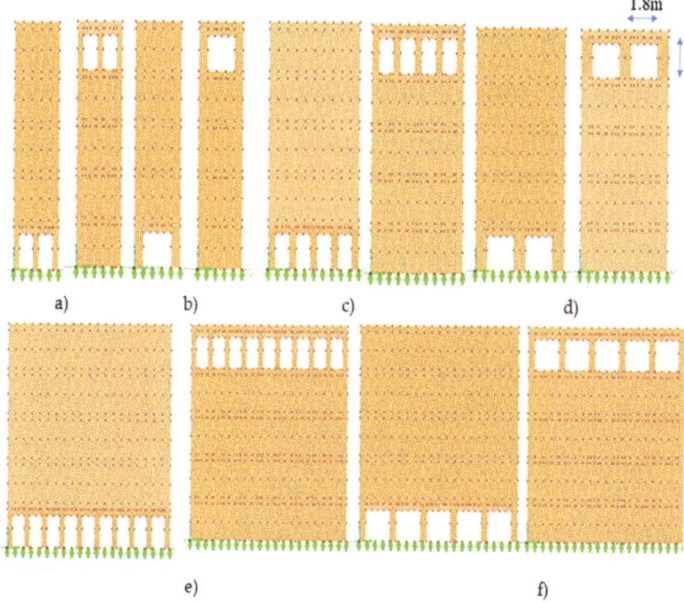

Figure 4. Location and size of horizontal system openings of three different widths of wooden frame walls. The walls 3 m wide (W1) with openings 0.9 × 2 m (Op1) (**a**) and openings 1.8 × 2 m (Op2) (**b**); the walls 6 m wide (W2) with openings 0.9 × 2 m (Op1) (**c**) and openings 1.8 × 2 m (Op2) (**d**); the walls 12 m wide (W3) with openings 0.9 × 2 m (Op1) (**e**) and openings 1.8 × 2 m (Op2) (**f**).

For the vertical opening system, the openings were located one above another on every floor with prime displacement equal to 0.5 m from the left edge of the wall. Then, the edge distance was increased by 0.5 m in every next step until the holes approached the opposite edge of the wall. An example of this type for Op1 and Op2 for W2 is shown in Figure 5; the same scenarios have been performed for W1 and W3.

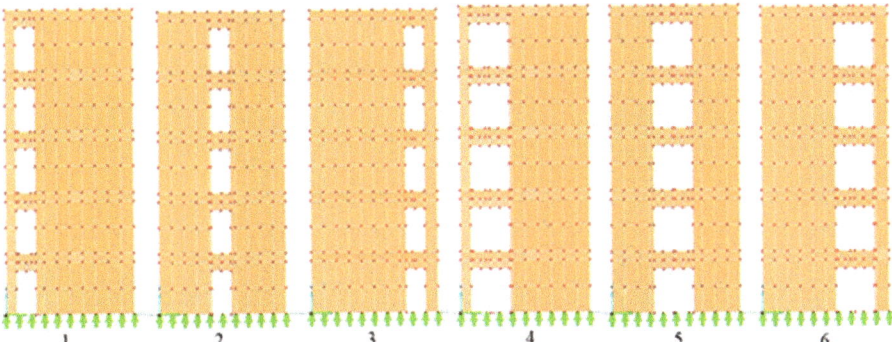

Figure 5. Locations of vertical opening system of W2. 1–3 are the different positions of the openings with dimensions 0.9 × 2 m. 4–6 are the different positions of the openings with dimensions 1.8 × 2 m.

3. Modal Analysis

The analysis was made by the finite elements method in Dlubal RFEM 5.17 software (Dlubal Software GmbH, Tiefenbach, Germany). The analysis shows that the systems have three natural vibrations; this is due to the fact that these kinds of frequencies appear in real five floor buildings [31]. The influence of location and size of openings was estimated by calculating the relative change of natural frequencies using the equation below:

$$d_i = \frac{f_{oi} - f_i}{f_i} * 100, i = 1:3 \qquad (1)$$

where d_i is the relative change of ith natural frequency, f_i is ith the natural frequency for the wall without openings, f_{oi} is ith the natural frequency for wall with openings.

Table 2 shows the response of W1, W2, and W3 without openings to the dynamic loading. The data from the table declares that the lowest frequency appears from the first natural frequency (1st NF) for the three walls, while the middle size wall recorded the lowest 1st NF among all.

Table 2. Tree NF forms of the walls without openings.

Wall Width (m)	Frequency (Hz)		
	Form 1	Form 2	Form 3
3	0.412	2.525	3.552
6	0.408	2.248	2.496
12	0.415	1.311	2.530

In Figure 6, the responses of W1 without and with openings to dynamic loading have been shown; Figure 6a is the 1st form of vibration for the wall without openings. While Figure 6b exemplifies the 1st NF of the wall with Op1 in the right bottom of the wall, Figure 6c is the 3rd NF of the opened wall in the bottom and the top by 1.8 × 2 m. The analysis has been performed for the three walls with different sizes and locations of the openings; the results were tabulated below.

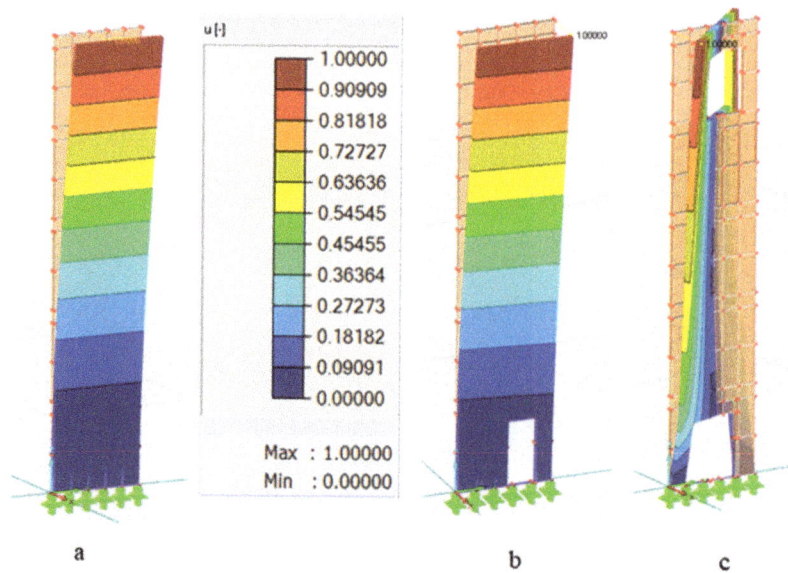

Figure 6. NF of W1: (**a**) 1st NF without openings, (**b**) 1st NF with Op1 at 1st position, and (**c**) 3rd NF with Op2 at 6th position.

The three natural forms of vibration after applying the dynamic load for Op1 and Op2 in six different locations of three various width walls were presented in Table 3. It can be clearly seen that during the 1st and 2nd NF, the lowest value is recorded when the opening is in the right bottom of the walls, while the highest value is recorded for the three forms of NF when the opening is located in the top right of the walls. However, the situation was different regarding the lowest value for the 3rd NF; the smallest NF was obtained when there were three openings in the right diagonal. Numerically speaking, the highest and lowest values were obtained for W1 with Op2. The former recorded 4.048 Hz, which was obtained from the 3rd NF for Op2 in the right top, and the latter recorded 0.29 Hz, which was attained from the 1st NF for the same size of the opening, but its location was in the right bottom.

Table 3. NF for W1, W2, and W3 with Op1 and Op2 shown in Figure 3.

Position	Frequency (Hz)																	
	3 m Wall						6 m Wall						12 m Wall					
	Opening Sizes																	
	0.9 × 2 m			1.8 × 2 m			0.9 × 2 m			1.8 × 2 m			0.9 × 2 m			1.8 × 2 m		
	Forms																	
	1	2	3	1	2	3	1	2	3	1	2	3	1	2	3	1	2	3
1	0.361	2.363	3.387	0.290	2.213	3.211	0.391	2.181	2.463	0.364	2.098	2.417	0.413	1.301	2.501	0.391	1.249	2.461
2	0.433	2.543	4.039	0.461	2.557	4.048	0.425	2.332	2.543	0.437	2.346	2.563	0.420	1.336	2.536	0.426	1.356	2.540
3	0.407	2.468	3.760	0.389	2.334	3.467	0.412	2.221	2.504	0.407	2.171	2.455	0.414	1.296	2.519	0.412	1.281	2.497
4	0.405	2.567	3.792	0.325	2.271	3.297	0.4	2.215	2.495	0.384	2.126	2.49	0.409	1.298	2.500	0.402	1.287	2.479
5	0.376	2.326	3.531	0.321	2.035	3.048	0.398	2.194	2.439	0.379	2.106	2.332	0.407	1.284	2.484	0.398	1.265	2.427
6	0.378	2.39	3.547	0.325	2.276	3.297	0.4	2.243	2.465	0.383	2.195	2.414	0.408	1.300	2.497	0.400	1.295	2.465

Table 4 shows the three forms of vibration for Op1 and Op2 in three different scenarios for the three size walls for the horizontal openings. It could be said that when the openings were in the top story, high natural frequencies were recorded, whereas when they were located in the bottom, low values of NF were obtained. To be more specific, the highest frequency was gained for the 3rd NF of the top Op1, while the smallest value was obtained for the 1st NF of the same opening size but in the bottom floor (see Figure 4).

Table 4. NF for W1, W2, and W3 with Op1 and Op2 of the horizontal system shown in Figure 4.

Floor	Frequency (Hz)																	
	3 m Wall						6 m Wall						12 m Wall					
	Opening Sizes																	
	0.9 × 2 m			1.8 × 2 m			0.9 × 2 m			1.8 × 2 m			0.9 × 2 m			1.8 × 2 m		
	Forms																	
	1	2	3	1	2	3	1	2	3	1	2	3	1	2	3	1	2	3
1	0.260	2.111	2.964	0.303	2.258	3.319	0.282	1.892	2.187	0.299	2.030	2.250	0.241	1.054	2.085	0.242	1.088	2.136
2	0.349	2.513	3.225	0.346	2.522	3.244	0.350	1.947	2.518	0.345	1.965	2.516	0.335	1.163	2.514	0.308	1.152	2.493
3	0.387	2.312	3.476	0.389	2.335	3.470	0.389	2.055	2.320	0.387	2.066	2.307	0.382	1.195	2.264	0.367	1.190	2.178
4	0.419	2.268	3.799	0.427	2.269	3.779	0.420	2.231	2.265	0.424	2.227	2.252	0.422	1.281	2.207	0.423	1.286	2.088
5	0.451	2.530	4.140	0.465	2.560	4.059	0.454	2.389	2.541	0.459	2.369	2.544	0.461	1.383	2.539	0.475	1.404	2.531

Table 5 shows three forms of natural frequencies of the walls for the vertical system openings. The table demonstrates that the values of the 1st and 3rd NF almost remain unchanged by changing the size and location of the openings, while the 2nd NF became higher by decreasing the size of the wall and the openings. The table clearly shows that the highest frequency gained for W2 from the 3rd NF was when Op1 had 0.5 m distance from the left (see Figure 5(1)), while the smallest value found for the same size wall was from the 1st NF when Op2 was located 0.5 m away from the left (see Figure 5(4)).

Table 5. NF for W1, W2, and W3 with Op1 and Op2 of the vertical system shown in Figure 5.

Edge Distance (m)	Frequency (Hz)																	
	3 m Wall						6 m Wall						12 m Wall					
	Opening Size																	
	0.9 × 2 m			1.8 × 2 m			0.9 × 2 m			1.8 × 2 m			0.9 × 2m			1.8 × 2 m		
	Forms																	
	1	2	3	1	2	3	1	2	3	1	2	3	1	2	3	1	2	3
0.5	0.371	2.299	3.371	0.316	1.979	2.624	0.396	2.213	2.429	0.378	2.101	2.323	0.407	1.304	2.482	0.399	1.303	2.433
1	0.372	2.307	3.225	-	-	-	0.397	2.158	2.432	0.379	2.011	2.330	0.407	1.296	2.484	0.400	1.286	2.438
1.5	0.380	2.347	3.342	-	-	-	0.400	2.105	2.448	0.380	1.961	2.332	0.408	1.288	2.493	0.400	1.271	2.440
2	-	-	-	-	-	-	0.399	2.106	2.449	0.379	1.943	2.330	0.408	1.286	2.493	0.400	1.259	2.439
2.5	-	-	-	-	-	-	0.399	2.097	2.449	0.379	1.952	2.330	0.408	1.280	2.493	0.400	1.249	2.440
3	-	-	-	-	-	-	0.400	2.083	2.448	0.383	1.969	2.344	0.408	1.274	2.493	0.400	1.236	2.448
3.5	-	-	-	-	-	-	0.396	2.116	2.432	0.379	2.072	2.327	0.407	1.271	2.493	0.400	1.234	2.441
4	-	-	-	-	-	-	0.396	2.147	2.431	-	-	-	0.407	1.268	2.485	0.400	1.229	2.442
4.5	-	-	-	-	-	-	0.400	2.226	2.449	-	-	-	0.408	1.262	2.493	0.400	1.226	2.443
5	-	-	-	-	-	-	-	-	-	-	-	-	0.408	1.266	2.493	0.400	1.225	2.443
5.5	-	-	-	-	-	-	-	-	-	-	-	-	0.408	1.265	2.493	0.402	1.227	2.454
6	-	-	-	-	-	-	-	-	-	-	-	-	0.408	1.261	2.493	0.401	1.222	2.449
6.5	-	-	-	-	-	-	-	-	-	-	-	-	0.407	1.266	2.485	0.400	1.223	2.441
7	-	-	-	-	-	-	-	-	-	-	-	-	0.407	1.268	2.485	0.400	1.238	2.442
7.5	-	-	-	-	-	-	-	-	-	-	-	-	0.408	1.268	2.493	0.400	1.245	2.442
8	-	-	-	-	-	-	-	-	-	-	-	-	0.408	1.275	2.493	0.400	1.255	2.440
8.5	-	-	-	-	-	-	-	-	-	-	-	-	0.408	1.279	2.493	0.400	1.266	2.438

Table 5. Cont.

Edge Distance (m)	Frequency (Hz)																	
	3 m Wall						6 m Wall						12 m Wall					
	Opening Size																	
	0.9 × 2 m			1.8 × 2 m			0.9 × 2 m			1.8 × 2 m			0.9 × 2m			1.8 × 2 m		
	Forms																	
	1	2	3	1	2	3	1	2	3	1	2	3	1	2	3	1	2	3
9	-	-	-	-	-	-	-	-	-	-	-	-	0.408	1.280	2.492	0.401	1.281	2.446
9.5	-	-	-	-	-	-	-	-	-	-	-	-	0.407	1.288	2.484	0.400	1.299	2.436
10	-	-	-	-	-	-	-	-	-	-	-	-	0.407	1.294	2.484	-	-	-
10.5	-	-	-	-	-	-	-	-	-	-	-	-	0.409	1.313	2.493	-	-	-

4. Relative Change of Natural Frequencies

The relative changes have been derived between the natural frequencies of the walls with and without openings. Furthermore, the figures illustrate the relative change of three forms of the natural frequencies for different sizes of openings, namely Op1 and Op2 in different locations that are shown in Figures 3–5. According to Equation (2), f is directly proportional with K and inversely proportional with M.; hence, when an opening is performed in a wall, K and M decrease. However, the diminishing of each depends on the size and the location of the opening. If K/M of the opened wall is greater than that of the non-opened wall, it means f is enhanced. Thus, the figures clearly illustrate the positivity and negativity of the relative change of the size and locations of openings of each wall size.

$$f = \sqrt{\frac{K}{M}} \qquad (2)$$

where K and M are the stiffness and the mass of the wall.

Figures 7–9 represent the relative changes (d1, d2, and d3) of the natural frequencies of W1, W2, and W3, respectively. One can see from Figure 7, the maximum relative changes obtained for Op1 in the 3rd NF and Op2 in the 1st and 3rd NF in location 2, shown in Figure 3, that the relative change was about 15%. On the other hand, the least relative change gained for Op2 in the 1st NF in location 1, shown in Figure 3, was 30%. In general, the charts in the three figures show that the relative changes of all three forms of NF for both opening sizes were positive when the opening was located in the 2nd position, whereas the relative changes in the majority of the other cases were negative when the openings were located other than in position 2, which is the top right of the wall.

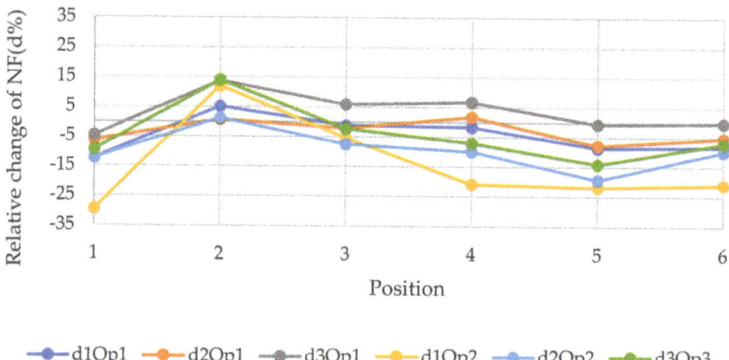

Figure 7. Relative changes of the natural frequencies for W1.

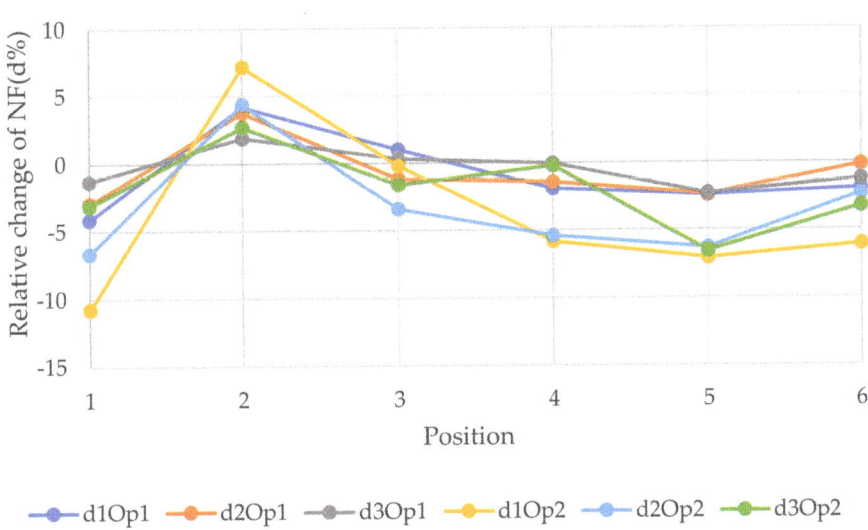

Figure 8. Relative changes of the natural frequencies for W2.

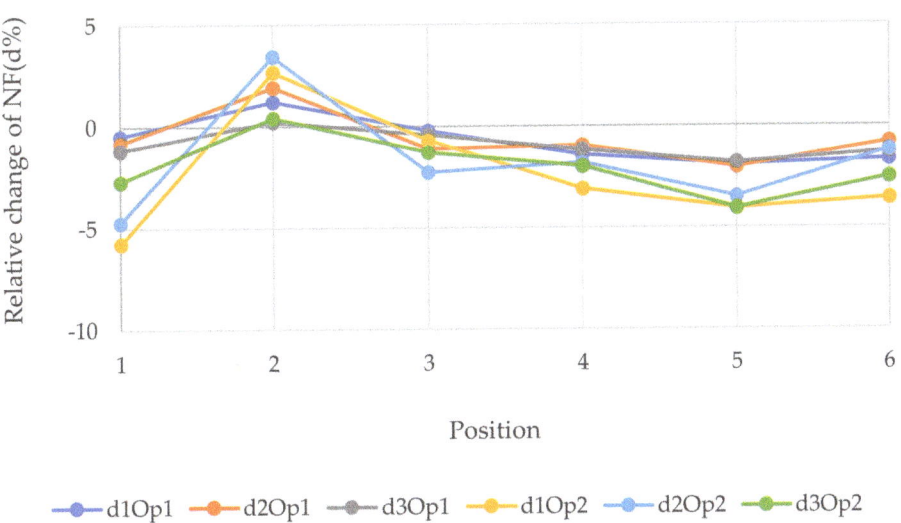

Figure 9. Relative changes of the natural frequencies for W3.

Figures 10–12 epitomize the relative changes (d1, d2, and d3) of the natural frequencies of the walls with the width 3, 6, and 12 m, respectively. As can be seen, the most considerable positive change of the 3rd natural frequency was for the 3 m wide wall with 1.8 × 2 m openings located in position 5, which is the fifth floor, and it was just over 15%. The reason for this could be that when the openings are performed at the bottom of the wall, the structure will be vulnerable. In contrast, when the openings are operated at the top level, the dynamic load will be least effective on the structure. Usually, for W2 and W3, the changes were similar for a particular position. However, there was a rapid increase in relative change due to a change in opening location in the 1st NF for both size openings for all size walls. Generally, the positive relative changes gained when the opening were located on the top floor, in contrast to the negative relative changes gained when the openings were on the 1st floor.

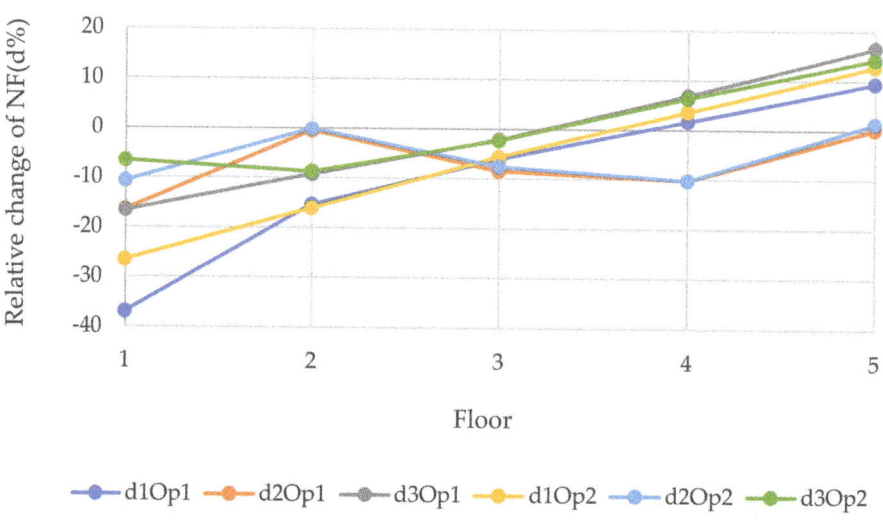

Figure 10. Relative changes of the natural frequencies for W1.

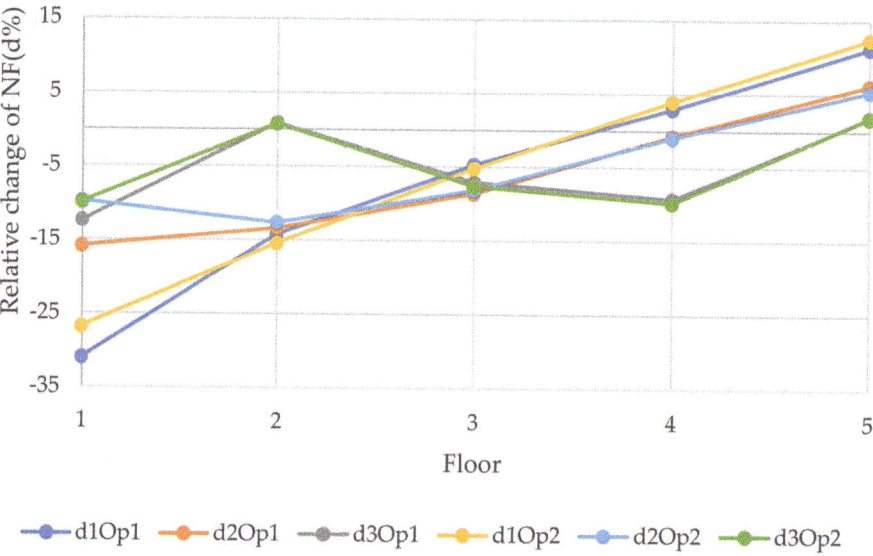

Figure 11. Relative changes of the natural frequencies for W2.

Figures 13–15 show the relative changes (d1, d2, and d3) of the natural frequencies of the walls with the change of the distance of the vertical openings to the left edge of the walls (see Figure 6). As can be seen from the figures, for the 1st and 3rd relative NF for both sizes of openings for 6 and 12 m walls, the lines remained almost flat with the change of the distance of the openings with the left edge of the wall. However, the 2nd RNF has the highest value when the opening is at the edges, and it declines when the opening is close to the middle of the wall and records the lowest value when the opening is at the middle of the wall. Although, for the 3 m wide wall the situation is different; for the 0.9 × 2 m opening, change is decreasing from the beginning, and for the 1.8 × 2 m opening, there is only one value because the hole consists of almost the whole width of the wall and, in this position, the relative change of the 2nd, 1st, and 3rd forms recorded the lowest values, −21.6, −23.3, and −26.1%, respectively. It could be said that the openings in the middle of the wall for the vertical system are considered to be vulnerable.

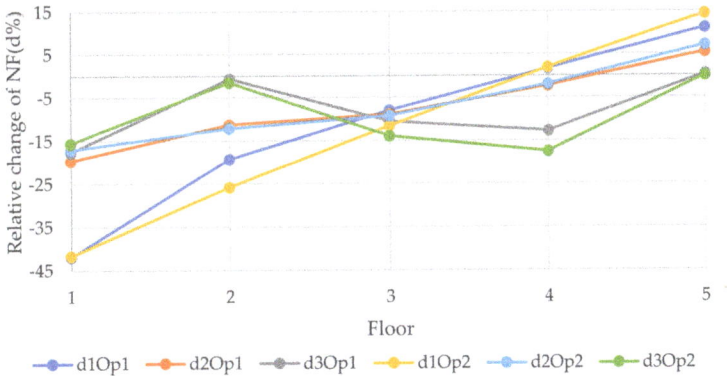

Figure 12. Relative changes of the natural frequencies for W3.

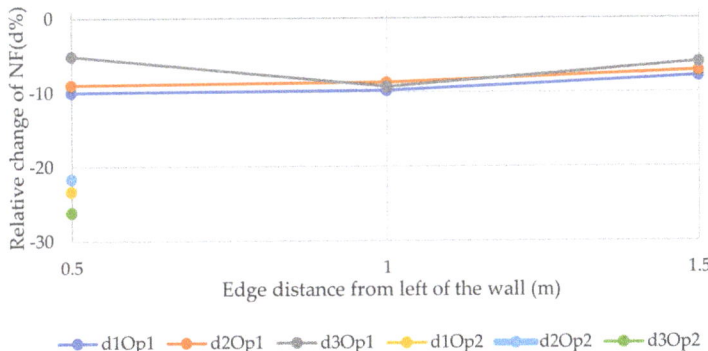

Figure 13. Relative changes of the natural frequencies for W1.

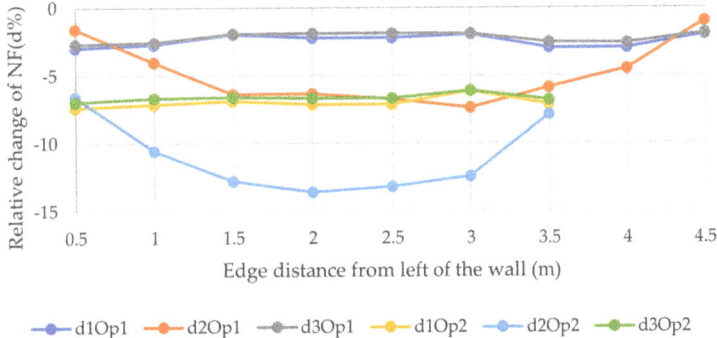

Figure 14. Relative changes of the natural frequencies for W2.

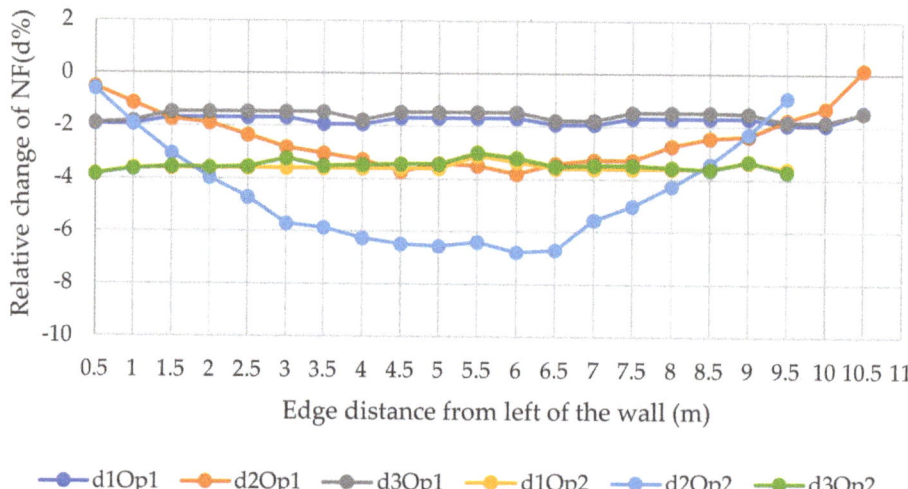

Figure 15. Relative changes of the natural frequencies for W3.

5. Conclusions

In this paper, three walls with different widths without and with two different sizes of openings in several locations have been tested to monitor the change in their natural frequencies under dynamic loads. It has been concluded that:

- Designing new openings with different sizes and locations can significantly affect the change of the values of the natural frequencies.
    - Small openings give better results than the big openings.
    - Vertical openings are the most unfavorable system of openings.
- For the 1st six positions of openings, the highest and lowest natural frequencies were recorded when the openings were located in the top right and bottom right of the walls, respectively. The relative change of the natural frequencies also gave the same outcomes.
- For the horizontal system of openings, the most suitable position for performing the openings is the top floor, while the 1st floor is considered as the worst scenario.
- Regarding the vertical openings, the appropriate case is when the openings are located at the edges, while the inappropriate case scenario has openings in the middle wall.
- According to the research, the appropriate size and location of the openings mitigate the impact of the seismic excitations to the timber frame walls.
- Further study can be conducted through dealing with the effect of the openings on a whole structure.

Author Contributions: Conceptualization, A.M. and M.S.; methodology, M.S.; software, N.S.; formal analysis, M.S. and N.S.; writing—original draft preparation, A.M., M.S. and D.C.; writing—review and editing, A.M., M.S. and D.C. All authors have read and agreed to the published version of the manuscript.

Funding: This research was financially supported by Gdańsk University of Technology under the DEC-1/2020/IDUB/II.1 grant under the Hydrogenium Supporting Membership In International Networks—'Excellence Initiative—Research University' program.

Institutional Review Board Statement: Not applicable.

Informed Consent Statement: Not applicable.

Data Availability Statement: The data presented in this study are available upon request from the corresponding author.

Conflicts of Interest: The authors declare no conflict of interest.

References

1. Sandanayake, M.; Lokuge, W.; Zhang, G.; Setunge, S.; Thushar, Q. Greenhouse gas emissions during timber and concrete building construction—A scenario based comparative case study. *Sustain. Cities Soc.* **2018**, *38*, 91–97. [CrossRef]
2. Vilčeková, S.; Harčárová, K.; Moňoková, A.; Burdová, E.K. Life Cycle Assessment and indoor environmental quality of wooden family houses. *Sustainability* **2020**, *12*, 10557. [CrossRef]
3. Boardman, C.R.; Glass, S.V. Improving the accuracy of a hygrothermal model for wood-frame walls: A cold-climate study. *Buildings* **2020**, *10*, 236. [CrossRef]
4. Burawska-Kupniewska, I.; Krzosek, S.; Mańkowski, P. Efficiency of visual and machine strength grading of sawn timber with respect to log type. *Forests* **2021**, *12*, 1467. [CrossRef]
5. Lunguleasa, A.; Dumitrascu, A.-E.; Ciobanu, V.-D. Comparative studies on two types of osb boards obtained from mixed resinous and fast-growing hard wood. *Appl. Sci.* **2020**, *10*, 6634. [CrossRef]
6. Igaz, R.; Krišťák, Ľ.; Ružiak, I.; Gajtanska, M.; Kučerka, M. Thermophysical properties of OSB boards versus equilibrium moisture content. *Bioresources* **2017**, *12*, 8106–8118. [CrossRef]
7. Di, J.; Zuo, H. Experimental and numerical investigation of light-wood-framed shear walls strengthened with parallel strand bamboo panels. *Coatings* **2021**, *11*, 1447. [CrossRef]
8. Ailenei, E.C.; Ionesi, S.D.; Dulgheriu, I.; Loghin, M.C.; Isopescu, D.N.; Maxineasa, S.G.; Baciu, I.-R. New waste-based composite material for construction applications. *Materials* **2021**, *14*, 6079. [CrossRef] [PubMed]
9. Jivkov, V.; Simeonova, R.; Antov, P.; Marinova, A.; Petrova, B.; Kristak, L. Structural application of lightweight panels made of waste cardboard and beech veneer. *Materials* **2021**, *14*, 5064. [CrossRef]
10. Sudoł, E.; Kozikowska, E. Mechanical properties of polyurethane adhesive bonds in a mineral wool-based external thermal insulation composite system for timber frame buildings. *Materials* **2021**, *14*, 2527. [CrossRef]
11. Antov, P.; Savov, V.; Neykov, N. Possibilities for manufacturing insulation boards with participation of recycled lignocellulosic fibres. *Manag. Sustain. Dev.* **2019**, *75*, 72–76.
12. Tudor, E.M.; Dettendorfer, A.; Kain, G.; Barbu, M.C.; Réh, R.; Krišťák, Ľ. Sound-absorption coefficient of bark-based insulation panels. *Polymers* **2020**, *12*, 1012. [CrossRef] [PubMed]
13. Migda, W.; Szczepański, M.; Jankowski, R. Increasing the seismic resistance of wood-frame buildings by applying PU foam as thermal insulation. *Period. Polytech. Civ. Eng.* **2019**, *63*, 480–488. [CrossRef]
14. Szczepański, M.; Migda, W.J.S. Analysis of validation and simplification of timber-frame structure design stage with PU-foam insulation. *Sustainability* **2020**, *12*, 5990. [CrossRef]
15. Szczepański, M.; Migda, W.; Jankowski, R. Modal analysis of real timber frame houses with different insulation materials. *Adv. Sci. Technol. Res. J.* **2016**, *10*, 215–221. [CrossRef]
16. Mohammadabadi, M.; Yadama, V.; Dolan, J.D. Evaluation of wood composite sandwich panels as a promising renewable building material. *Materials* **2021**, *14*, 2083. [CrossRef]
17. Latif, E.; Ciupala, M.A.; Tucker, S.; Wijeyesekera, D.C.; Newport, D.J.J.B. Hygrothermal performance of wood-hemp insulation in timber frame wall panels with and without a vapour barrier. *Build. Environ.* **2015**, *92*, 122–134. [CrossRef]
18. Filiatrault, A. Static and dynamic analysis of timber shear walls. *Can. J. Civ. Eng.* **1990**, *17*, 643–651. [CrossRef]
19. Porcu, M.C.; Bosu, C.; Gavrić, I. Non-linear dynamic analysis to assess the seismic performance of cross-laminated timber structures. *J. Build. Eng.* **2018**, *19*, 480–493. [CrossRef]
20. Popovski, M.; Schneider, J.; Schweinsteiger, M. Lateral load resistance of cross-laminated wood panels. In Proceedings of the World Conference on Timber Engineering, Trento, Italy, 20–24 June 2010; p. 24.
21. Schieweck, A. Very volatile organic compounds (VVOC) as emissions from wooden materials and in indoor air of new prefabricated wooden houses. *Build. Environ.* **2021**, *190*, 107537. [CrossRef]
22. He, M.; Lam, F.; Foschi, R.O. Modeling three-dimensional timber light-frame buildings. *J. Str. Eng.* **2001**, *127*, 901–913. [CrossRef]
23. Šušteršič, I.; Dujič, B. Simplified cross-laminated timber wall modelling for linear-elastic seismic analysis. *CIB-W18* **2012**, *45*, 45-15.
24. Yasumura, M.; Sugiyama, H. Shear properties of plywood-sheathed wall panels with opening. *Trans. Archit. Instit. Jpn.* **1984**, *338*, 88–98. [CrossRef]
25. Dujič, B.; Klobčar, S.; Žarnić, R. Influence of openings on shear capacity of wooden walls. *N. Z. Timber Des. J.* **2008**, *16*, 5–17.
26. Dujič, B.; Klobčar, S.; Žarnić, R. Shear capacity of cross-laminated wooden walls. In Proceedings of the 10th World Conference on Timber Engineering, Miyazaki, Japan, 2–5 June 2008.
27. EN 338. *Timber Structures—Strength Classes*; European Committee for Standarization (CEN): Brussels, Belgium, 2016.
28. Isopescu, D.; Stanila, O.; Astanei, I. Analysis of wood bending properties on standardized samples and structural size beams tests. *Bul. Instit. Politeh. Iasi. Sectia Constr. Arhit.* **2012**, *58*, 65.

29. Szczepański, M.; Migda, W.; Jankowski, R. Experimental study on dynamics of wooden house wall panels with different thermal isolation. *Appl. Sci.* **2019**, *9*, 4387. [CrossRef]
30. European Committee for Standardization (CEN). *Eurocode 5—Design of Timber Structures—Part 1-1: General Rules and Rules for Buildings*; European Committee for Standardization: Brussels, Belgium, 2004.
31. Ciesielski, R.; Kuźniar, K.; Maciąg, E.; Tatara, T. Empirical formulae for fundamental natural periods of buildings with load bearing walls. *Arch. Civ. Eng.* **1992**, *38*, 291–299.

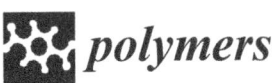

Article

Thermal Modification of Spruce and Maple Wood for Special Wood Products

Anna Danihelová [1,*], Zuzana Vidholdová [2], Tomáš Gergeľ [3], Lucia Spišiaková Kružlicová [4] and Michal Pástor [3]

[1] Department of Fire Protection, Faculty of Wood Sciences and Technology, Technical University in Zvolen, T. G. Masaryka 24, 960 01 Zvolen, Slovakia

[2] Department of Wood Technology, Faculty of Wood Sciences and Technology, Technical University in Zvolen, TU in Zvolen, Masaryka 24, 960 53 Zvolen, Slovakia; zuzana.vidholdova@tuzvo.sk

[3] National Forest Centre, Forest Research Institute, T. G. Masaryka 22, 960 01 Zvolen, Slovakia; tomas.gergel@nlcsk.org (T.G.); michal.pastor@nlcsk.org (M.P.)

[4] Department of Furniture Design and Interior, Faculty of Wood Sciences and Technology, Technical University in Zvolen, TU in Zvolen, Masaryka 24, 960 53 Zvolen, Slovakia; kruzlicovalucia@gmail.com

* Correspondence: danihelova@acoustics.sk

Abstract: This article presents a proposal of thermal modification of Norway spruce and sycamore maple for special wood products, mainly for musical instruments. Selected physical and acoustical characteristics (PACHs), including the density (ρ), dynamic modulus of elasticity along the wood grain (E_L), specific modulus (E_{sp}), speed of sound along the wood grain (c_L), resonant frequency (f_r) and acoustic constant (A), logarithmic decrement (ϑ), loss coefficient (η), acoustic conversion efficiency (ACE), sound quality factor (Q), and the timbre of sound, were evaluated. These two wood species were chosen regarding their use in the production or repair of musical instruments. For the thermal modification, a similar process to the ThermoWood process was chosen. Thermal modification was performed at the temperatures 135 °C, 160 °C and 185 °C. The resonant dynamic method was used to obtain the PACHs. Fast Fourier transform (FFT) was used to analyze the sound produced. The changes in the observed wood properties depended on the treatment temperature. Based on our results of all properties, the different temperature modified wood could find uses in the making of musical instruments or where the specific values of these wood characteristics are required. The mild thermal modification resulted in a decrease in mass, density, and increased speed of sound and dynamic modulus of elasticity at all temperatures of modification. The thermally modified wood showed higher sound radiation and lower loss coefficients than unmodified wood. The modification also influenced the timbre of sound of both wood species.

Keywords: thermal modification; spruce and maple wood; physical and acoustical characteristics; timbre of sound; musical instruments

Citation: Danihelová, A.; Vidholdová, Z.; Gergeľ, T.; Spišiaková Kružlicová, L.; Pástor, M. Thermal Modification of Spruce and Maple Wood for Special Wood Products. *Polymers* **2022**, *14*, 2813. https://doi.org/10.3390/polym14142813

Academic Editor: Antonios N. Papadopoulos

Received: 22 June 2022
Accepted: 8 July 2022
Published: 10 July 2022

Publisher's Note: MDPI stays neutral with regard to jurisdictional claims in published maps and institutional affiliations.

Copyright: © 2022 by the authors. Licensee MDPI, Basel, Switzerland. This article is an open access article distributed under the terms and conditions of the Creative Commons Attribution (CC BY) license (https://creativecommons.org/licenses/by/4.0/).

1. Introduction

The modification of wood properties using heat has been used for many years in the manufacture of musical instruments, where the technique is used to replicate the highly desirable tones and inherent stability of aged guitars and violins. String musical instruments have come a long way since their inception. The development has reached a stage where it is possible to move materially and structurally only within certain limits (e.g., due to prestressing, the volume and mass of the board can be reduced or the center of gravity of the board can be shifted). It is, therefore, desirable to focus on the material from which the musical instrument is made.

Musical instrument manufacturers have been using wood that is naturally dried for more than 5 years. The aging of wood can be defined as slow chemical reactions in wood that occur over time under variations in climatic conditions [1]. These chemical changes, as well as the changes in the microstructure of wood, are the cause of the changes in the physical, acoustical, and mechanical properties of wood [2,3]. Long-term natural drying reduces the hygroscopicity of wood. The reduced hygroscopicity improves the dimensional stability of wood, as well as stabilizing its mechanical and acoustic properties. The speed of sound increases and the internal damping decreases with a reduction in moisture content; therefore, the musical instrument often sounds noticeably brighter [4,5].

Obtaining long-term aging wood is becoming increasingly difficult; therefore, manufacturers of musical instruments are becoming increasingly interested in various methods that allow the simulation of the effects of long-term natural drying wood. The aim of wood modification is to achieve the required properties of wood for musical instruments. Mechanical modification (densification) of wood, thermal modification, or their combination, appear to be suitable modifications of wood for musical instruments [6–12]. Similarly, the biological modification of wood with wood-staining fungi or wood-destroying fungi can be used to modify the relevant properties [13–15]. These modifications of wood improve its properties, producing material that when disposed at the end of the product life cycle, does not present an environmental hazard any greater than unmodified wood [16]. Heat treatment as a wood modification process is based on chemical degradation of wood by heat transfer [17]. In the heat treatment process, wood is heated to temperatures ranging from 180 °C to 260 °C, where lower temperatures cause minor changes in wood components and higher temperatures cause severe changes [16]. Thermal modification causes the number of hydroxyl groups of cellulose and hemicelluloses to decrease, thereby resulting in the decrease in the absorption of water [18,19]. Thermally modified wood is dimensionally more stable, which is beneficial to wind instruments (due to changes in relative humidity, the instrument does not lose tuning. The increase in dimensional stability [20] reduces the formation of cracks and improves the efficiency of paint coating systems. Higher dimensional stability can give greater tuning stability, which means that instruments will need to be tuned less [7]. Thermally modifying the wood also changes other important wood properties, such as biological durability, hardness, and UV-stability [18,21,22]. This is very important in wooden organ pipe making. The thermal modification also induces a darker coloration of the wood. The darkness intensity is dependent upon the treatment time and temperature [23]. Different wood species will turn a different darker shade during the process, due to their natural characteristics. The darker shade of wood color is welcome for some kinds of musical instruments (marimba, xylophone, guitar, etc.); however, this is provided that the wood has the required physical and acoustic properties in addition to color [6,24,25].

The aim of thermal modification of wood for musical instruments making is to maximize sonic benefits and minimize degradation (in guitar-making, this process is also known as "thermo curing", "wood torrefaction" or "roasting" [25]). Guitar manufacturers have begun using acoustic sound boards and electric guitar fretboards that are thermally treated to help prevent the warping and cracking that often occurs. As a secondary benefit, the acoustic guitars (also violins) tend to sound similar to well-broken-in aged instruments much sooner than instruments made from thermally unmodified wood [24].

Based on the above-mentioned properties of long-term stored wood and thermally modified wood, it can be stated that relatively mild thermal treatment will accelerate changes in the structure, and thus also in the physical and acoustic properties of the wood [8,26]. The quality of a string musical instrument strongly depends on that of its soundboard. It is generally known that a high acoustic constant, low density, and a large degree of anisotropy are required for an excellent soundboard [27,28]. The resonant wood (for string musical instruments) should have a low density ρ (around 430 kg·m^{-3}), a low logarithmic damping decrement ϑ (around 0.022) and a high acoustic constant A (around 13 m^4·kg^{-1}·s^{-1}). On the other hand, maple wood should have a high density (around 600 kg·m^{-3}), a low acoustic constant A (around 6.5 m^4·kg^{-1}·s^{-1}) and logarithmic damping decrement ϑ (around 0.05). These relevant properties of used wood are important because flexible wood tends to shape well and projects the highest quality of sound [27,29–31].

The main objective of this study was to propose the appropriate regime of thermally modifying spruce and maple wood (for producing musical instruments) to improve their physical and acoustical characteristics (density, dynamic modulus of elasticity, acoustic constant, speed of sound, logarithmic damping decrement, etc.). The second objective was to determine the effect of thermal modification on the sound.

2. Material and Methods

Wood specimens were prepared from Norway spruce (*Picea abies* (L.) Karst.) and Sycamore maple (*Acer pseudoplatanus*, L.) growing in Slovakia. The specimens were prepared after natural drying the wood outdoors. The specimens from spruce and maple wood were sawn radially into the shape of a bar with dimensions (10 × 10 × 400) mm. The specimens from spruce wood had an annual ring width, which is typical for the bass to discant zone of pianos [32]. The specimens made from maple wood included curly maple and maple with straight grains. The specimens were divided into three sets for thermal modification at temperatures 135 °C, 160 °C and 185 °C. Each set consisted of 30 specimens.

The experiment was divided into three parts. In the first part of the experiment, the specimens were conditioned to 8% moisture at (20 ± 2) °C and (45 ± 5)% relative humidity to avoid splitting during thermal modification. After conditioning, the physical and acoustical characteristics (density ρ, modulus of elasticity along the grain E_L, acoustic constant A, speed of sound c_L, logarithmic damping decrement ϑ, loss coefficient η, acoustic conversion efficiency (ACE) and sound quality factor Q), the dimensions, volume, mass and moisture content of natural spruce and maple wood were determined. The moisture content of the samples after conditioning was in the intervals (8.0 ± 0.2)%.

The second part of the experiment was thermally modifying the specimens. The thermal modification used in this experiment used the ThermoWood process [33]. This method enables the modifying of the wood at lower temperatures (smaller decomposition/depolymerization of the construction polymers of wood) [34]. This is a "dry technology" wood that is modified in air, and the oil is not used to increase the interaction temperature, as there is no need to create an environment of steam typical for the PlatoWood process [35].

The thermal modification of wood consisted of the following three steps (Figure 1): high temperature drying (at 100 °C for 90 min and then the steadily increasing to 130 °C for 2.5 h); heat treatment at t = 135 °C, 160 °C and 185 °C for a 50 min period, cooling and moisture conditioning. The wood was cooled down in a controlled way so that its temperature reached 80 °C. At the end of the thermal modification, the kiln was turned off and the specimens were kept inside. The specimens were allowed to cool naturally until their temperature reached 20 °C.

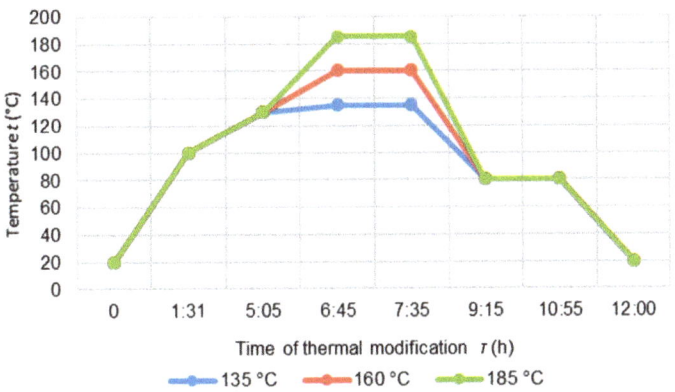

Figure 1. Diagram of the thermal modification process.

After thermal modification, all specimens underwent conditioning under the same conditions as before the thermal modification to stabilize wood moisture content. After the stabilization of moisture content of wood (i.e., after reaching the equilibrium humidity of the wood) the physical and acoustical characteristics (density ρ, modulus of elasticity along the grain E_L, acoustic constant A, speed of sound c_L, logarithmic damping decrement ϑ, loss coefficient η, acoustic conversion efficiency (ACE), sound quality factor Q), the dimensions, volume, mass, and moisture content) as well as dimensions, volume and mass of specimens were measured again.

The moisture content of spruce and maple species were $7.0 \pm 0.2\%$ and $5.8 \pm 0.2\%$, respectively, because of thermal modification at 135 °C and 185 °C temperatures. After modifying at 160 °C, the moisture contents (MC) of the spruce and maple wood specimens were different. MC of the spruce wood was $6.4 \pm 0.2\%$, but MC of maple wood reached only $5.8 \pm 0.2\%$.

For obtaining the relevant physical and acoustical characteristics (PACH) before and after modification, the resonant dynamic method was used. The principle of this method searches for the resonant frequency fr of a vibrating body (usually of bar shape), for which the dynamic modulus of elasticity along the grain E_L can be calculated. The measuring device MEARFA (Figure 2) was used for the measurements of relevant characteristics.

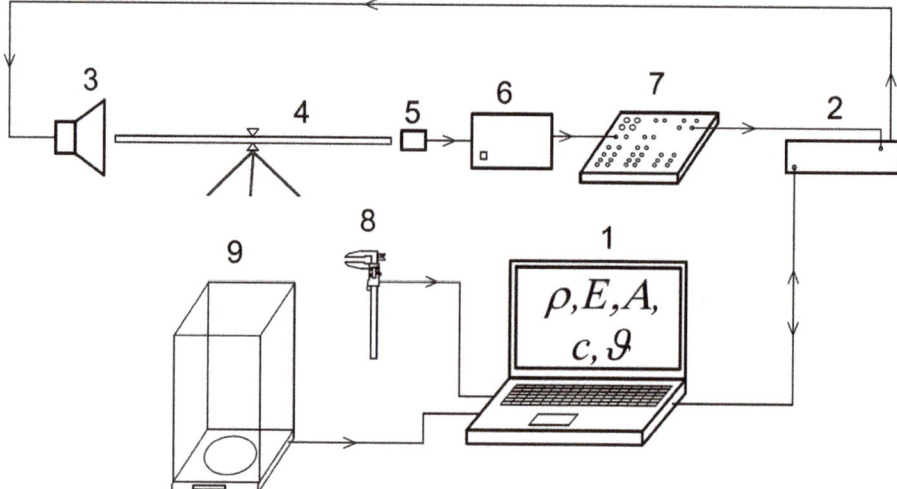

Figure 2. Measuring device MEARFA [36].

The device consists of a computer (1); generator of sinusoidal signal (in the range of acoustic frequencies) and response detector (2); exciter-loudspeaker (3); test specimen (4); electromagnetic detector (5); high-pass filter (6); preamplifier (7); digital gauger, Mitutoyo (8) and electronic scale (9). The fundamental resonance frequency fr, as well as frequencies f_1 and f_2, were determined. These were stored in the computer for further processing and evaluation using the computer software. The physical and acoustical characteristics (density ρ, modulus of elasticity along the grain E_L, acoustic constant A, speed of sound c_L and logarithmic damping decrement ϑ) were calculated according to the Equations (1)–(4).

The dynamic modulus of elasticity along the wood grain E_L (Pa) of the specimen in the shape of bar was calculated [4,27] using the following equation:

$$E_L = 4.l.f_r^2.\rho \qquad (1)$$

where ℓ (m) is the length of the specimen, ρ (kg·m^{-3}) is the density and fr (Hz) is the resonant frequency. The longitudinal specific modulus of elasticity E_{sp} (m^2·s^{-2}) was calculated with the equation

$$E_{sp} = \frac{E_L}{\rho} \quad (2)$$

The acoustic constant (also well known as sound radiation coefficient—R) A (m^4·kg^{-1}·s^{-1}) was calculated [27,29] using the equation

$$A = \sqrt{\frac{E_L}{\rho^3}} = \frac{c_L}{\rho} \quad (3)$$

where E_L (Pa) is the modulus of elasticity along the wood grain; c (m·s^{-1}) is the speed of sound. For calculating the speed of sound in wood [29], the following equation was used:

$$c_L = \sqrt{\frac{E_L}{\rho}} \quad (4)$$

where E_L (Pa) is the modulus of elasticity along the wood grain; ρ (kg·m^{-3}) is the density. The logarithmic damping decrement (expresses the absorption of acoustic energy by the material) was calculated [27] by the equation

$$\vartheta = \frac{\pi}{\sqrt{3}} \cdot \frac{f_2 - f_1}{f_r} \quad (5)$$

where f_1 and f_2 (Hz) are frequencies at which the amplitude of the vibrations is half the maximum amplitude at resonant frequency fr.

Given the importance of vibration damping in musical instruments, the internal friction is also relevant. This is an intrinsic material property, unlike other loss mechanisms, such as the radiation of acoustic energy. An indicator of internal friction is the loss coefficient η. It was measured through the logarithmic decrement and calculated with the equation

$$\eta = \frac{\vartheta}{\pi} \quad (6)$$

Combining A (m^4·kg^{-1}·s^{-1}) and η, the acoustic conversion efficiency (ACE) (m^4·kg^{-1}·s^{-1}) was calculated according to the equation

$$ACE = \frac{A}{\eta} \quad (7)$$

The ACE is useful to show group effects, e.g., the effect of internal friction and sound radiation together [37,38]. The η is related to the sound quality factor Q, which represents the mechanical gain of a structure at a resonant frequency. The Q is a descriptor of the wood sound quality because it provides information about the resonance sharpness. The Q was calculated using the equation

$$Q = \frac{1}{\eta} \quad (8)$$

The characteristic acoustic impedance z (MPa·s·m^{-1}) represents the resistance of wood to sound wave propagation. It is important for musical instruments because it describes vibration transfer. A high value of characteristic acoustic impedance presents a high rate of reflection of sound wave by the medium with which it travels in. At the boundary between the media of different acoustic impedances, some of the wave energy is reflected and some is transmitted. Two materials with a large difference in values of characteristic acoustic impedance give much larger reflections as the transmissions. The characteristic acoustic impedance z was calculated using the equation

$$z = \rho \cdot c_L \tag{9}$$

The physical and acoustical characteristics were evaluated using Statistica software (Statsoft Inc., Version 7, Prague, Czech Republic), the two-way analysis of variance (ANOVA) as well as correlation analyses with the coefficient of determination (R2) and Duncan's multiple range test to determine the significance of the variation at a 0.05 significance level (α).

The third part of experiment was fast Fourier transform (FFT), which is a digital implementation of the Fourier transform. The musical instruments in fact produce sound as a result of the vibration of strings, bars, or plates. Vibration causes a periodic variation in air pressure that is heard as a sound and these sound waves can be analyzed using Fourier series. The sound produced by musical instruments of the same kind differs, because they have different timbre sounds. The reason for this is that the energy in each of the harmonics is different. The Fourier analysis provides mathematical evidence of the dissimilarities of musical timbre sound.

The output of the frequency analysis is the frequency response, showing the sound intensity levels of the individual frequency components of the signal. The impulse excitation technique was used (described by [39]) to excite the bar vibrations. For our experiment, the experimental device (shown in Figure 3) was designed to simulate the conditions resembling those in which a musician would play a xylophone. The important part of device is the mechanical system of impulse excitation. This system ensures the same conditions of excitation for all specimens. The mechanical system consists of a pendulum that hangs on a metal rod embedded in the plates of plexiglass. A pendulum consists of a nylon cord ($\ell = 0.3$ m) and a metal ball ($d = 14$ mm, $m = 12$ g). The pendulum was set in motion to trigger a vibration in the wood bar by hitting the end with the metal ball. The system is designed to trigger a vibration in the wood bar by hitting it at the bottom of the bar at a distance of 10 cm from its end.

Figure 3. Experimental device for FFT analysis.

The test specimen was fastened in the node line of the 4th (2, 0) mode of vibration. The position of the nodes was determined as the excitation of the 4th mode of vibration before FFT analysis. The vibration frequency of the supporting system was low (below 10 Hz). The microphone was placed on a separate holder above the specimen at a distance of 25 mm from the other bar end to measure the acoustic pressure that radiates at impact. The sounds were produced and recorded in the laboratory in a free field.

The FFT analysis of sound before and after thermal modification of spruce and maple wood was performed using Adobe Audition program 1.5.

3. Results and Discussion

The mean values of the physical and acoustical characteristics (density ρ, dynamic modulus of elasticity E_L, acoustic constant A, speed of sound c_L, resonant frequency f_r, logarithmic damping decrement ϑ, loss coefficient η, acoustic conversion efficiency (ACE) and sound quality factor Q) of spruce and maple specimens before thermal modification and after thermal modification at the temperatures 135 °C, 160 °C and 185 °C are given in Table 1.

Table 1. PACH (density ρ, dynamic modulus of elasticity E_L, acoustic constant A, speed of sound c_L, resonant frequency f_r, logarithmic damping decrement ϑ, loss coefficient η, characteristic acoustic impedance z, acoustic conversion efficiency (ACE) and sound quality factor Q) of spruce and maple wood before and after thermal modification.

PACH		Norway Spruce				Sycamore Maple			
		Unmod.	135 °C	160 °C	185 °C	Unmod.	135 °C	160 °C	185 °C
ρ (kg·m^{-3})	MV	425	420	413	415	609	601	603	604
	SD	32.8	21.6	13.8	17.9	38.6	24.2	19.6	20.5
		y = 0.0671x + 426.64 R² = 0.8274				y = −0.037x + 608.87 R² = 0.628			
E_L (GPa)	MV	12.48	13.12	12.99	12.63	10.70	11.35	11.43	11.02
	SD	0.025	0.018	0.020	0.017	0.034	0.021	0.028	0.0361
		y = 0.0021x + 12.544 R² = 0.257				y = 0.0032x + 10.723 R² = 0.4911			
E_{sp} (10^6·m^2·s^{-2})	MV	29.36	31.24	31.45	30.43	17.57	18.88	18.95	18.24
	SD	0.031	0.023	0.019	0.13	0.33	0.24	0.25	0.28
		y = 0.0098x + 29.396 R² = 0.5668				y = 0.0064x + 17.614 R² = 0.519			
c_L (m·s^{-1})	MV	5419	5589	5608	5517	4192	4346	4353	4271
	SD	390	228	263	281	285	223	231	197
		y = 0.089x + 5422 R² = 0.5739				y = 0.747x + 4197 R² = 0.5215			
A (m^4·kg^{-1}·s^{-1})	MV	12.75	13.31	13.58	13.29	6.88	7.23	7.22	7.07
	SD	1.15	0.88	0.96	0.86	0.82	0.63	0.68	0.65
		y = 0.0042x + 12.707 R² = 0.7767				y = 0.0017x + 6.89 R² = 0.5587			
f_r (Hz)	MV	6025	6211	6238	6192	5240	5410	5382	5328
	SD	527	462	428	483	535	436	323	341
		y = 0.0042x + 12.105 R² = 0.8241				y = 0.0042x + 12.105 R² = 0.8241			
ϑ (-)	MV	0.036	0.030	0.031	0.033	0.055	0.052	0.053	0.051
	SD	0.00369	0.00312	0.00279	0.00264	0.00427	0.00381	0.00363	0.00356
		y = −3·10^{-5}x + 0.0358 R² = 0.5142				y = −2·10^{-5}x + 0.0554 R² = 0.8283			
η (-)	MV	0.0115	0.0095	0.0099	0.0105	0.0175	0.0165	0.0168	0.0162
	SD	0.0018	0.0009	0.0011	0.0012	0.0021	0.0019	0.0017	0.0019
		y = −9·10^{-6}x + 0.0114 R² = 0.5108				y = −7·10^{-6}x + 0.0176 R² = 0.8532			
z (MPa·s·m^{-1})	MV	2303	2347	2316	2290	2553	2612	2625	2580
	SD	12.84	4.79	3.63	5.03	11.01	5.35	4.53	4.39
		y = 0.0072x + 2313 R² = 0.0005				y = 0.2966x + 2555 R² = 0.4449			
Q	MV	86.96	105.26	101.01	95.23	56.82	60.61	59.52	61.73
	SD	5.8	4.1	3.7	3.9	5.2	4.6	4.2	3.8
		y = 0.0734x + 87.938 R² = 0.4568				y = 0.0268x + 56.324 R² = 0.8613			
ACE (m^4·kg^{-1}·s^{-1})	MV	1109	1401	1372	1266	393.1	438.2	429.8	436.4
	SD	183	165	162	157	108	98	96	93
		y = 1.3508x + 1118 R² = 0.556				y = 0.2714x + 390.45 R² = 0.8767			

3.1. Effect of Thermal Modification on Density

Table 1 shows that thermal treatment caused a decrease in spruce wood (SW) density at each temperature used in our experiments. The largest density decrease (on average around 2.8%) was recorded at 160 °C. The mean values of maple wood (MW) density reduction varied between 0.8 and 1.3%, in which the highest reduction was found after the treatment at the temperature of 135 °C. According to the ANOVA test, there is no statistically significant difference between the density of thermal modified and unmodified wood (F = 2.28, p > 0.05). Positive correlation between the temperature of modification and density was observed for both wood species (R^2 = 0.83 for SW; R^2 = 0.63 for MW).

The density reduction was mainly due to a reduction in mass (SW varied from 2.1% to 3.6%; MW from 2.5% to 4.6%) and a decrease in volume (SW from 1.8% to 2.9%; MW 1.2% to 3.3%), which was lower than the corresponding mass reduction. Our results showed that the mass loss of both wood species increased with temperature, which agrees with earlier results for spruce [40] and maple wood [35]. The mass loss values were similar to the results found in other studies of thermally modified softwoods and hardwoods [41,42]. The extent depends upon temperature and treatment time [43–45].

Thermal modification at lower temperatures led to a lower mass reduction associated mainly with the loss of volatiles and bound water. Higher mass loss was observed when the samples were treated at temperatures above 150 °C, which is the effect of the partial decomposition of hemicellulose and cellulose, but also other changes in the chemical structure of the thermally modified wood [35,46]. The intensity of thermic degradation (mass loss) maple wood was higher than spruce wood. It can be explained through the differences in chemical composition of hardwoods and softwoods because hardwoods contain slightly more hemicelluloses (25%) compared to softwoods (20%), which are the least resistant to thermal degradation [47,48].

3.2. Effect of Thermal Modification on Modulus of Elasticity

Mean values of modulus of elasticity (MOE) of spruce wood after modification increased by 5.1% (at 135 °C), by 4.1% (at 160 °C), and only by 1.2% after modification at 185 °C. This is in agreement with Millett and Gerhards [48]. Navickas et al. [49] reported slightly higher MOE of spruce wood (by 0.4%) than it was before heating (at 190 °C). Likewise, Kubojima et al. [50] found that heat treatment at 160 °C resulted in increased dynamic MOE. The modulus of elasticity maple wood increased by 6.0%, 6.8% and 3.0% (at 135 °C, 160 °C and 185 °C, respectively) compared to unmodified maple wood. The values of the R^2 show a slight dependence speed of sound on temperature (R^2 = 0.51 for SW; R^2 = 0.49 for MW). The above-mentioned changes in the moduli of elasticity after thermal modification are not statistically significant (F = 0.93, p > 0.05) and the coefficient of determination was also low (for SW, it was 0.26 and for MW 0.49), respectively. This result corresponds with the results of Bekhta and Niemz [23]. Rusche [51] presents that irrespective of treatment and wood species, the decrease in MOE was significant when the mass loss exceeded 8%.

3.3. Effect of Thermal Modification on Acoustic Constant

The mean value of the acoustic constant of spruce wood after thermal modification increased by 4.4%, 6.5 and 4.2% and the acoustic constant of maple wood increased by 6.1, 6.8 and 3.0%. Stronger correlation was found for spruce wood (R^2 = 0.78) than for maple wood (R^2 = 0.56). The effect of thermal modification on the acoustic constant was significant (F = 3.68, p < 0.05). The sound radiation coefficient describes how much the vibration of a body is damped due to sound radiation. Particularly in the case of idiophones, such as xylophones and soundboards, a large sound radiation coefficient of the material is desirable [27,29]. An increase in acoustic constant due to thermal modification of spruce is, therefore, welcome because soundboards prefer values over 10 $m^4 \cdot kg^{-1} \cdot s^{-1}$ [27,29]. In terms of the use on the back plate string instruments, this change is negative. On the other hand, this increase is welcome, for example, in making bassoons and others woodwind instruments and drums kits as they are favored for their bright sound. The acoustic constant for these instruments can be in the range 4 to 8 $m^4 \cdot kg^{-1} \cdot s^{-1}$ [52].

3.4. Effect of Thermal Modification on Speed of Sound

The speed of sound in a material depends on how quickly vibrational energy can be transferred through the medium. Therefore, the speed of sound in a material depends on the material and on the state of the material. A high speed of sound (over 5000 m·s^{-1}) is preferred for soundboards [27,52]. Our results showed that the sound waves move faster in thermally modified wood species. The speed of sound has increased after the thermal modification in both wood species, although the density slightly has decreased. Speed of sound of spruce wood increased (0.8 to 1.8%) and ranged from 5589 to 5608 m·s^{-1}, i.e., this modification is appropriate for the soundboards of string musical instruments [53]. A higher increase was found for maple (1.9 to 3.8%). These results are consistent with an increase in the modulus of elasticity after thermal modification. The similar results were published by Puszynski and Warda [7], Pfriem [8] as well as by Esteves and Pereira [54]. Positive correlation between the temperature of modification and speed of sound was observed for spruce ($R^2 = 0.57$) and for maple wood ($R^2 = 0.52$). These changes are not statistically significant ($F = 2.14$, $p > 0.05$).

3.5. Effect of Thermal Modification on Specific Modulus of Elasticity

Specific modulus of elasticity E_{sp} represents the stiffness to mass ratio and is also known as specific stiffness. It is one of the most important parameters of resonant wood. E_{sp} is directly proportional to the cell-wall substances [55]. Hardwoods, in contrast to softwoods, include a diverse range of cell types, e.g., fibers, vessels, and parenchyma [56]. Softwood lignin differs from hardwood lignin in its structural structure and content. Lignin content is higher in softwood (25 to 36%); hardwood contains less lignin (15 to 25%). Lignin gives wood strength by joining individual fibers into a compact whole. Thermal modification even at lower temperatures (below 190 °C) causes degradation of cellulose and hemicelluloses. As a result, the lignin content increases, increases mechanical strength and, conversely, decreases permeability. Our results confirm this. Specific modulus of elasticity E_{sp} of spruce increased by 6.0%, 6.8% and 3.0% and E_{sp} of maple increased by 7.4%, 7.8% and 7.8% (at 135 °C, 160 °C and 185 °C, respectively) compared to unmodified wood. Lower density gives a higher modulus of elasticity. Lower density also gives a higher E_{sp} [7], which means a positive impact on most of the acoustic properties of wood.

3.6. Effect of Thermal Modification on Logarithmic Damping Decrement

The logarithmic damping decrement (ϑ) is another important acoustic characteristic, and it is preferred to be small in wood with excellent acoustic quality [57]. Lower logarithmic damping decrements result in higher material acoustic values [58]. This characteristic (ϑ) is related to internal friction, and it cannot be correlated to the mass of wood or the volume of wood, but it closely correlates to their structure.

The logarithmic damping decrement (ϑ) was varied depending on the temperature of modification. The lowest reduction in logarithmic damping decrement (16%) was observed after modification at 185 °C in the case of spruce wood (ϑ drops 8.3%). The higher effect of thermal modification was observed when spruce wood was modified at temperatures 130 °C (ϑ drops 16.3%) and 160 °C (ϑ drops 13.8%). The lowest reduction in damping of maple wood was recorded at 160 °C (ϑ drops 3.6%) and the highest drops ϑ (=7.3%) at 185 °C. R^2 values in the case of ϑ were ($R^2 = 0.51$ for SW and $R^2 = 0.83$ for MW). Duncan's test showed that the changes in logarithmic damping decrement caused by the thermal modification (of both species) are not statistically significant ($F = 0.91$, $p > 0.05$). Our results showed a high dependence between E_{sp} and ϑ ($R^2 = 0.97$ for SW and $R^2 = 0.69$ for MW).

The vibrational properties (E_{sp} and logarithmic damping decrement ϑ) depend on the cell-wall structure. The transformation of lignin and the increase in the concentration contributes to the reduced water adsorption (lower equilibrium moisture content). The lower equilibrium moisture and, hence, reduced swelling and shrinkage, is characteristic of the thermally modified wood, i.e., the microfibrils' angle in the S2 layer is smaller. The orientation of cellulose microfibrils in the S2 layer affects these properties the most.

While E_{sp} is decreasing, damping ϑ is increasing, with an increasing microfibril angle. The microfibrils' angle in S2 and the deflection of the fibers show significant correlations with the vibrational properties in the longitudinal direction but not perpendicular to the wood grains. The collective effect of the microfibril angle and grain angle on both E_{sp} and ϑ is higher than their individual effects on these properties [59–61].

3.7. Effect of Thermal Modification on Loss Coefficient and Characteristic Acoustic Impedance

A measured logarithmic damping decrement was used to calculate the internal friction-loss coefficient (Equation (6)). Loss coefficient is important because it shows the amount of absorbed acoustic energy. It increases with higher moisture content, and it is reduced by thermal treatment [51,62]. Our results are consistent with this claim. The loss coefficient of spruce wood decreased by 17.4% (at 135 °C), 14.8% (at 160 °C) and 9% (at 185 °C); for maple, the decrease was lower at each modification temperature, i.e., 5.7%, 4% and 7.6%. Since low damping is required for the top plate (spruce) and higher damping for the back plate (maple), it can be stated that the heat treatment seems to be suitable in this aspect.

The characteristic acoustic impedance z of spruce wood is very similar to maple (around 2300 MPa·s·m^{-1} for SW and 2550 MPa·s·m^{-1} for MW). Due to thermal modification, z increased but not significantly. The lowest increase was recorded at 185 °C for both types of wood (around 1%). A low z value is required for the top plate of string instruments (between 1200 and 3392 MPa·s·m^{-1}) and high value z (between 1680 and 5760 MPa·s·m^{-1}) is required for percussion and woodwind instruments [29,63] because the sound should decay slowly.

3.8. Effect of Thermal Modification on Acoustic Conversion Efficiency (ACE) and on Sound Quality Factor Q

The acoustic conversion efficiency (ACE) reflects the ability of the material to convert vibration energy into sound energy [55,64]. ACE combines internal friction and sound radiation together. The highest ACE value was measured after thermal modification at 135 °C for both wood species (ACE_{SW} = 1401 m^4·kg^{-1}·s^{-1}; ACE_{MW} = 438 m^4·kg^{-1}·s^{-1}). Similarly, Q, as a wood sound quality descriptor, was highest at 135 °C (Q_{SW} = 105.3 and Q_{MW} = 60, 61). The ACE increase compared to untreated wood was 8.3% for SW and 11% for MW. The increase in Q of SW was considerably higher, up to 21%, while at MW, it was 7%. A high ACE value of spruce wood is especially required for soundboards because such a board should show the maximum ratio of the sound radiation to internal friction, i.e., a high ACE and low loss coefficient η is preferred [54].

3.9. Effect of Thermal Modification on Timber of Sound

The FFT analysis demonstrated a positive effect of the thermal modification on the timbre of sound of the maple specimens, as well as of the spruce wood specimens (Figures 4 and 5). Influence of thermal modification on the acoustic constant as well as on the frequency spectrum (FFT analysis), which both investigated the wood species, were identical. The higher value of the acoustic constant is characterized by the increasing of acoustic energy radiation, as confirmed by increasing the sound pressure levels (Figures 4 and 5).

The results of the FFT analysis (time course of signal, sound spectrum) of the bars of spruce and maple wood before (blue curve) and after modification (red curve) are presented in Figures 4 and 5. The mean ratios of aliquot frequencies in maple wood after thermal modification were as follows: 1:1.6:2.4 at 135 °C (1:1.7:2.6 before modification), 1:1.7:2.9 at 160 °C (1:1.8:2.8 before), 1:1.7:2.7 at 185 °C (1:1.6:2.6 before). Figure 4 shows that the fundamental resonant frequency of maple slightly increased after each modification and the sound pressure level reduced the thermal modification effect.

The sound pressure level of low frequencies decreased and of higher frequencies increased, which means that the lower frequencies are repressed, and the higher frequencies are highlighted, so the timber of sound will be sharper [65].

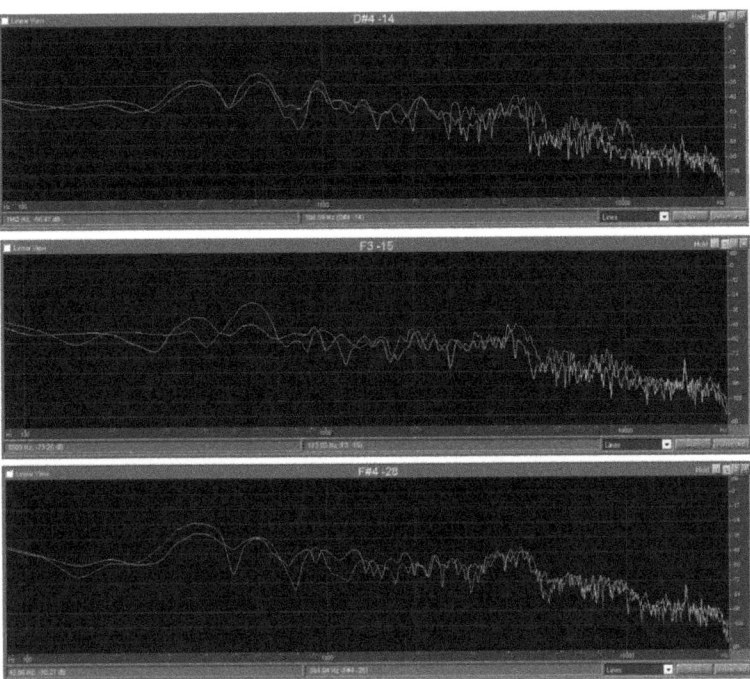

Figure 4. FFT analysis maple before and after thermal modification at 135 °C, 160 °C and 185 °C.

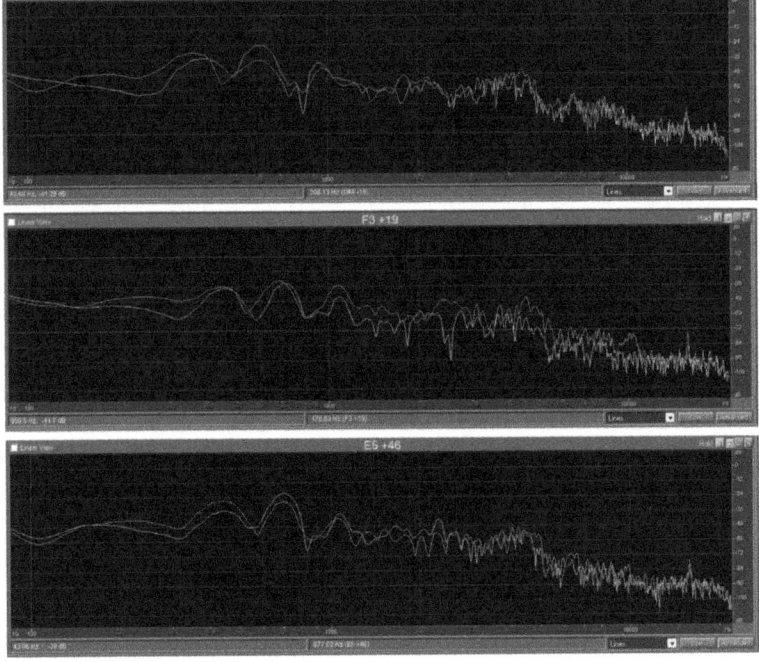

Figure 5. FFT analysis spruce before and after thermal modification at 135 °C, 160 °C and 185 °C.

4. Conclusions

As indicated by our experiments, we can improve the properties of wood for musical instruments by means of thermal modification at lower temperatures (135 °C to 185 °C). This modification causes a decrease in the mass and density but increase in the speed of sound and MOE, as well as the acoustic constant. These changes are not statistically significant, except for the acoustic constant.

Thermal modification leads to changes in vibrational properties by increasing the specific modulus of elasticity E_{sp} and decreasing the loss coefficient η. Our results showed a high dependence between E_{sp} and ϑ.

The characteristic acoustic impedance z of spruce and maple slightly increased.

The thermal modification process is the cause of not only the changes in the wood properties but also the changes in timbre and sound quality.

Author Contributions: Conceptualization, A.D. and Z.V.; methodology, A.D. and L.S.K.; software T.G.; validation, M.P. and L.S.K.; formal analysis, Z.V. and M.P.; investigation, A.D. and Z.V.; resources, T.G. and M.P.; data curation, A.D.; writing—original draft preparation, A.D.; writing—review and editing, A.D. and T.G. All authors have read and agreed to the published version of the manuscript.

Funding: This scientific work was supported by grant APVV-17-0583 "Construction and decorative materials based on recycled and modified wood", as well as by grant APVV-20-0326 "Research possibilities of the utilisation of black walnut (*Juglans nigra* L.) and chestnut (*Castanea sativa* Mill.) from a production-ecological aspect in agroforestry systems in Slovakia" and APVV-21-0032 "Certification of log quality based on internal wood defects". This publication is one of the results of the project implementation of Centrum excelentnosti lesnícko-drevárskeho komplexu LignoSilva (Centre of Excellence of Forest-based Industry, ITMS: 313011S735) supported by the Research and Development Operational Programme funded by the ERDF.

Institutional Review Board Statement: Not applicable.

Informed Consent Statement: Not applicable.

Data Availability Statement: Not applicable.

Conflicts of Interest: The authors declare no conflict of interest.

References

1. Froidevaux, J.; Parviz, N. Aging law of spruce wood. *Wood Mater. Sci. Eng.* **2013**, *8*, 46–52. [CrossRef]
2. Kránitz, K.; Sonderegger, W.; Bues, C.T. Effects of aging on wood: A literature review. *Wood Sci. Technol.* **2016**, *50*, 7–22. [CrossRef]
3. Borgin, K.; Faix, O.; Schweers, W. The effect of aging on lignins of wood. *Wood Sci. Technol.* **1975**, *9*, 207–211. [CrossRef]
4. Bucur, V. Ageing of wood. In *Handbook of Materials for String Musical Instruments*; Springer: Cham, Switzerland, 2016; Volume 1, pp. 283–323.
5. Obataya, E. Effects of natural and artificial ageing on the physical and acoustic properties of wood in musical instruments. *J. Cult. Herit.* **2017**, *27*, S63–S69. [CrossRef]
6. Danihelová, A.; Čulík, M.; Němec, M.; Gejdoš, M.; Danihelová, Z. Modified wood of black locust—Alternative to Honduran rosewood in the production of xylophones. *Acta Phys. Pol. A* **2015**, *127*, 106–109. [CrossRef]
7. Puszyński, J.; Warda, M. Possibilities of using the thermally modified wood in the electric string instruments. *For. Wood Technol.* **2014**, *85*, 200–204.
8. Pfriem, A. Thermally Modified Wood for Use in Musical Instruments. *Drv. Ind.* **2015**, *66*, 251–253. [CrossRef]
9. Mania, P.; Gąsiorek, M. Acoustic Properties of Resonant Spruce Wood Modified Using Oil-Heat Treatment (OHT). *Materials* **2020**, *13*, 1962. [CrossRef]
10. Mania, P.; Moliński, W.; Roszyk, E.; Górska, M. Optimization of Spruce (*Picea abies* L.) Wood Thermal Treatment Temperature to Improve Its Acoustic Properties. *Bioresources* **2019**, *15*, 505–516. [CrossRef]
11. Rapp, A.O.; Sailer, M. Oil-heat-treatment of wood—Process and properties. *Drv. Ind.* **2001**, *52*, 63–70.
12. Zatloukal, P.; Suchomelová, P.; Dömény, J.; Doskočil, T.; Manzo, G.; Tippner, J. Possibilities of Decreasing Hygroscopicity of ResonanceWood Used in Piano Soundboards Using Thermal Treatment. *Appl. Sci.* **2021**, *11*, 475. [CrossRef]
13. Danihelová, A.; Spišiak, D.; Reinprecht, L.; Gergeľ, T.; Vidholdová, Z.; Ondrejka, V. Acoustic Properties of Norway Spruce Wood Modified with Staining Fungus (*Sydowia polyspora*). *BioResources* **2019**, *18*, 3432–3444. [CrossRef]
14. Schwarze, F.W.M.R.; Spycher, M.; Fink, S. Superior wood for violins–wood decay fungi as a substitute for cold climate. *New Phytol.* **2008**, *179*, 1095–1104. [CrossRef]

15. Spycher, M. The Application of Wood Decay Fungi to Improve the Acoustic Properties of Resonance Wood for Violins. Ph.D. Thesis, Albert-Ludwigs-Universität, Freiburg im Breisgau, Germany, 2008. Available online: https://d-nb.info/98820701X/34 (accessed on 20 June 2022).
16. Hill, C. *Wood Modification: Chemical, Thermal and Other Processes*; John Wiley & Sons, Ldt.: West Sussex, UK, 2006; p. 264.
17. Candelier, K.; Thevenon, M.F.; Petrissans, A. Control of wood thermal treatment and its effects on decay resistance. *Ann. For. Sci.* **2016**, *73*, 571–583. [CrossRef]
18. Korkut, S.; Budakci, M. The effects of high temperature heat—Treatment on physical properties and surface roughness of rowan (Sorbus Aucuparia, L.) wood. *Wood Res.* **2010**, *55*, 67–78. [CrossRef]
19. Mitsui, K.; Inagaki, T.; Tsuchikawa, S. Monitoring of hydroxyl groups in wood during heat treatment using NIR spectroscopy. *Biomacromolecules* **2008**, *9*, 286–288. [CrossRef]
20. Čermák, P.; Rautkari, L.; Horáček, P.; Saake, B.; Rademacher, P.; Sablík, P. Analysis of Dimensional Stability of Thermally Modified Wood Affected by Re-Wetting Cycles. *BioResources* **2015**, *10*, 3242–3253. [CrossRef]
21. Reinprecht, L. *Wood Deterioration, Protection and Maintenance*; Wiley-Blackwell: Hoboken, NJ, USA, 2016; p. 376.
22. Tjeerdsma, B.; Stevens, M.; Militz, H. Durability aspects of (hydro) thermal treated wood. In Proceedings of the International Research Group on Wood Preservation, Kona, Hawaii, 14–19 May 2000; IRG-WP: Stockholm, Sweden, 2000.
23. Bekhta, P.; Niemz, P. Effect of High Temperature on the Change in Color, Dimensional Stability and Mechanical Properties of Spruce Wood. *Holzforschung* **2003**, *57*, 539–546. [CrossRef]
24. Nieminen, P.; Nieminen, R. *Thermally Aged Tonewood*; MES—The Finnish Music Foundation: Helsinki, Finland, 2020; p. 35.
25. Huss and Dalton TD R 20th Anniversary Thermo-Cured Red Spruce and Bubinga. Available online: https://reverb.com/item/2834248-huss-and-dalton-td-r-20th-anniversary-guitar-thermo-cured-red-spruce-and-bubinga-pre-owned-2015 (accessed on 8 November 2021).
26. Wagenführ, A.; Pfriem, A.; Grothe, T.; Eichelberger, K. Untersuchungen zur vergleichenden Charakterisierung von thermisch modifi zierter Fichte für Resonanzdecken von Gitarren. *Holz als Roh- Und Werkstoff* **2006**, *64*, 313–316. [CrossRef]
27. Rajčan, E. Application of acoustic to some problems of material science related to the making of musical instruments. *Acustica* **1998**, *84*, 122–128.
28. Obataya, E.; Ono, T.; Norimoto, M. Vibrational properties of wood along the grain. *J. Mater. Sci.* **2000**, *35*, 2993–3001. [CrossRef]
29. Wegst, U.G.K. Wood for Sound. *Am. J. Bot.* **2006**, *93*, 1439–1448. [CrossRef] [PubMed]
30. Čulík, M.; Danihelová, A.; Danihelová, Z. Wood for musical instruments. *Akustika* **2016**, *25*, 66–72.
31. Gejdoš, M.; Suchomel, J. Potential of wood for musical instruments in Slovakia. *Akustika* **2013**, *20*, 16–23.
32. Čulík, M. *Drevo a Jeho Využitie vo Výrobe Hudobných Nástrojov [Wood and Its Use in the Manufacture of Musical Instruments]*; TU in Zvolen: Zvolen, Slovakia, 2013; p. 93.
33. Tetri, T. *ThermoWood® Handbook*; International ThermoWood Association: Helsinki, Finland, 2003; p. 56.
34. Korošec, R.C.; Lavrič, B.; Pohleven, F.; Bukovec, P. Thermogravimetry as a possible tool for determining modification degree of thermally treated Norway spruce wood. *J. Therm. Anal. Calorim.* **2009**, *98*, 189–195. [CrossRef]
35. Reinprecht, L.; Vidholdová, Z. *Termodrevo—Príprava, Vlastnosti a Aplikácie [Thermowood—Its Preparation, Properties and Application]*; Technical University in Zvolen: Zvolen, Slovakia, 2009; p. 89.
36. Danihelová, A.; Čulík, M.; Danihelová, Z. The most popular wood for making Slovak folk flutes with open ends. *Akustika* **2012**, *17*, 6–9.
37. Obataya, E.F.; Tanaka, M.; Norimoto, B.T. Hygroscopicity of heat-treated wood 1. Effects of after-treatments on the hygroscopicity of heat-treated wood. *J. Jpn. Wood Res. Soc.* **2000**, *46*, 77–87.
38. Abdolahian Sohi, A.M.; Khademi-Eslam, H.; Hemmasi, A.H.; Roohnia, M.; Talaiepour, M. Nondestructive detection of the effect of drilling on acoustic performance of wood. *BioResources* **2011**, *6*, 2632–2646. [CrossRef]
39. Bismarck, G. Sharpness as an Attribute of the Timbre of Steady Sounds. *Acta Acust. United Acust.* **1974**, *30*, 159–172.
40. Alén, R.; Kotilainen, R.; Zaman, A. Thermochemical behavior of Norway spruce (Picea abies) at 180–225 °C. *Wood Sci. Technol.* **2002**, *36*, 163–171. [CrossRef]
41. Esteves, B.; Marques, A.V.; Domongos, I.; Pereira, H. Influence of steam heating on the properties of pine (*Pinus pinaster*) and eucalypt (*Eucalyptus globulus*) wood. *Wood Sci. Technol.* **2007**, *41*, 193–207. [CrossRef]
42. Esteves, B.M.; Domingos, I.J.; Pereira, H.M. Pine wood modification by heat treatment in air. *Bioresources* **2008**, *3*, 142–154. [CrossRef]
43. Stamm, A.J. Thermal degradation of wood and cellulose. *Ind. Eng. Chem.* **1956**, *48*, 413–417. [CrossRef]
44. Fung, D.P.C.; Stevenson, J.A.; Shields, J.K. The effect of heat and $NH_4H_2PO_4$ on the dimensional and anatomical properties of Douglas-fir. *Wood Sci.* **1974**, *7*, 13–20.
45. Zauer, M.; Kowalewski, A.; Sproßmann, R.; Stonjek, H.; Wagenführ, A. Thermal modification of European beech at relatively mild temperatures for the use in electric bass guitars. *Eur. J. Wood Wood Prod.* **2016**, *74*, 43–48. [CrossRef]
46. Čabalová, I.; Výbohová, E.; Igaz, R.; Kristak, L.; Kačík, F.; Antov, P.; Papadopoulos, A.N. Effect of oxidizing thermal modification on the chemical properties and thermal conductivity of Norway spruce (*Picea abies* L.) wood. *Wood Mater. Sci. Eng.* **2021**. [CrossRef]
47. Alén, R.; Oesch, P.; Kuoppala, E. Py-GC/AED studies on thermochemical behavior of softwood. *J. Anal. Appl. Pyrolysis* **1995**, *35*, 259–265. [CrossRef]

48. Millett, M.A.; Gerhards, G.C. Accelerated aging: Residual weight and flexural properties of wood heated in air at 115 °C to 175 °C. *Wood Sci. Technol.* **1972**, *4*, 193–201.
49. Navickas, P.; Karpavičiūtė, S.; Albrektas, D. Effect of heat treatment on wettability and MOE of pine and spruce wood. *Mater. Sci.* **2015**, *21*, 400–404. [CrossRef]
50. Kubojima, Y.; Okano, T.; Ohta, M. Bending strength and toughness of heat-treated wood. *J. Wood Sci.* **2000**, *46*, 8–15. [CrossRef]
51. Rusche, H. Thermal degradation of wood at temperatures up to 200 deg C. I. Strength properties of wood after heat treatment. *Holz Als Roh-Und Werkst.* **1973**, *31*, 273–281. [CrossRef]
52. Wegst, U.G.K. Bamboo and Wood in Musical Instruments. *Ann. Rev. Mater. Res.* **2008**, *38*, 323–349. [CrossRef]
53. Bucur, V. *Acoustics of Wood*, 2nd ed.; Timell, T.E., Wimmer, R., Eds.; Birkhäuser: Berlin, Germany, 2006; p. 403.
54. Esteves, B.; Pereira, H. Wood modification by heat treatment: A review. *BioResources* **2008**, *4*, 370–404. [CrossRef]
55. Brémaud, I. Acoustical properties of wood in string instruments soundboards and tuned idiophones: Biological and cultural diversity. *J. Acoust. Soc. Am.* **2012**, *131*, 807–818. [CrossRef]
56. Wu, Y.; Wu, X.; Yang, F.; Zhang, H.; Feng, X.; Zhang, J. Effect of Thermal Modification on the Nano-Mechanical Properties of the Wood Cell Wall and Waterborne Polyacrylic Coating. *Forests* **2020**, *11*, 1247. [CrossRef]
57. Zhang, X.; Li, L.; Xu, F. Chemical Characteristics of Wood Cell Wall with an Emphasis on Ultrastructure: A Mini-Review. *Forests* **2022**, *13*, 439. [CrossRef]
58. Roohnia, M.; Kohantorabi, M.; Tajdini, A. Maple wood extraction for a better acoustical performance. *Eur. J. Wood Wood Prod.* **2015**, *73*, 139–142. [CrossRef]
59. Karlinasari, L.; Baihaqi, H.; Maddu, A.; Mardikanto, T.R. The Acoustical Properties of Indonesian Hardwood Species. *Makara J. Sci.* **2012**, *16*, 110–114. [CrossRef]
60. Brémaud, I.; El Kaïm, Y.; Guibal, D.; Minato, K.; Thibaut, B.; Gril, J. Characterisation and categorisation of the diversity in viscoelastic vibrational properties between 98 wood types. *Ann. For. Sci.* **2012**, *69*, 373–386. [CrossRef]
61. Alkadri, A.; Carlier, C.; Wahyudi, I.; Gril, J.; Langbour, P.; Brémaud, I. Relationships between anatomical and vibrational properties of wavy sycamore maple. *IAWA J.* **2018**, *39*, 63–86. [CrossRef]
62. Zhu, L.; Liu, Y.; Liu, Z. Effect of high-temperature heat treatment on the acoustic-vibration performance of picea jezoensis. *BioResources* **2016**, *11*, 4921–4934. [CrossRef]
63. Hilde, C.; Woodward, R.; Avramidis, S.; Hartley, I.D. The Acoustic Properties of Water Submerged Lodgepole Pine (*Pinus contorta*) and Spruce (*Picea* spp.) Wood and Their Suitability for Use as Musical Instruments. *Materials* **2014**, *7*, 5688–5699. [CrossRef] [PubMed]
64. Barlow, C.Y. Materials selection for musical instruments. *Proc. Inst. Acoust.* **1997**, *19*, 69–78.
65. Ilmoniemi, M.; Välimäki, V.; Huotilainen, M. Subjective evaluation of musical instrument timbre modifications. In Proceedings of the Joint Baltic-Nordic Acoustics Meeting, Mariehamn, Aland Islands, 8–10 June 2004.

Article

Flammability Characteristics of Thermally Modified Meranti Wood Treated with Natural and Synthetic Fire Retardants

Milan Gaff [1,*], Hana Čekovská [1], Jiří Bouček [1], Danica Kačíková [2], Ivan Kubovský [2], Tereza Tribulová [1], Lingfeng Zhang [3], Salvio Marino [1] and František Kačík [2]

1. Department of Wood Processing and Biomaterials, Faculty of Forestry and Wood Sciences, Czech University of Life Sciences Prague, Kamýcká 1176, Praha 6—Suchdol, 16521 Prague, Czech Republic; hanacekovska@gmail.com (H.Č.); jboucek@fld.czu.cz (J.B.); tereza.tribulova@gmail.com (T.T.); salviomarino@yahoo.it (S.M.)
2. Faculty of Wood Sciences and Technology, Technical University in Zvolen, T.G.Masaryka 24, 960 01 Zvolen, Slovakia; kacikova@tuzvo.sk (D.K.); kubovsky@tuzvo.sk (I.K.); kacik@tuzvo.sk (F.K.)
3. School of Civil Engineering, Southeast University, Nanjing 211189, China; lfzhang@yzu.edu.cn
* Correspondence: gaffmilan@gmail.com

Citation: Gaff, M.; Čekovská, H.; Bouček, J.; Kačíková, D.; Kubovský, I.; Tribulová, T.; Zhang, L.; Marino, S.; Kačík, F. Flammability Characteristics of Thermally Modified Meranti Wood Treated with Natural and Synthetic Fire Retardants. *Polymers* 2021, 13, 2160. https://doi.org/10.3390/polym13132160

Academic Editors: Antonios N. Papadopoulos and Ignazio Blanco

Received: 17 May 2021
Accepted: 24 June 2021
Published: 30 June 2021

Publisher's Note: MDPI stays neutral with regard to jurisdictional claims in published maps and institutional affiliations.

Copyright: © 2021 by the authors. Licensee MDPI, Basel, Switzerland. This article is an open access article distributed under the terms and conditions of the Creative Commons Attribution (CC BY) license (https://creativecommons.org/licenses/by/4.0/).

Abstract: This paper deals with the effect of synthetic and natural flame retardants on flammability characteristics and chemical changes in thermally treated meranti wood (*Shorea* spp.). The basic chemical composition (extractives, lignin, holocellulose, cellulose, and hemicelluloses) was evaluated to clarify the relationships of temperature modifications (160 °C, 180 °C, and 210 °C) and incineration for 600 s. Weight loss, burning speed, the maximum burning rate, and the time to reach the maximum burning rate were evaluated. Relationships between flammable properties and chemical changes in thermally modified wood were evaluated with the Spearman correlation. The thermal modification did not confirm a positive contribution to the flammability and combustion properties of meranti wood. The effect of the synthetic retardant on all combustion properties was significantly higher compared to that of the natural retardant.

Keywords: thermal modification; fire retardant; flammability characteristics; meranti

1. Introduction

Meranti (*Shorea* spp.) is one of the most widely used tropical hardwoods. It is relatively easy to process, and it has a course, fibrous structure with open pores. With a straight-grain consistency, meranti trees produce long, straight pieces of lumber. It is used for molding, structural elements, furniture, cabinets, window and door trim, and veneers for plywood. Meranti is one of the more affordable hardwoods, due in part to numerous subspecies, prolific growing characteristics, and availability. Similar to Teak (*Tectona grandis*), meranti and other hardwoods are resistant to damage from insects, fungus, and moisture decay. The wood is dimensionally stable and resistant to warping and twisting [1,2]. However, additional treatments are necessary to increase its fire resistance.

A very popular and ecological method is thermal modification of wood [3–8]. The purpose of thermal modification is to decrease the content of flammable substances present in the wood. Thermal treatments can also be a great alternative to improve physical and aesthetic properties, increase market value, and, consequently, the number of applications of wood from meranti [9,10]. Thermal treatment is normally performed at temperatures between 180 and 260 °C. Temperatures lower than 140 °C do not significantly affect the structure of the wood. Volatile and vapor substances are released under higher temperature. The process has to be very carefully optimized in order to not degrade the wood. Temperatures higher than 260 °C cause undesirable degradation [11,12].

According to the literature [13–15], the temperature is a key parameter of the process responsible for the highest effect of modification in the properties of thermally treated wood. Certified technologies that are used with commercial significance not only differ in applied temperature for a certain time, but also in ambient humidity, the use of a gaseous (air or nitrogen) or liquid (vegetable oils) environment, etc. All aforementioned factors, i.e., temperature, duration of thermal modification, and environment, have a certain impact on the results [16].

Surface treatment of wood with a flame-retardant finish was also applied in this research. With regard to environmental aspects of the thermal degradation of wood to retard the burning process, it should be noted that chemical treatment of these materials is approached on the basis of the "lesser risk" theory [17]. Most commercial fire-retardant formulations are not environmentally friendly [18–20].

For this reason, arabinogalactan was tested as a natural compound with a potential fire-retardant effect. It is well known that arabinogalactan is a biopolymer consisting of arabinose and galactose monosaccharides. In plants, it is a major component of many gums and it is often found attached to proteins. The resulting arabinogalactan protein functions as both an intercellular signaling molecule and a glue to seal plant wounds [21]. The big advantage is that it can be extracted from plant cells in its natural form using only water [22]. It is currently used in many ways, e.g., as an additive in food, animal feed, cosmetics, pharmacy, construction, pulp production, oil production, plant growth, etc. It generally plays an essential role in environmental preservation. With regard to this, and based on its properties [21–23], arabinogalactan is destined to become a suitable and effective flame retardant. This assessment was based on a comparison with the same previously thermally modified samples, treated with the synthetic fire-retardant system Flamgard Transparent. The main active ingredient of this commercial formulation is one of the most common fire-retardant inorganic chemicals—mono ammonium phosphate. Phosphates are some of the oldest known fire-retardant systems and they are usually included in most proprietary systems used for wood [24,25].

This article deals with the investigation of the thermal stability of thermally modified meranti wood that was subsequently treated with synthetic and natural fire retardants. The aim was to expand knowledge of thermal processes, disclose the possibilities of using arabinogalactan as an active ingredient in fire retardant formulations in gaining new facts on chemical reactions of thermal degradation, and to compare the efficiency of this environmentally friendly treatment with the application of a synthetic two-component formulation based on ammonium phosphates and an acrylic topcoat. The results of this study provide main flammability characteristics of modified meranti wood samples at temperatures of 160 °C, 180 °C, and 210 °C, namely, weight loss, burn rate, maximum burning rate, and ratio of the maxim um burning rate to time to reach the maximum burning rate.

2. Materials and Methods

2.1. Wood

We used meranti (*Shorea* spp.) wood for our research. Place of origin: Southeast Asia. Tree dimensions: height, 30 m; stem diameter, 1.5 m; dry wood density, 710 kg·m^{-3}. Color: dark red-brown to red-brown. Wood hardness: 7.73 MPa. Samples were radially cut into 20 mm × 100 mm × 200 mm specimens. All wood samples were conditioned under specific conditions (relative humidity of 65% ± 3% and a temperature of 20 °C ± 2 °C) to achieve an equilibrium of moisture content of 12%. The conditioned samples were divided into three basic groups according to the fire-retardant application:

1. Thermal modification (TM);
2. Thermal modification + synthetic fire retardant (SFR);
3. Thermal modification + natural fire retardant (NFR).

2.2. Synthetic Fire Retardant

The flame retardant was Flamgard Transparent, which consists of two components. The first component is a flammable water-borne coating composition (a mixture composed of ammonium phosphates, foam-forming agent, oxalic and acetic acid, and fire-retardant additives), which is applied in three layers so that the total coat is at least 500 g·m^{-2}. The second component is acrylic topcoat space 1818, which is only applied in one layer with a minimum coating of 80 g·m^{-2}.

Both components were applied according to the manufacturer's specifications. The curing of the individual layers took place at an ambient temperature of 20 °C.

2.3. Natural Fire Retardant

Macromolecules containing arabinose and galactose have been found in most plant tissues. In some situations, they are isolated as polysaccharides free from associated protein; in other situations, they occur in covalent association with protein, either as proteoglycans, in which the protein component carries polysaccharide substituents [26,27], or as glycoproteins, in which the protein component is substituted by one or more oligosaccharide residues [28,29]. Arabinogalactans are a class of polysaccharides found in a wide range of plants; however, they are most abundant in plants of the genus Larix [30]. They have a highly branched structure. Their main chain consists of β-(1→3) linked galactose units. Approximately one-half of the side chains consist of β-(1→6)-linked dimers of galactopyranose; galactopyranose monomers account for about a quarter, and the remainder contains a major part of the polysaccharide's arabinose in aggregates of two or more monomers [31–33]. Arabinose fragments mostly occur as side chains consisting of 3-O-substituted β-L-arabinofuranose residues and terminal residues of β-L-arabinopyranose, β-D-arabinofuranose, and α-L-arabinofuranose. However, arabinose fragments were also found in the main chain. Arabinogalactan has good solubility in cold water and the uniquely low viscosity of concentrated aqueous solutions [32,33].

2.4. Methods

2.4.1. Thermal Modification

The thermal treatment was based on the Thermo Wood principle, developed by VTT (Espoo, Finland). The wood samples were treated in an S400/03 type thermal chamber (LAC Ltd., Rajhrad, Czech Republic). Three final temperatures of 160 °C, 180 °C, and 210 °C were set, and the treatment was carried out in a protective atmosphere (water steam) to prevent overheating and burning. The thermal modification was carried out in three basic stages (Table 1). All samples were dried in a ULM 400 oven (Memmert GmbH & amp; Co. KG, Schwabach, Germany) at T = 103 ± 2 °C to dry state. This condition facilitates the thermal modification of one group and also generates similar moisture for samples without thermal treatment, which is important from a comparative perspective.

Table 1. Conditions and Parameters of Thermal Modification.

Thermal Modification Parameters			
Temperatures (°C)	160	180	210
Heating (h)	13.5	13.6	18.7
Thermal treatment (h)	3.0	3.0	3.0
Cooling (h)	9.6	11.1	11.2
Total modification time (h)	26.1	27.7	32.9

The thermal modification of meranti wood was performed in the same way, according to the method described in publications on the wood species teak, padauk, and oak [34–36]. The density of the sample before and after the thermal modification is shown in Table 2.

Table 2. The density of Shorea Samples.

Temperature	Thermal Modification Temperature			
	Unmodifield	160 °C	180 °C	210 °C
Density before TM (kg·m^{-3})	626 (9.3)	662 (11.4)	634 (10.8)	626 (8.5)
Density after TM (kg·m^{-3})		625 (9.2)	608 (10.4)	650 (7.2)

Numbers in parentheses represent coefficients of variation (CV) in %.

2.4.2. Physical Properties

For the sorting of sample quality, auxiliary criteria such as moisture content and density values can be used. These physical parameters should be sorted so that they do not affect the values of the basic assessment criteria. These two parameters were determined according to ISO 13061-2 [37] and ISO 13061-1 [38].

2.4.3. Determination of Flammability Characteristics

The method of simulating actual fire conditions was performed according to the previously published procedure [38]. The heat source in this experiment was an open flame, patented, USBEC 1011/1 propane gas burner, supplied and regulated flame source (DIN-DVGW reg. NG-2211AN0133), 1.7 kW. Mass was measured by Mettler Toledo (MS1602S/MO1, Mettler Toledo, Geneva, Switzerland). Weight changes were recorded according to evaluation criteria performed with BalanceLink 4.2.0.1 (Mettler Toledo, Switzerland).

2.4.4. Calculation of Given Characteristics for the Experiment

All flammability factors were evaluated in Statistica 13 (Statsoft Inc., Tulsa, OK, USA) using the two-way analysis of variance (ANOVA) and Spearman's correlation. The sample density of meranti according to ISO 13061-2 (2014) [33] was determined together, and the moisture content of all samples according to ISO 13061-1 (2014) [39] was measured together before and after the thermal treatment.

2.4.5. Chemical Analyses

All samples were mechanically decomposed into sawdust and a fraction size 0.5–1.0 mm was extracted according to ASTM D1107-96 [40]; the lignin content was determined according to Sluiter [41], holocellulose was determined using the method of [42], cellulose by the Seifert method [43], and the hemicelluloses content was calculated as the difference between holocellulose and cellulose. All measurements were performed on four replicates per specimen. The data were presented as percentages of the oven-dry weight of wood (odw) per unextracted wood.

The obtained values of the burning characteristics and percentage content of chemical wood components were evaluated in Statistica 13 (Statsoft Inc., Tulsa, OK, USA) with an analysis of variance (ANOVA) and Spearman's correlation.

3. Results and Discussion

The average values of the chemical components of wood are shown in Table 3.

Table 4 shows the statistical significance of the effects of flammability properties on thermal modification and retardant treatment, including their interaction on meranti wood.

Based on the significance level of "P", it is conclusive that both the flame retardant and the thermal modification or interaction of both factors have a strong statistical effect on the above characteristics.

The thermal modification in interaction with the retardant had a significant statistical effect on the burning characteristics (WL., BR., RMBR. and TRMBR.). However, for the rest of the characteristics (BR. and MBR.), thermal modification had a negligible effect.

Table 3. Chemical analyses of the main wood components of meranti (%) and the main results of a way ANOVA analysis evaluating the effect of the thermal modification temperature on the values of interest.

Temperature (°C)	Extractives (%)	Fisher's F-Test Level of Significance	Lignin (%)	Fisher's F-Test Level of Significance	Cellulose (%)	Fisher's F-Test Level of Significance
Unmodified	2.51 (2.37)		32.42 (0.25)		53.04 (0.89)	
160	3.22 (1.94)	297.04	36.67 (0.26)	1468	51.02 (0.19)	543.5
180	3.20 (1.47)	***	36.29 (0.24)	***	52.04 (0.31)	***
210	3.92 (2.32)		35.33 (0.37)		59.06 (0.60)	

Temperature (°C)	Holocellulose (%)	Fisher's F-Test Level of Significance	Hemicelluloses (%)	Fisher's F-Test Level of Significance
Unmodified	71.53 (0.58)		18.49 (4.65)	
160	66.51 (1.07)	259.2	15.50 (5.14)	367.925
180	66.12 (0.23)	***	14.08 (2.16)	***
210	62.71 (0.54)		3.65 (6.13)	

The data represent the mean percentages of oven-dry weight (odw), numbers in parentheses represent coefficients of variation (CV) in %, $n = 4$, NS—not significant, ***—significant, $P < 0.05$.

Table 4. Basic Statistical Characteristics Evaluation of the Effect of Thermal Treatment and Retardants on the Values of the Monitored Characteristics.

Influence	Sum of Squares	Degrees of Freedom	Variance	Fisher's F-Test	Significance Level P
		WL.—600 s (%)			
Intercept	1495.404	1	1495.404	1430.428	***
Retardant	374.391	2	187.196	179.062	***
TM	14.324	3	4.775	4.567	***
Retardant × TM	17.222	6	2.870	2.746	***
Error	50.180	48	1.045		
	The model corresponds to approximately 89.0% of the total sum of squares.				
		BR.—600 s (%·s^{-1} × 10^{-5})			
Intercept	17.66190	1	17.66190	174.4060	***
Retardant	1.38970	2	0.69485	6.8615	***
TM	0.13538	3	0.04513	0.4456	NS
R × TM	0.79910	6	0.13318	1.3151	NS
Error	4.86091	48	0.10127		
	The model corresponds to approximately 32.3% of the total sum of squares.				
		MBR. (%·s^{-1})			
Intercept	531.7951	1	531.7951	1244.976	***
Retardant	50.8410	2	25.4205	59.511	***
TM	2.3821	3	0.7940	1.859	NS
Retardant × TM	4.6048	6	0.7675	1.797	NS
Error	20.5033	48	0.4272		
	The model corresponds to approximately 73.8% of the total sum of squares.				
		RMBR. (%)			
Intercept	0.456095	1	0.456095	558.1834	***
Retardant	0.036244	2	0.018122	22.1784	***
TM	0.019321	3	0.006440	7.8817	***
Retardant × TM	0.026381	6	0.004397	5.3810	***
Error	0.039221	48	0.000817		
	The model corresponds to approximately 67.6% of the total sum of squares.				

Table 4. Cont.

Influence	Sum of Squares	Degrees of Freedom	Variance	Fisher's F-Test	Significance Level P
		TRMBR.			
Intercept	108,800.4	1	108,800.4	464.2151	***
Retardant	21,450.8	2	10,725.4	45.7618	***
TM	544.6	3	181.5	0.7745	NS
Retardant × TM	4079.2	6	679.9	2.9007	***
Error	11,250.0	48	234.4		

The model corresponds to approximately 69.9% of the total sum of squares.

NS—not significant, ***—significant, $P < 0.05$, TM.—Thermal modification, WL.—Weight loss, BR.—Burn rate, MBR.—Maximum burn rate, RMBR.—Ratio of the maximum burn rate, TRMBR.—Time to reach the maximum burn rate.

In similar research, other authors also who dealt with this topic [40] showed that the degree of thermal treatment has no significant effect on any of the monitored characteristics in the case of spruce wood, i.e., WL.—600 s and 900 s (%), BR.—600 s and 900 s (%·s^{-1} × 10^{-5}), RMBR. (%·s^{-1}), (%) TRMBR. (s). These results show that none of the monitored characteristics are affected by thermal treatment of the wood and thermally modified wood is therefore suitable for use in timber structures just like wood that has not undergone thermal modification. The same team of authors in another study reported that thermal treatment only had a statistically significant effect on the values of monitored characteristics for weight loss—600 s (%) in the case of Teak wood.

Gašparík et al. [44] evaluated the effect of thermal modification and fire-retardant treatment on the flammability characteristics of oak wood. They concluded that thermal modification has an insignificant effect on the burn rate. Interaction of fire-retardant treatment and thermal modification has an insignificant effect on the MBR. Last but not least, fire-retardant treatment, thermal modification, and their interaction have an insignificant effect on the TRMBR.

The significant effect of the flame retardant on weight loss is evident from the diagram shown in (Figure 1). The lowest average weight loss values were measured when a synthetic flame retardant was applied. The highest weight loss values were measured on samples treated with a natural flame retardant [44]. A similar observation in these trends was found for thermally modified oak wood and oak wood treated with a synthetic fire retardant [45]. The highest weight loss values for thermally treated wood were also found at 160 and 180 °C, and using a synthetic flame retardant at 160 °C. In the case of wood modified with a synthetic flame retardant, there were also minor differences in the decrease in weight loss at individual temperatures of the thermal modification.

In the case of burn rate characteristics—600 s (Figure 2), both the thermal modification and the flame retardant, as well as their interactions have an insignificant effect (however, the retardant is significant according to the table). In the above-mentioned research, the effect of thermal modification on the burn rate was not unambiguous [41]. Generally, the highest burn rate of thermally modified wood spruce, teak, and oak was observed at 180 °C, declining again at higher temperatures [44,45]. The expected effect of reduced burn rate with the application of flame retardants has not been confirmed in our case. It can be argued that, in almost all cases, the burn rate values of wood treated with flame retardants were equal to or higher than that of only thermally modified wood. Flame retardants reduce the flammability of materials by physical or chemical means: they may reduce or extinguish flame using an endothermic chemical reaction (occurs temperature limitation), or the pyrolysis modification process may reduce the amount of flammable volatile matter and increase production of fewer flammable substances that act as a barrier to protect others material. Retardants may act before ignition substances change during the pyrolysis process, they can also react in a flame and reduce the flammability of the substance during combustion, whether they can prevent the access of oxygen or heat to the hearth [46]. Synthetic and natural flame retardants exhibit similar behaviour, but only at 160 °C; the synthetic flame retardant provides better results, i.e., a lower burn rate.

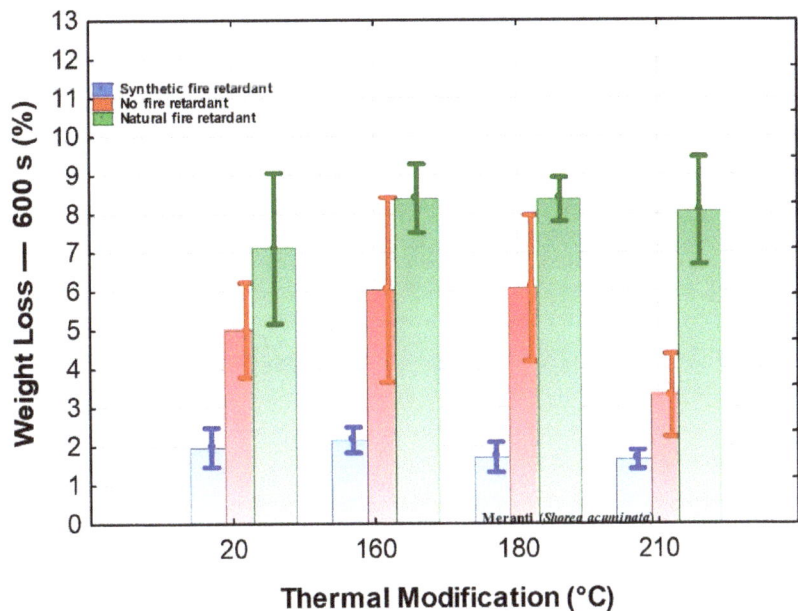

Figure 1. The effect of thermal modification and fire retardant on weight loss.

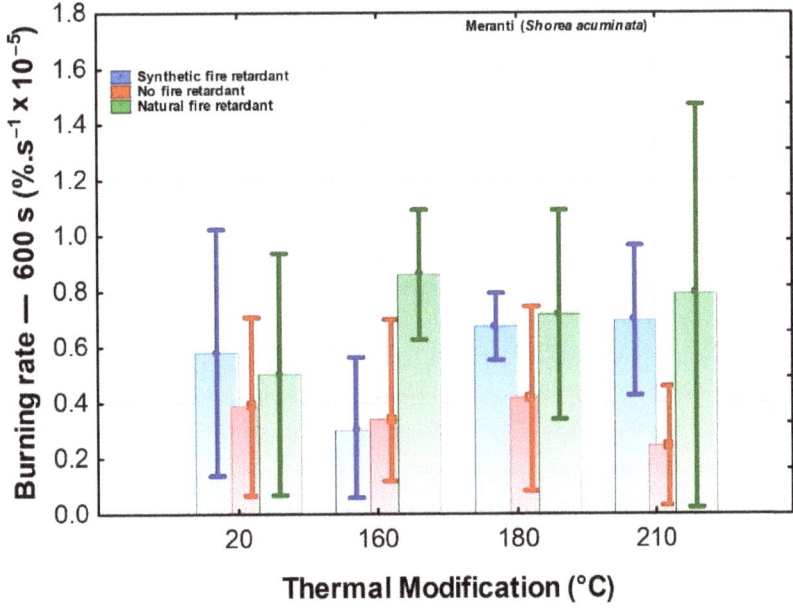

Figure 2. The effect of thermal modification and fire retardant on the burning rate.

The burning rate is significantly affected by the application of a flame retardant (Figure 3). A more significant effect of the thermal modification can be seen at higher temperatures (180 and 210 °C) of thermal modification. The application of a synthetic flame retardant significantly reduces the values of this characteristic compared to only thermally modified wood, especially at temperatures of 20 °C, 160 °C, and 180 °C. The application of a natural flame retardant at lower temperatures has a very similar effect to that of thermally modified wood, deteriorating significantly at 180 and 210 °C, i.e., the maximum burn rate values increase. While, in this case, the effect of the thermal modification proved to be statistically insignificant (as in published works [44,45,47]), the effect of thermal modification of oak wood on the maximum burn rate was found to be statistically significant.

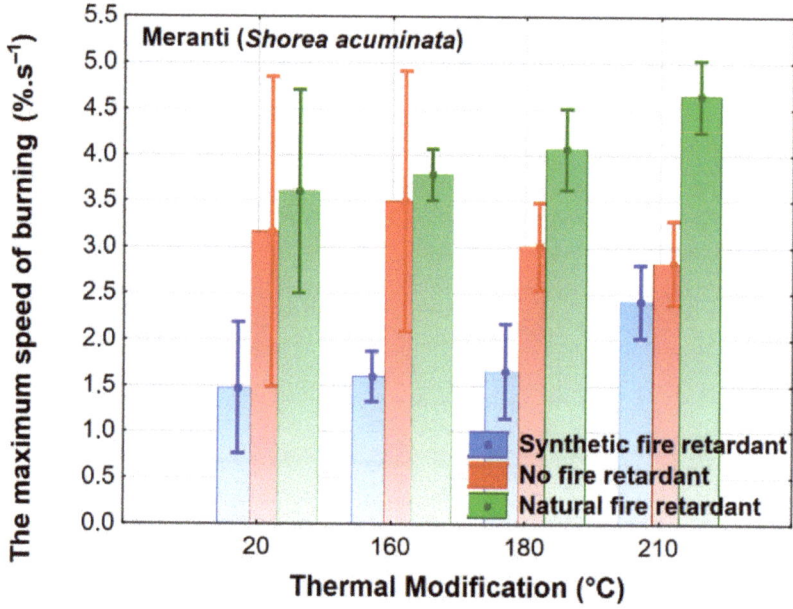

Figure 3. The effect of thermal modification and fire retardant on the maximum burn rate.

The ratio of the maximum burn rate is significantly affected by the temperature of the thermal modification and the flame retardant and, therefore, also by the mutual interaction of these effects. The differences between thermally modified wood and wood treated with flame retardants only become apparent under higher thermal modification temperatures (Figure 4). From 180 °C, this characteristic has a downward trend in thermally modified wood, while, in wood treated with a flame retardant, it has a rising trend. The differences in the use of different types of flame retardants are insignificant in this case, the greatest relative differences are observed at a temperature of 160 °C, when the natural flame retardant causes a lower ratio of the maximum burn rate of the examined wood.

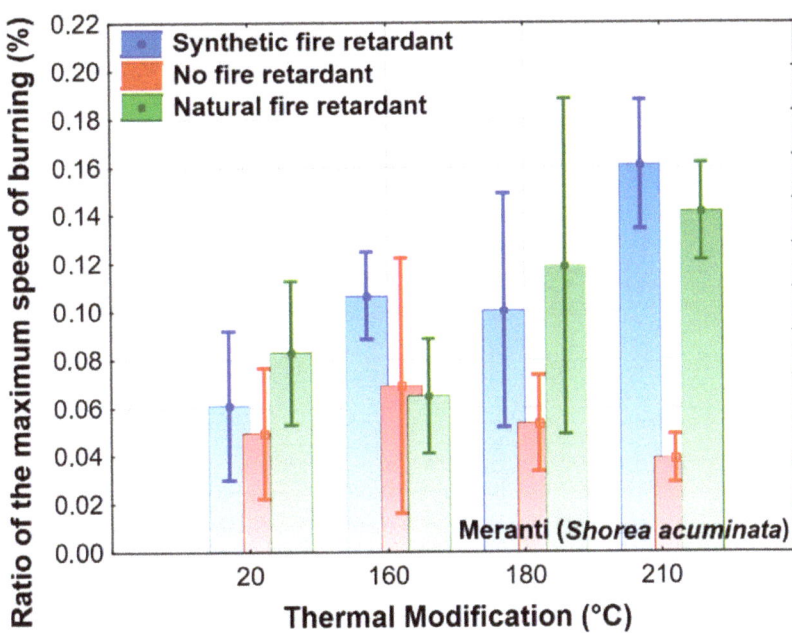

Figure 4. The effect of thermal modification and fire retardant on the ratio of the maximum burning rate.

In our research, between the thermal modification and the time to reach the maximum burning rate, the synthetic retardant had a very statistically significant effect on the maximum burn rate, up to two times lower values compared to treatment a with natural retardant. The observed characteristics for both retardants are recorded in (Figure 5), where they are shown on the blue and green curves we can see a similar trend for this characteristic at temperatures of 160 °C and 180 °C.

The results evaluating the effect of the observed factors on combustion characteristics were also evaluated using Duncan's tests. The test results fully confirm the results of Figures 1–5. Because of the large-scale work, we do not report these results.

During thermal treatment, the content of extractives increases (Table 3). Some extractives disappear or degrade at higher temperatures; however, new components are formed due to the degradation of main wood components, especially hemicelluloses. The rise in the content of extractives is due to the degradation of lignin and polysaccharides [46]. The increase in lignin content at a temperature of 160 °C is caused by condensation reactions with degradation products of polysaccharides and further cross-linking, thus increasing the apparent lignin content, i.e., pseudo-lignin [48,49]. A similar trend was observed in thermally treated teak and iroko [49]; however, at higher temperatures (180 °C and 210 °C), the lignin content declines, probably due to predominant degradation reactions.

Hemicelluloses are the most affected by higher temperatures, and their content decreases by 80% at a temperature of 210 °C when compared to untreated samples. The sharp drop in hemicelluloses between 180 °C and 210 °C can be attributed to the decomposition of xylan, the most thermally unstable hemicellulose [50]. Xylan is the dominant hemicellulose in meranti wood, and it can be quantitatively converted to xylose by dilute acid hydrolysis [51]. The increase in cellulose content is about 12%, probably because of its crystalline nature [51]. It is worth noting that tropical tree hemicelluloses are more thermally unstable compared to temperate tree hemicelluloses. The decrease in hemicellulose content in oak and spruce wood is 58% and 37%, respectively; on the other hand, the drop in hemicellulose content in teak, meranti, and merbau wood is 67%, 80%, and 90%, respectively [44,45,52]. This phenomenon may be due to the different structure of hemicelluloses in tropical and temperate wood species, e.g., molecular weight, branching, etc.

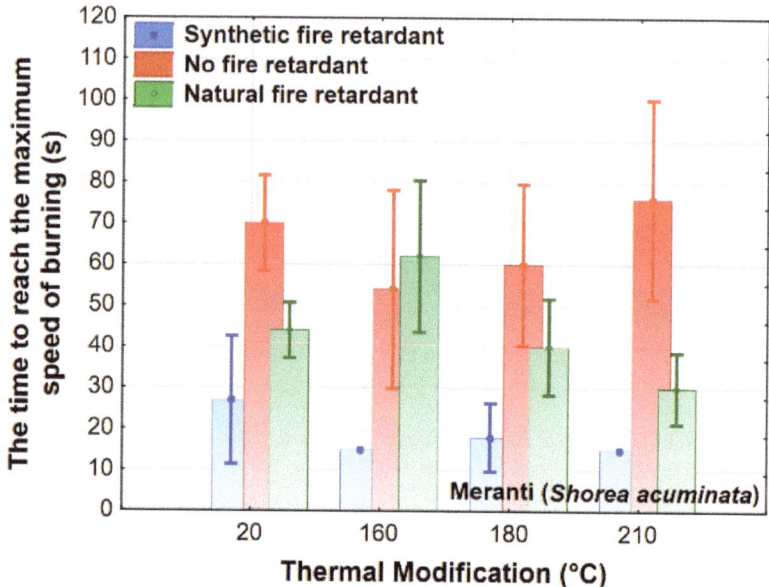

Figure 5. The effect of thermal modification and fire retardant on the time to reach the maximum burning rate.

The results of the Spearman correlation (Table 5) showed a high degree of dependence between the burning properties and the chemical components of the wood. The chemical components of the wood and the temperature of the thermal modification have reached very high values of a degree of dependence of more than 99%. The combustion characteristics had a moderate degree of dependence in interaction with the chemical components of the wood. The dependency ranged from −31% to 28%.

Table 5. Spearman's correlation between burning characteristics and chemical wood components.

Variable	FR	TM	WL	BR	MBR	RMBR	TRMBR	Ex.	Li.	Hol.	Cell.
FR											
TM.	0										
WL.	90	−4									
BR.	14	10	23								
MBR.	81	17	86	24							
RMBR.	−5	32	3	36	19						
TRMBR.	48	−14	43	−17	40	−79					
Ex.	0	84	−3	3	18	28	−9				
Li.	0	19	13	6	2	11	−8	34			
Hol.	0	−97	4	−11	−17	−32	14	−84	−19		
Cell.	0	39	−16	−1	8	9	−1	28	−77	−41	
Hemicell.	0	−97	4	−11	−17	−31	13	−83	−19	99	−42

Legend: FR.—Retardant, TM.—Thermal modification, WL.—Weight loss, BR.—Burning rate, MBR.—Maximum burning rate, RMBR.—The ratio of the maximum burning rate, TRMBR.—Time to reach the maximum burning rate, Ex.—Extractives, Li.—Lignin, Hol.—Holocellulose, Cell.—Cellulose, Hemicell.—Hemicelluloses.

4. Conclusions

The effect of synthetic fire retardant proved to be a statistically very significant agent, influencing the combustion process in all monitored characteristics. These findings are comparable with the results of other works.

Weight loss values can be reduced by impregnating the wood matrix with synthetic flame retardant (ammonium phosphate). In contrast, no significant effect on weight loss values was observed related to the reference sample treated with a natural flame retardant (arabinogalactan).

The effect of the temperature modification on the burning rate had a similar course as the weight loss. The thermal modification had a significant effect on the burning rate with all values increased. There was a similar course to the previous characteristic, and there was no statistically significant difference between the temperatures. A significant positive effect on the reduction of the combustion rate at all recorded temperatures for samples treated with synthetic flame retardant compared with non-treated samples was observed again. A statistically significant effect at higher temperatures (180 °C and 210 °C), with a decrease of the burning rate values of the treated wood, was observed.

A significant reduction of burning rate for synthetic retardant treated samples was observed. After the application of the natural retardant, there was a decrease in the values of the maximum burning rate, however not so significant related to those synthetic.

Heat-treated samples treated with synthetic retardant also had a significant statistical effect on the maximum burning rate ratio. After application of the natural retarder or between the temperatures of the heat treatment itself, there was an increased resistance of the synthetic retardant compared to the natural thermal modification at 160 and 210 °C, however, this effect was not statistically proven. The differences between the effects of synthetic and natural flame retardant on the burning properties of heat-treated meranti wood were studied. Due to the heterogeneity of the wood material and better interpretation of the data in further research, we will choose more repetitions for measurement in order not to achieve such a high coefficient of variation.

The relative content of the extractives, lignin, and cellulose after the heat treatment was increased, whereas the values of holocellulose and especially hemicelluloses were decreased significantly. The positive effect of thermal modification on the flammability characteristics of meranti wood was not confirmed.

Positive effects on all flame characteristics for samples treated with synthetic retardant compared to those treated with the natural one were described. In the case of arabinogalactan application and its maximum potential positive retarding effect, it is necessary to optimize the conditions of the experiments for its purpose.

Author Contributions: M.G., D.K. and F.K. conceived and designed the experiments; M.G. and D.K. performed flammability experiments; H.Č. measure fire properties; I.K., J.B. and F.K. performed chemical analyses; D.K., I.K. and F.K. calculated dependencies from obtained data; S.M., L.Z., T.T., D.K., I.K., M.G. and F.K. wrote the paper. All authors have read and agreed to the published version of the manuscript.

Funding: The authors are grateful for the support of "Advanced research supporting the forestry and wood-processing sector's adaptation to global change and the 4th industrial revolution, No. CZ.02.1.01/0.0/0.0/16_019/0000803 financed by OP RDE" and the authors are grateful for the support of the Internal Grant Agency (IGA) of the Faculty of Forestry and Wood Sciences, project No. IGA B 20/02. This work was also supported by the Slovak Research and Development Agency under contracts No. APVV-16-0326, and No. APVV-17-0005, and by the VEGA Agency of Ministry of Education, Science, Research and Sport of the Slovak Republic No. 1/0387/18.

Data Availability Statement: The data presented in this study are available on request from the corresponding author.

Conflicts of Interest: The authors declare no conflict of interest. The funders had no role in the design of the study; in the collection, analyses, or interpretation of data; in the writing of the manuscript, and in the decision to publish the results.

References

1. Baral, S.; Gaire, N.P.; Aryal, S.; Pandey, M.; Rayamajhi, S.; Vacik, H. Growth Ring Measurements of *Shorea robusta* Reveal Responses to Climatic Variation. *Forests* **2019**, *10*, 466. [CrossRef]
2. Baral, S.; Neumann, M.; Basnyat, B.; Gauli, K.; Gautam, S.; Bhandari, S.K.; Vacik, H. Form Factors of an Economically Valuable Sal Tree (*Shorea robusta*) of Nepal. *Forests* **2020**, *11*, 754. [CrossRef]
3. Hill, C.A.S. *Wood Modification: Chemical Thermal, and Other Processes*; John Wiley & Sons: Hoboken, NJ, USA, 2006; p. 260, ISBN 978-0_470-02172-9.
4. Rowell, R.M. *Handbook of Wood Chemistry and Wood Composites*, 2nd ed.; CRC Press: Boca Raton, FL, USA, 2012; p. 703, ISBN 9781439853801.
5. Militz, H.; Altgen, M. Processes and Properties of Thermally Modified Wood Manufactured in Europe. In *Deterioration and Protection of Sustainable Biomaterials*; ACS Symposium Series Vol. 1158, Chapter 16; American Chemical Society: Washington, DC, USA, 2014; pp. 269–285, ISBN 13: 9780841230040.
6. Rosu, L.; Mustata, F.; Varganici, C.D.; Rosu, D.; Rusu, T.; Rosca, I. Thermal behaviour and fungi resistance of composites based on wood and natural and synthetic epoxy resins cured with maleopimaric acid. *Polym. Degrad. Stabil.* **2019**, *160*, 148–161. [CrossRef]
7. Bakar, B.F.A.; Hiziroglu, S.; Tahir, P.M. Properties of some thermally modified wood species. *Mater Des.* **2013**, *43*, 348–355. [CrossRef]
8. Gérardin, V. New alternatives for wood preservation based on thermal and chemical modification of wood—A review. *Ann. For. Sci.* **2016**, *73*, 559–570. [CrossRef]
9. Čekovská, H.; Gaff, M.; Osvald, A.; Kačík, F.; Kubš, J.; Kaplan, L. Fire Resistance of Thermally Modified Spruce Wood. *BioResources* **2017**, *12*, 947–959. [CrossRef]
10. Baysal, E.; Kart, S.; Toker, H.; Degirmentepe, S. Some physical characteristics of thermally modified oriental-beech wood. *Maderas Cienc. Tecnol.* **2014**, *16*, 291–298. [CrossRef]
11. LeVan, S.L. Chemistry of fire retardancy. In *The Chemistry of Solid Wood. Advances in Chemistry Series 207*; Rowell, R.M., Ed.; American Chemical Society: Washington, DC, USA, 1984; Chapter 14; pp. 531–574.
12. Poletto, M.; Zattera, A.J.; Forte, M.C.; Mariam, M.C.; Santana, R. Thermal decomposition of wood: Influence of wood components and cellulose crystallite size. *Bioresour. Technol.* **2012**, *109*, 148–153. [CrossRef] [PubMed]
13. Mitchell, P.H. Irreversible property changes of small loblolly pine specimens heated in air, nitrogen, or oxygen. *Wood Fiber Sci.* **1988**, *20*, 320–355.
14. Korkut, D.S.; Guller, B. The effects of heat treatment on physical properties and surface roughness of red-bud maple (*Acer trautvetteri* Med.) wood. *Bioresour. Technol.* **2008**, *99*, 2846–2851. [CrossRef]
15. Cademartori, P.H.G.; Missio, A.L.; Mattos, B.D.; Gatto, D.A. Effect of thermal treatments on the technological properties of wood from two Eucalyptus species. *An. Acad. Bras. Cienc.* **2015**, *87*, 471–481. [CrossRef]
16. Esteves, B.M.; Pereira, H.M. Wood modification by heat treatment: A review. *BioResources* **2009**, *4*, 370–404. [CrossRef]
17. Šimkovic, I. Trends in thermal stability study of chemically modified lignocellulose materials. In *Polymer Degradation and Stability Research Developments*; Albertov, L.B., Ed.; Nova Science Publishers, Inc.: New York, NY, USA, 2007; pp. 217–236, ISBN 978-1-60021-827-9.
18. Harper, C.A. *Handbook of Building Materials for Fire Protection*; McGraw-Hill Handbooks: New York, NY, USA, 2003; p. 800, ISBN 13: 978-0071388917.
19. Alaee, M.; Wenning, R.J. The significance of brominated flame retardants in the environment: Current understanding, issues, and challenges. *Chemosphere* **2002**, *46*, 579–582. [CrossRef]
20. Östman, B.; Voss, A.; Hughes, A.; Hovde, P.J.; Grexa, O. The durability of Fire Retardant Treated Wood Products at Humid and Exterior Conditions Review of Literature. *Fire Mater* **2001**, *25*, 95–104. [CrossRef]
21. Nothnagel, E.A.; Bacic, A.; Clarke, A.E. *Cell and Developmental Biology of Arabinogalactan-Proteins*; Kluwer Academic/Plenum Publishers: Amsterdam, The Netherlands, 2000; 301 Seiten; p. 301, ISBN 978-0-306-46469-0.
22. Spiridon, I.; Popa, V.I. Hemicelluloses: Major sources, properties, and applications. In *Monomers, Polymers, and Composites from Renewable Resources*; Belgacem, M.N., Gandini, A., Eds.; Elsevier: Amsterdam, The Netherlands, 2008; Chapter 13; pp. 289–304, ISBN 978-0-08-045316-3.
23. Karaseva, V.; Bergeret, A.; Lacoste, C.; Ferry, L.; Fulcrand, H. Influence of Extraction Conditions on Chemical Composition and Thermal Properties of Chestnut Wood Extracts as Tannin Feedstock. *ACS Sustain. Chem. Eng.* **2019**, *7*, 17047–17054. [CrossRef]
24. LeVan, S.L.; Winandy, J.E. Effect of fire retardant treatments of wood strength: A review. *Wood Fiber. Sci.* **1990**, *22*, 113–131.
25. Unger, A.; Schniewind, A.; Unger, W. *Conservation of Wood Artifacts: A Handbook*; Springer: Berlin/Heidelberg, Germany; Cham, Switzerland, 2001; p. 578, ISBN 3540415807.
26. Gottschalk, A. *Glycoproteins: Their Composition, Structure, and Function*, 2nd ed.; Elsevier: Amsterdam, The Netherlands, 1972; p. 1378, ISBN 13: 9780444409492.
27. Reid, R.; Clamp, J.R. The biochemical and histochemical nomenclature of mucus. *Br. Med. Bull* **1978**, *34*, 5–8. [CrossRef] [PubMed]
28. Kornfeld, R.; Kornfeld, S. Comparative Aspects of Glycoprotein Structure. *Annu. Rev. Biochem.* **1976**, *45*, 217–238. [CrossRef]
29. Marshall, R.D. Glycoproteins. *Annu. Rev. Biochem.* **1972**, *41*, 673–702. [CrossRef]
30. D'Adamo, P. Larch arabinogalactan. *J. Naturopath. Med.* **1996**, *4*, 32–39.

31. Goellner, E.M.; Utermoehlen, J.; Kramer, R.; Classen, B. Structure of arabinogalactan from Larix laricina and its reactivity with antibodies directed against type-II-arabinogalactans. *Carbohydr. Polym.* **2011**, *86*, 1739–1744. [CrossRef]
32. Grube, B.; Stier, H.; Riede, L.; Gruenwald, L. Tolerability of a proprietary larch arabinogalactan extract: A randomized, double-blind, placebo-controlled clinical trial in healthy subjects. *Food Nutr. Sci.* **2012**, *3*, 1533–1538. [CrossRef]
33. ISO 13061-2. *Wood-Determination of Density for Physical and Mechanical Tests*; International Organization for Standardization: Geneva, Switzerland, 2014.
34. Gaff, M.; Kačík, F.; Gašparík, M.; Todaro, L.; Jones, D.; Corleto, R.; Osvaldová, L.M.; Čekovská, H. The effect of synthetic and natural fire-retardants on burning and chemical characteristics of thermally modified teak (*Tectona grandis* L. f.) wood. *Constr. Build. Mater.* **2019**, *200*, 551–558. [CrossRef]
35. Aydin, T.Y. Ultrasonic evaluation of time and temperature-dependent orthotropic compression properties of oak wood. *J. Mater. Res. Technol.* **2020**, *9*, 6028–6036. [CrossRef]
36. Corleto, R.; Gaff, M.; Niemz, P.; Sethy, A.K.; Todaro, L.; Ditommaso, G.; Macků, J. Effect of thermal modification on properties and milling behaviour of African padauk (*Pterocarpus soyauxii* Taub.) wood. *J. Mater. Res. Technol.* **2020**, *9*, 9315–9327. [CrossRef]
37. ISO 13061-1. *Wood-Determination of Moisture Content for Physical and Mechanical Tests*; International Organization for Standardization: Geneva, Switzerland, 2014.
38. ČSN 73 0862/B-2. *Determining the Degree of Flammability of Construction Materials—Amemedment B-2*; Czech Standards Institute: Prague, Czech Republic, 1991.
39. ISO 13061-1. *Physical and Mechanical Properties of Wood—Test Methods for Small Clear Wood Specimens—Part 1: Determination of Moisture Content for Physical and Mechanical Tests*; International Organization for Standardization: Geneva, Switzerland, 2014.
40. ASTM D1107-96. *Standard Test Method for Ethanol-Toluene Solubility of Wood*; ASTM International: West Conshohocken, PA, USA, 2013.
41. Sluiter, A.; Hames, B.; Ruiz, R.; Scarlata, C.; Sluiter, J.; Templeton, D.; Crocker, D. Determination of Structural Carbohydrates and Lignin in Biomass. In *Laboratory Analytical Procedure (LAP)*; NREL/TP-510-42618; National Renewable Energy Laboratory: Golden, CO, USA, 2012.
42. Wise, L.E.; Murphy, M.; D'Addieco, A.A. Chlorite holocellulose, its fractionation and bearing on summative wood analysis and studies on the hemicelluloses. *Paper Trade J.* **1946**, *122*, 35–43.
43. Seifert, K. Uber ein Neues Verfahren Zur Schnellbestimmung der Rein- Cellulose. *Papier* **1956**, *10*, 301–306.
44. Gašparík, M.; Osvaldová, L.; Čekovská, H.; Potůček, D. Flammability characteristics of thermally modified oak wood treated with fire-retardant. *BioResources* **2017**, *12*, 8451–8467.
45. Sikora, A.; Kačík, F.; Gaff, M.; Vondrová, V.; Bubeníková, T.; Kubovský, I. Impact of thermal modification on color and chemical changes of spruce and oak wood. *J. Wood. Sci.* **2018**, *64*, 406–416. [CrossRef]
46. Esteves, B.; Videira, R.; Pereira, H. Chemistry and ecotoxicity of heat-treated pine wood extractives. *Wood Sci. Technol.* **2011**, *45*, 661–676. [CrossRef]
47. Kačíková, D.; Kubovský, I.; Gaff, M.; Kačík, F. Changes of Meranti, Padauk and Merbau Wood Lignin during ThermoWood Process. *Polymers* **2021**, *13*, 993. [CrossRef]
48. Mohamed, A.L.; Hassabo, A.G. Flame Retardant of Cellulosic Materials and Their Composites. In *Flame Retardants*; Engineering Materials; Visakh, P., Arao, Y., Eds.; Springer: Cham, Switzerland, 2015.
49. Hu, F.; Jung, S.; Ragauskas, A. Pseudo-lignin formation and its impact on enzymatic hydrolysis. *Bioresour. Technol.* **2012**, *117*, 7–12. [CrossRef] [PubMed]
50. Werner, K.; Pommer, L.; Broström, M. Thermal decomposition of hemicelluloses. *J. Anal. Appl. Pyrol.* **2014**, *110*, 130–137. [CrossRef]
51. Rafiqul, I.S.M.; Sakinah, A.M.M.; Karim, M.R. Production of Xylose from Meranti Wood Sawdust by Dilute Acid Hydrolysis. *Appl. Biochem. Biotechnol.* **2014**, *174*, 542–555. [CrossRef] [PubMed]
52. Shinde, S.D.; Meng, X.; Kumar, R.; Ragauskas, A.J. Recent advances in understanding the pseudolignin formation in a lignocellulosic biorefinery. *Green Chem.* **2018**, *20*, 2192–2205. [CrossRef]

Article

Effect of Natural Aging on Oak Wood Fire Resistance

Martin Zachar [1], Iveta Čabalová [2,*], Danica Kačíková [1] and Tereza Jurczyková [3]

[1] Department of Fire Protection, Faculty of Wood Sciences and Technology, Technical University in Zvolen, T. G. Masaryka 24, 960 53 Zvolen, Slovakia; zachar@tuzvo.sk (M.Z.); kacikova@tuzvo.sk (D.K.)
[2] Department of Chemistry and Chemical Technologies, Faculty of Wood Sciences and Technology, Technical University in Zvolen, T. G. Masaryka 24, 960 53 Zvolen, Slovakia
[3] Department of Wood Processing, Czech University of Life Sciences in Prague, Kamýcká 1176, 16521 Praha 6-Suchdol, Czech Republic; jurczykova@fld.czu.cz
* Correspondence: cabalova@tuzvo.sk; Tel.: +421-4-5520-6375

Citation: Zachar, M.; Čabalová, I.; Kačíková, D.; Jurczyková, T. Effect of Natural Aging on Oak Wood Fire Resistance. *Polymers* **2021**, *13*, 2059. https://doi.org/10.3390/polym13132059

Academic Editor: Nicolas Brosse

Received: 4 June 2021
Accepted: 22 June 2021
Published: 23 June 2021

Publisher's Note: MDPI stays neutral with regard to jurisdictional claims in published maps and institutional affiliations.

Copyright: © 2021 by the authors. Licensee MDPI, Basel, Switzerland. This article is an open access article distributed under the terms and conditions of the Creative Commons Attribution (CC BY) license (https://creativecommons.org/licenses/by/4.0/).

Abstract: The paper deals with the assessment of the age of oak wood (0, 10, 40, 80 and 120 years) on its fire resistance. Chemical composition of wood (extractives, cellulose, holocellulose, lignin) was determined by wet chemistry methods and elementary analysis was performed according to ISO standards. From the fire-technical properties, the flame ignition and the spontaneous ignition temperature (including calculated activation energy) and mass burning rate were evaluated. The lignin content does not change, the content of extractives and cellulose is higher and the content of holocellulose decreases with the higher age of wood. The elementary analysis shows the lowest proportion content of nitrogen, sulfur, phosphor and the highest content of carbon in the oldest wood. Values of flame ignition and spontaneous ignition temperature for individual samples were very similar. The activation energy ranged from 42.4 kJ·mol^{-1} (120-year-old) to 50.7 kJ·mol^{-1} (40-year-old), and the burning rate varied from 0.2992%·s^{-1} (80-year-old) to 0.4965%·s^{-1} (10-year-old). The difference among the values of spontaneous ignition activation energy is clear evidence of higher resistance to initiation of older wood (40- and 80-year-old) in comparison with the younger oak wood (0- and 10-year-old). The oldest sample is the least thermally resistant due to the different chemical composition compared to the younger wood.

Keywords: oak wood; historical wood; chemical composition; flame ignition temperature; spontaneous ignition temperature; activation energy; mass burning rate

1. Introduction

When using wood as a structural element, especially as a part of the building structure, it is necessary to assess it from the point of view of fire safety of buildings. It can be realized by using Eurocodes [1]. An important part of ensuring the fire safety of buildings is knowledge of the burning process and forecasting the dynamics of the development of internal fire. Knowledge of fire dynamics is an important starting point, for example, in building design, controlled evacuation, physicochemical and mathematical description of fire–materials and fire–human organism interactions, in the process of determining the causes of fire in finding possible scenarios of fire origin and development, and determining the most similar cause of fire [2].

A significant factor affecting burning rate is chemical composition [3–8], primarily the lignin content [3,7,9,10], the species of the wood [3,4,11–13], density, moisture content, permeability, anatomy [3] and, last but not least, the aging process [14–16]. According to Reinprecht [17], wood decomposition is caused by a complex of chemical reactions associated with mass and heat transfer processes. The wood is ignited when sufficient initial thermal energy is supplied (approximately 104 W·m^{-2}), while the specific temperature at which the wood ignites is between 250 and 400 °C.

Activation energy has a great influence on the combustion process. There are several methods to calculate the activation energy. Simple methods (e.g., Arrhenius equation)

to much more complex methods (e.g., Ozawa–Flynn–Wall, Kissinger–Akahira–Sunos or Friedman, ASTM E698–18) requiring the results of thermogravimetry (TG), differential thermogravimetry (DTG) and differential scanning calorimetry (DSC), and other progressive instruments and methodologies result in calculating the activation energy and the combustion process kinetics in a complex and precise way. The progressive approaches to the analysis of biomass combustion kinetics are also evident in the study of Majlingová et al. [18]. Based on the flash ignition temperature and spontaneous ignition temperature, according to the ISO 871:2006, the relative comparison of a material resistance against ignition can be carried out.

Several authors studied the thermal properties of wood. They preferred the Arrhenius equation to calculate the activation energy of woody biomass. Martinka et al. [19] focused on the influence of spruce wood form on ignition activation energy and they also investigated the impact of heat flux on fire risk of the selected samples. Rantuch et al. [20] used the Arrhenius equation to calculate the ignition activation energy of materials based on polyamide. In another study, Martinka et al. [21] carried out research on initiatory parameters of poplar wood (*Populus tremula* L.). These initiatory parameters (the critical heat flux density and the surface temperature at the time of initiation) were set on a conical calorimeter using a testing procedure in accordance with ISO 5660-1:2015. Luptáková et al. [22] clarified a comparison of activation energies of thermal degradation of heat-sterilized silver fir wood samples to larval frass in terms of fire safety. The ignition activation energies of wood samples using the Arrhenius equation were also calculated. Zachar et al. [23] published the results of an analysis focusing on the activation energy required for spontaneous ignition and flash point of Norway spruce and thermowood specimens. Karlsson and Quintere [24] stated that for flash ignition temperature phenomenon to occur, the temperature in the fire compartment should reach 500 to 600 °C, or the radiant flux on the floor should be 15 to 20 $kW \cdot m^{-2}$. These temperatures are significantly higher than the spontaneous ignition temperatures of most lignocellulosic materials.

The fire parameters of wood species are well known [25], but only a few scientific studies have focused on the exact assessment of changes of the key fire parameters (i.e., activation energy and burning rate) of wood and wood-based materials due to their aging.

This paper deals with the assessment of the age (0, 10, 40, 80 and 120 years) of oak wood and its chemical composition on the burning rate for the purposes of determining the causes of fires. We also investigated the relative weight loss and duration of flame burning as important characteristics when assessing materials from the perspective of fire.

2. Experimental

2.1. Material

Samples of oak wood (*Quercus robur* L.) were prepared from beams (taken from the interior part of building) of different ages (40, 80 and 120 years). The beams were obtained from historic buildings in Slovakia during their reconstruction. The age was marked on the beams and verified in the historical records of building construction. Samples prepared from 0- and 10-year-old wood were taken from wood harvested in the Zvolen locality in Slovakia.

Samples with the dimensions of 20 × 20 × 10 mm (standard STN ISO 871) [26] were conditioned at a temperature of 23 ± 2 °C and relative air humidity of 50 ± 5% for at least 40 h to the final sample moisture content of 12% (standard STN EN ISO 291) [27].

The average density of wood samples ranged from 0.641 to 0.702 $g \cdot cm^{-3}$ (Table 1).

Table 1. The average density of oak wood samples.

Approximate Age of Oak Sample (years)	Density (g·cm^{-3})
recent	0.681 ± 0.03
10	0.641 ± 0.11
40	0.660 ± 0.28
80	0.688 ± 0.12
120	0.702 ± 0.21

2.2. Methods

2.2.1. Chemical Composition of Wood

Samples were disintegrated into sawdust, and the size fractions 0.5 to 1.0 mm were used for the chemical analyses. According to ASTM D1107-21 [28], the extractives content was determined in a Soxhlet apparatus with a mixture of ethanol and toluene (2:1). The lignin content was determined according to Sluiter et al. [29], the cellulose according to the method by Seifert [30], and the holocellulose according to the method by Wise et al. [31]. Hemicelluloses were calculated as a difference between the holocellulose and the cellulose content. Measurements were performed on four replicates per sample. The results were presented as oven-dry wood percentages.

Elementary analysis was accomplished as follows: carbon (C) was determined according to STN ISO 10,694 [32] by using elemental analysis with thermally conductive analysis; nitrogen (N) according to ISO 13,878 [33] by using elemental analysis with thermally conductive analysis; sulfur (S) according to ISO 15,178 [34] by using elemental analysis with thermally conductive analysis; and phosphor (P), calcium (Ca), magnesium (Mg) and potassium (K) according to ISO 11,885 [35] by using atomic emission spectrometry with inductively coupled plasma.

2.2.2. Flame Ignition Temperature and Spontaneous Ignition Temperatures

The flame ignition temperature (FIT) and spontaneous ignition temperature (SIT) were determined according to the STN ISO 871 standard [26]. Measurements were performed on twenty replicates per sample. The principle of the test is to heat the test material in a heating chamber at different temperatures. By positioning a small ignition flame impinging on the opening cover of the hot-air furnace, the released gases ignite and the FIT can be determined. SIT is determined the same way as the FIT, but without the igniting flame. The temperature profile in the furnace was measured using thermocouples (type K) with a diameter of 0.5 mm; the data logger ALMEMO®710 (Ahlborn Mess- und Regelungstechnik GmbH, Holzkirchen, Germany) was used for recording temperature. The lowest air temperature at which the sample was ignited within 10 min was recorded as the spontaneous ignition temperature. Subsequently, the induction time was found. Analysis of dependences between the induction time and the inverse values of thermodynamic temperature for the samples was performed using Statistica 12 software. The exponential equation was derived in the same software. The pre-exponential factor was further used to calculate the activation energy of spontaneous ignition. The calculation of the activation energy of spontaneous ignition (kJ·mol^{-1}) was performed according to Equation (1), which is analogous to the Arrhenius equation.

$$E = \ln\left(\frac{\varnothing}{A}\right) \times R \times T \qquad (1)$$

where:

τ—induction time of spontaneous ignition (s);
A—pre-exponential (frequency) factor (-);
E—activation energy of spontaneous ignition (J·mol^{-1});
R—gas constant (8.314 J·K^{-1}·mol^{-1});

T—ignition thermodynamic temperature (K).

2.2.3. The Mass Burning Rate

The reaction to fire tests was determined according to the ISO 11925-2 [36] standard. The mass burning rate was measured with an instrument consisting of an electronic balance with an accuracy of two decimal places, a weight protection unit, a metal sample holder, a metal loading frame for placing the radiant heat source and an infrared thermal heater with an input of 1000 W. The sample was placed into the holder at a distance of 30 mm from the heat source for a specific time of 600 s and the weight change recorded every 10 s. The heat flux of the infrared thermal heater was 30 kW·m^{-2}. Measurements were performed on twenty replicates per sample.

To determine the burning rate in the specified time interval, the absolute burning rate v (was calculated according to the relational Equation (2):

$$\vartheta = \frac{\delta(\tau) - \delta(\tau + \Delta\tau)}{\Delta\tau} \qquad (2)$$

where:

ϑ—absolute burning rate (%·s^{-1});
$\delta(\tau)$—specimen mass in the time (τ) (%);
$\delta(\tau + \Delta\tau)$—specimen mass in the time ($\tau + \Delta\tau$) (%);
$\Delta\tau$—time interval in which the mass values are recorded (s).

3. Results and Discussion

3.1. Wood Chemical Composition

The content of the main chemical compounds, except the lignin amount, differs when comparing older and recent wood (Table 2). The proportion of the lignin does not change considerably despite the different age of the wood samples. Several authors determined the lignin content between 14.78% and 28.8% depending on the type of oak wood, zone of wood, etc. [37–40]. Kolář and Rybíček [38] analyzed subfossil oak wood (*Quercus robur* L.). Based on their results, subfossil wood contains a greater amount of fibers. In the view of chemical composition, the cellulose content does not change despite the different age of the trunk, but the lignin content is higher in subfossil wood. The cellulose and extractives comparison of our samples does display higher amounts of these components in older wood. Kačík et al. [41] observed that the cellulose in old fir wood beam samples (from 108- to 390-years-old) increased by 13%; both the lignin and holocellulose dropped by 4% compared to the recent fir wood. In contrast, the content of hemicelluloses in our samples decreased due to the aging process. Hemicelluloses are the most susceptible to degradation, even at relatively low temperatures [42–44]. Several authors studied wood with different natural aging time and the results confirm that the proportion of saccharides gradually decreases (mainly due to the hemicellulose degradation) and the content of lignin increases successively with increasing time [41,45,46]. According to Fengel and Wegener [47], chemical analyses of old woods show a decrease in polysaccharides and an increase in the nonhydrolyzable residues. Jebrane et al. [48] explain the degradation of polysaccharides with the presence of acetyl groups that are thermally labile and lead to the formation of acetic acid, thereby causing acid-catalyzed degradation of the polysaccharides. A more reliable indicator seems to be the cellulose to hemicelluloses ratio (C/H) [41]. In the samples of oak wood, the C/H ratio increased with age (Table 2) due to a lower stability of hemicelluloses towards cellulose.

Table 2. Relative content of the main chemical components of oak wood.

Age of Oak Sample (years)	Extractives (%)	Lignin (%)	Cellulose (%)	Holocellulose (%)	Hemicelluloses (%)	C/H Ratio
0	3.93 ± 0.04	23.04 ± 0.14	33.48 ± 0.06	73.03 ± 0.11	39.55 ± 0.17	0.84
10	3.97 ± 0.06	22.86 ± 0.04	33.79 ± 0.57	73.16 ± 0.02	39.37 ± 0.55	0.86
40	5.78 ± 0.04	22.14 ± 0.02	33.41 ± 0.11	72.08 ± 0.06	38.67 ± 0.23	0.86
80	6.62 ± 0.03	22.31 ± 0.02	36.22 ± 0.01	71.08 ± 0.04	34.85 ± 0.03	1.04
120	7.34 ± 0.11	22.91 ± 0.04	36.40 ± 0.08	69.75 ± 0.15	33.35 ± 0.23	1.09

Historic wood samples have different chemical compositions, primarily depending on the deposition conditions. The conditions determine the mechanisms and rates of wood degradation [49,50]. According to Krutul and Kocoń [51], time plays only a secondary role in the destroying process. Factors influencing wood aging include UV radiation, which has sufficient energy for photochemical degradation of structural polymer components of wood (lignin, cellulose and hemicelluloses) [52].

The results in Table 3 show the differences in elementary composition of oak wood. The 120-year-old oak wood contains the lowest amount of N, S and P, and the largest amount of C elements due to the initial carbonization process [53]. We determined a calcium content of 0.7 g·kg^{-2} in the recent wood samples, which is in accordance with the results of Krutul [54]. The amount of this element decreased with the age of wood.

Table 3. Elementary analysis of oak wood.

Age of Oak Sample (years)	Carbon (g·kg^{-2})	Nitrogen (g·kg^{-2})	Sulfur (mg·kg^{-2})	Phosphor (g·kg^{-2})	Calcium (g·kg^{-2})	Magnesium (g·kg^{-2})	Potassium (g·kg^{-2})
0	489	1.32	299	0.151	0.702	0.044	0.621
10	491	1.07	252	0.155	0.495	0.134	0.921
40	486	1.35	291	0.145	0.296	0.029	0.834
80	493	1.29	122	0.14	0.244	0.008	0.435
120	501	0.98	94	0.118	0.293	0.022	0.544

Cárdenas-Gutiérrez [39] performed ash microanalysis of various oak woods with an X-ray spectrometer (*Quercus candicans*—Qc; *Quercus laurina*—Ql; *Quercus rugosa*—Qr). Based on their results, sample Qc contains 19.91% of Mg, 8.27% of P, 1.28% of S, 38.33% of K and 24.83% of Ca; sample Ql contains 5.41% of Mg, 1.22% of P, 0.22% of S, 8.93% of K and 77.11% of Ca, and sample Qr contains 27.71% of Mg, 10.34% of P, 0.97% of S, 27.07% of K and 33.28% of Ca. Kovář and Rybíček [38] conducted the analysis of the inorganic elements of subfossil wood and determined the highest content of Ca. Based on the results they assume that the samples have gone through a process of calcification.

3.2. Fire-Technical Properties

The spontaneous ignition temperatures together with the temperature recalculated values (inverse value of the temperature in °C to the thermodynamic temperature in K, necessary for the calculation of the activation energy) are shown in Table 4.

Based on the results given in Table 4, the sample of 10-year-old oak wood has the highest fire resistance in the view of FIT, because we marked its average temperature of 434.14 °C in a time of 257.4 s. From the point of view of fire resistance, it is important to record the highest temperature and the longest ignition time.

Table 4. Flame ignition temperature and spontaneous ignition temperature of oak wood samples.

Age of Oak Sample (years)	Thermal Loading	Average Time τ (s)	Average Temperature t (°C)	Average Temperature T (K)	Inverse Value 1/T (K^{-1})
0	FIT	219.6 ± 21.38	436.02 ± 21.05	709.17	0.0014111
	SIT	336.4 ± 18.49	374.96 ± 23.09	648.11	0.0015448
10	FIT	257.4 ± 13.33	434.14 ± 23.14	706.19	0.0014153
	SIT	336.1 ± 19.51	375.43 ± 24.30	648.58	0.0015436
40	FIT	248.2 ± 25.28	432.43 ± 17.62	705.58	0.0014182
	SIT	376.1 ± 20.29	371.24 ± 26.72	644.39	0.0015542
80	FIT	221.6 ± 17.14	429.52 ± 16.24	702.67	0.0014274
	SIT	379.2 ± 17.58	361.23 ± 22.13	633.33	0.0015777
120	FIT	229.5 ± 24.30	430.48 ± 18.12	703.63	0.0014219
	SIT	391.2 ± 19.54	363.37 ± 21.89	636.52	0.0015727

The results of the calculated activation energy using the Arrhenius equation are introduced. Table 5 shows the values of pre-exponential factor (A) representing the regression coefficient in the correlation equation calculated between the spontaneous ignition temperature and the induction time values of the individual samples. These data were used for the calculation of the activation energy of individual samples, including the average values of the spontaneous ignition temperature and the induction time.

Table 5. The activation energy values of individual test samples.

Age of Oak Sample (years)	Exponential Equation	τ (s)	A	Activation Energy (kJ·mol^{-1})
0	$y = 0.067 \times e^{5715.9x}$	219.6	0.0672	47.710
10	$y = 0.108 \times e^{5472x}$	257.4	0.1080	45.727
40	$y = 0.044 \times e^{6077x}$	248.2	0.0439	50.684
80	$y = 0.045 \times e^{5941x}$	221.6	0.0450	49.535
120	$y = 0.162 \times e^{5090.9x}$	229.5	0.1621	42.444

The activation energy of oak samples (Table 5) with different age ranged from 42.4 kJ·mol^{-1} (120-year-old) to 50.7 kJ·mol^{-1} (40-year-old). The value of the activation energy of a 40-year-old sample was obtained in an average time of 248.2 s from the beginning of the thermal loading. Based on these results, we can state that this sample has the higher resistance against this kind of thermal loading. The data indicate that the activation energy of spontaneous ignition may become a suitable tool for a more accurate comparison of the thermal resistance of materials. The difference in spontaneous ignition temperature between these materials is minimal (according to ISO 871:2006, the ignition temperature is measured with an accuracy of 10 °C), but the burning time achieved was different. The difference in the activation energy of spontaneous ignition is very significant.

The lowest thermal resistance of oak wood was observed in the oldest wood (120-year-old), which could cause its degradation and decomposition of the main chemical components. The higher content of extractives in this sample indicates the decomposition of lignin as the most stable macromolecule [55,56]. This sample also has the lowest proportion of saccharides, mainly due to the degradation of hemicelluloses (C/H ratio = 1.09). Several authors describe the decrease of polysaccharides during natural aging of wood [45,57]. Therefore, older wood poses a greater risk of ignition compared to the younger oak wood.

In general, the values of the induction time (τ) decrease with increasing values of the temperatures (t) [23].

The scientific work of Tureková and Balog [58] showed that the activation energy of spontaneous ignition, even if the sample weight is up to one gram, significantly depends on the weight of the sample. The solution for obtaining the spontaneous ignition activation energy from these external factors is the measure the activation energy of spontaneous ignition for a particular sample and the specific conditions (e.g., modeling the dynamics of development of a particular fire, investigation of a specific fire, etc.).

The results of the mass burning rate are shown in Figure 1, where the maximum values of burning rate and the time needed to reach the maximum values of burning rate are given.

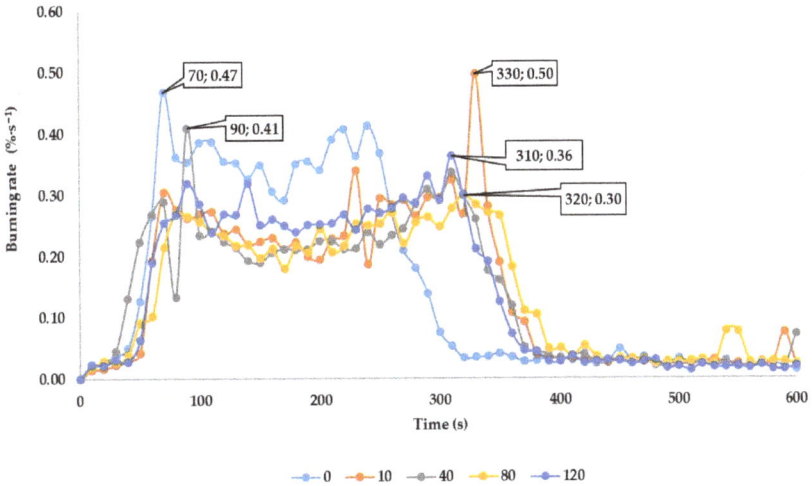

Figure 1. The absolute burning rate of the oak wood.

The maximum values of burning rate and the time needed to reach the maximum values of burning rate are presented in Figure 1.

Based on the changes in the mass burning rate curves shown in Figure 1, it can be stated that the maximum burning rates can be observed over the entire time interval during which the samples were subjected to the thermal loading.

The maximum burning rate of $0.4965\% \cdot s^{-1}$ was reached in 330 s for 10-year-old samples. This was followed by 0-year-old samples, where the maximum burning rate was $0.4678\% \cdot s^{-1}$ that was reached in 70 s; 40-year-old samples with the burning rate of $0.4093\% \cdot s^{-1}$ that was reached in 90 s and 120-year-old samples with the burning rate of $0.3625\% \cdot s^{-1}$ that was reached in 320 s. The lowest burning rate of $0.2992\% \cdot s^{-1}$ was reached by 80-year-old samples in 320 s from the beginning of the thermal load. Based on the results of burning rates, no clear conclusion can be drawn about the influence of the age of the samples on the burning rate. The results show that the initiation and the subsequent thermal degradation of oak wood occur in time interval from 30 to 90 s from the beginning of the thermal loading. The impact of the age of the samples on initiation and subsequent development of the fire cannot be confirmed.

The reason for a lower rate values was the difference in the initial phase of sample burning. Kačíková a Makovická [59] compared coniferous species. The lowest average burning rate, $0.090\% \cdot s^{-1}$, was reached for spruce wood. The maximum burning rate for spruce wood reached $0.187\% \cdot s^{-1}$ in 180 s from the beginning of the thermal load.

4. Conclusions

The content of extractives and cellulose is higher and hemicelluloses lower with increasing age of oak wood. The proportion of lignin does not change.

The 120-year-old samples contain the lowest relative content of hemicelluloses and the highest relative content of both lignin and extractives due to the degradation processes.

From the perspective of elementary analysis, the 120-year-old oak wood sample contains the highest amount of carbon due to the initial carbonization process and the lowest amount of nitrogen, sulfur and phosphor.

Samples of 10-year-old oak wood have the greatest fire resistance in the view of FIT.

In terms of fire-technical properties, especially the highest activation energy reached, 50.684 kJ·mol^{-1}, we can state that the 40-year-old oak wood samples have the highest thermal resistance.

The maximal burning rate values are range from 0.2992%·s^{-1} (80-year-old oak wood samples) to 0.4965%·s^{-1} (10-year-old oak wood samples).

Comparing the activation energy (49.535 kJ·mol^{-1}) and the mass burning rate (0.2992%·s^{-1}), the 80-year-old samples have the greatest fire resistance. Samples of 120-year-old oak wood have the lowest values of these parameters; these samples also contain the lowest amount of N, S, P and the largest amount of C elements due to the initial carbonization process.

Author Contributions: M.Z. and I.Č. conceived and designed the experiments; M.Z., I.Č. and T.J. carried out the laboratory experiments; M.Z., I.Č. and D.K. analyzed the data, interpreted the results, prepared figures and wrote the manuscript. All authors have read and agreed to the published version of the manuscript.

Funding: This research received no external funding.

Acknowledgments: This work was supported by the Slovak Research and Development Agency under the contract No. APVV-17-0005 (100%).

Conflicts of Interest: The authors declare no conflict of interest.

References

1. EN 1995-1-1 + A1. *Eurocode 5. Design of Wooden Structures. Depending on the Required Fire Resistance of the Building Structure, the Minimum Cross-Sectional Dimensions of Load-Bearing Wooden Elements in Wooden Structures Must Be Designed*; European Committee for Standardization: Brussels, Belgium, 2008.
2. Kačíková, D.; Majlingová, A.; Veľková, V.; Zachar, M. *Modelling of Internal Fires Using the Results of Progressive Methods of Fire Engineering*, 1st ed.; Technical University: Zvolen, Slovakia, 2017; p. 147.
3. Friquin, K.L. Material properties and external factors influencing the charring rate of solid wood and glue-laminated timber. *Fire Mater.* **2010**, *35*, 303–327. [CrossRef]
4. Cachim, P.B.; Franssen, J.-M. Assessment of Eurocode 5 Charring Rate Calculation Methods. *Fire Technol.* **2010**, *46*, 169–181. [CrossRef]
5. White, R.; Dietenberger, M. Wood Products: Thermal Degradation and Fire. In *Encyclopedia of Materials: Science and Technology*; Elsevier BV: Amsterdam, The Netherlands, 2001; pp. 9712–9716.
6. Bartlett, A.I.; Hadden, R.M.; Bisby, L.A. A Review of Factors Affecting the Burning Behaviour of Wood for Application to Tall Timber Construction. *Fire Technol.* **2019**, *55*, 1–49. [CrossRef]
7. Lau, P.W.C.; White, R.; Van Zeeland, I. Modelling the charring behaviour of structural lumber. *Fire Mater.* **1999**, *23*, 209–216. [CrossRef]
8. Očkajová, A.; Kučerka, M.; Kminiak, R.; Krišťák, Ľ.; Igaz, R.; Réh, R. Occupational exposure to dust produced when milling thermally modified wood. *Int. J. Environ. Res. Public Health* **2020**, *17*, 1478. [CrossRef]
9. Aristri, M.; Lubis, M.; Yadav, S.; Antov, P.; Papadopoulos, A.; Pizzi, A.; Fatriasari, W.; Ismayati, M.; Iswanto, A. Recent Developments in Lignin- and Tannin-Based Non-Isocyanate Polyurethane Resins for Wood Adhesives—A Review. *Appl. Sci.* **2021**, *11*, 4242. [CrossRef]
10. Kačíková, D.; Kubovský, I.; Ulbriková, N.; Kačík, F. the impact of thermal treatment on structural changes of teak and iroko wood lignins. *Appl. Sci.* **2020**, *10*, 5021. [CrossRef]
11. Frangi, A.; Fontana, M. Charring rates and temperature profiles of wood sections. *Fire Mater.* **2003**, *27*, 91–102. [CrossRef]
12. Njankouo, J.M.; Dotreppe, J.-C.; Franssen, J.-M. Experimental study of the charring rate of tropical hardwoods. *Fire Mater.* **2004**, *28*, 15–24. [CrossRef]
13. Schmid, J.; Just, A.; Klippel, M.; Fragiacomo, M. The Reduced Cross-Section Method for Evaluation of the Fire Resistance of Timber Members: Discussion and Determination of the Zero-Strength Layer. *Fire Technol.* **2015**, *51*, 1285–1309. [CrossRef]
14. Sonderegger, W.; Kránitz, K.; Bues, C.-T.; Niemz, P. Aging effects on physical and mechanical properties of spruce, fir and oak wood. *J. Cult. Herit.* **2015**, *16*, 883–889. [CrossRef]
15. Kránitz, K.; Sonderegger, W.; Bues, C.-T.; Niemz, P. Effects of aging on wood: A literature review. *Wood Sci. Technol.* **2016**, *50*, 7–22. [CrossRef]

16. Topaloglu, E.; Ustaomer, D.; Ozturk, M.; Pesman, E. Changes in wood properties of chestnut wood structural elements with natural aging. *Maderas Cienc. Tecnol.* **2021**, *23*, 23. [CrossRef]
17. Reinprecht, L. *Wood Deterioration, Protection and Maintenance*, 1st ed.; Wiley & Sons, Ltd.: Chichester, UK, 2016; pp. 145–217.
18. Majlingová, A.; Zachar, M.; Lieskovský, M.; Mitterová, I. The analysis of mass loss and activation energy of selected fast-growing tree species and energy crops using the Arrhenius equation. *Acta Fac. Xylologiae Zvolen* **2018**, *60*, 175–186.
19. Martinka, J.; Mózer, V.; Hroncová, E.; Ladomerský, J. Influence of spruce wood form on ignition activation energy. *Wood Res.* **2015**, *60*, 815–822.
20. Rantuch, P.; Wachter, I.; Hrušovský, I.; Balog, K. Ignition Activation Energy of Materials based on Polyamide 6. *Trans. VSB Tech. Univ. Ostrav. Saf. Eng. Ser.* **2016**, *11*, 27–31. [CrossRef]
21. Martinka, J.; Hroncová, E.; Kačíková, D.; Rantuch, P.; Balog, K.; Ladomerský, J. Ignition parameters of poplar wood. *Acta Fac. Xylologiae* **2017**, *59*, 85–95.
22. Luptáková, J.; Kačík, F.; Eštoková, A.; Kačíková, D.; Šmíra, P.; Nasswettrová, A.; Bubeníková, T. Comparison of activation energy of thermal degradation of heat sterilised silver fir wood to larval frass regarding fire safety. *Acta Fac. Xylologiae Zvolen* **2018**, *60*, 19–29.
23. Zachar, M.; Majlingová, A.; Šišulák, S.; Baksa, J. Comparison of the activation energy required for spontaneous ignition and flash point of the Norway spruce wood and thermowood specimens. *Acta Fac. Xylologiae Zvolen* **2017**, *59*, 79–90.
24. Karlsson, B.; Quintiere, J. *Enclosure Fire Dynamics*; Informa UK Limited: London, UK, 1999; p. 336.
25. Shi, L.; Chew, M.Y.L. Experimental study of woods under external heat flux by autoignition. *J. Therm. Anal. Calorim.* **2012**, *111*, 1399–1407. [CrossRef]
26. STN ISO 871. *Plastics. Determination of Ignition Temperature Using a Hot-Air Oven*; International Organization for Standardization: Geneva, Switzerland, 2010.
27. STN EN ISO 291. *Plastics. Standard Atmospheres for Conditioning and Testing*; International Organization for Standardization: Geneva, Switzerland, 2008.
28. ASTM D1107-21. *Standard Test Method for Ethanol-Toluene Solubility of Wood*; ASTM International: West Conshohocken, PA, USA, 2021.
29. Sluiter, A.; Hames, B.; Ruiz, R.; Scarlata, C.; Sluiter, J.; Templeton, D.; Crocker, D. *Determination of Structural Carbohydrates and Lignin in Biomass (NREL/TP-510-42618)*; National Renewable Energy Laboratory: Golden, CO, USA, 2012.
30. Seifert, V.K. About a new method for rapid determination of pure cellulose. *Das Pap.* **1956**, *10*, 301–306. (In German)
31. Wise, L.E.; Murphy, M.; D'addieco, A.A. Chlorite holocellulose, its fractionation and bearing on summative wood analysis and on studies on the hemicelluloses. *Pap. Trade J.* **1946**, *122*, 35–44.
32. ISO 10694. *Soil Quality. Determination of Organic and Total Carbon after Dry Combustion (Elementary Analysis)*; International Organization for Standardization: Geneva, Switzerland, 1995.
33. ISO 13878. *Soil Quality. Determination of Total Nitrogen Content by Dry Combustion (Elemental Analysis)*; International Organization for Standardization: Geneva, Switzerland, 1998.
34. ISO 15178. *Soil Quality. Determination of Total Sulfur by Dry Combustion*; International Organization for Standardization: Geneva, Switzerland, 2000.
35. ISO 11885. *Water Quality. Determination of Selected Elements by Inductively Coupled Plasma Optical Emission Spectrometry (ICP-OES)*; International Organization for Standardization: Geneva, Switzerland, 2007.
36. ISO 11925-2. *Reaction to Fire Tests—Ignitability of Products Subjected to Direct Impingement of Flame—Part 2: Single-Flame Source Test*; International Organization for Standardization: Geneva, Switzerland, 2020.
37. Santos, R.B.; Capanema, E.A.; Balakshin, M.Y.; Chang, H.; Jameel, H. Lignin structural variation in hardwood species. *J. Agric. Food Chem.* **2012**, *60*, 4923–4930. [CrossRef] [PubMed]
38. Kolář, T.; Rybníček, M. The changes in chemical composition and properties of subfossil oak deposited in holocene sediments. *Wood Res.* **2014**, *59*, 149–166.
39. Cárdenas-Gutiérrez, M.A.; Pedraza-Bucio, F.E.; López-Albarrán, P.; Rutiaga-Quiñones, J.G. Chemical components of the branches of six hardwood species. *Wood Res.* **2018**, *63*, 795–808.
40. Hrčka, R.; Kučerová, V.; Hýrošová, T. Correlations between oak wood properties. *BioResources* **2018**, *4*, 8885–8898. [CrossRef]
41. Kačík, F.; Šmíra, P.; Kačíková, D.; Reinprecht, L.; Nasswettrova, A. Chemical changes in fir wood from old buildings due to ageing. *Cellul. Chem. Technol.* **2014**, *48*, 79–88.
42. Kučerová, V.; Lagaňa, R.; Výbohová, E.; Hýrošová, T. The effect of chemical changes during heat treatment on the color and mechanical properties of fir wood. *BioResources* **2016**, *11*, 9079–9094. [CrossRef]
43. Kubovský, I.; Kačíková, D.; Kačík, F. Structural Changes of Oak Wood Main Components Caused by Thermal Modification. *Polymers* **2020**, *12*, 485. [CrossRef]
44. Čabalová, I.; Zachar, M.; Kačík, F.; Tribulová, T. Impact of thermal loading on selected chemsical and morphological properties of spruce ThermoWood. *BioResources* **2019**, *1*, 387–400.
45. Popescu, C.-M.; Hill, C.A.S. The water vapour adsorption–desorption behaviour of naturally aged *Tilia cordata* Mill. wood. *Polym. Degrad. Stab.* **2013**, *98*, 1804–1813. [CrossRef]
46. Zhao, C.; Zhang, X.; Liu, L.; Yu, Y.; Zheng, W.; Song, P. Probing Chemical Changes in Holocellulose and Lignin of Timbers in Ancient Buildings. *Polymers* **2019**, *11*, 809. [CrossRef] [PubMed]
47. Fengel, D.; Wegener, G. *Wood—Chemistry, Ultrastructure, Reactions*, 2nd ed.; Walter de Gruyter: Berlin, Germany, 1989; p. 613.

48. Jebrane, M.; Pockrandt, M.; Cuccui, I.; Allegretti, O.; Uetimane, E.; Terziev, N. Comparative Study of two softwood species industrially modified by Thermowood (R) and thermo-vacuum process. *BioResources* **2018**, *13*, 715–728. [CrossRef]
49. Passialis, C.N. Physico-chemical characteristics of waterlogged archaeological wood. *Holzforschung* **1977**, *51*, 111–113. [CrossRef]
50. Florian, M.L.E. Scope and history of archaeological wood. In *Archaeological Wood: Properties, Chemistry, and Preservation*; Rowell, R.M., Barbour, R.J., Eds.; Oxford University Press: Pxford, UK, 1990; pp. 3–32.
51. Krutul, D.; Kocoń, J. Inorganic constituents and scanning electron microscopic study of fossil oak wood (*Quercus* sp.). *Holzforsch. Holzverwendung* **1982**, *34*, 69–77.
52. Teaca, C.A.; Roşu, D.; Mustaţă, F.; Rusu, T.; Roşu, L.; Roşca, I.; Varganici, C.D. Natural bio-based products for wood coating and protection against degradation: A Review. *BioResources* **2019**, *14*, 4873–4901. [CrossRef]
53. Carrión, J.S. *Plant Evolution (Evolución Vegetal)*; Marin, D., Ed.; University of Murcia: Madrid, Spain, 2003; p. 497.
54. Krutul, D.; Radomski, A.; Zawadzki, J.; Zielenkiewicz, T.; Antczak, A. Comparison of the chemical composition of the fossil and recent oak wood. *Wood Res.* **2010**, *55*, 113–120.
55. Wikberg, H.; Maunu, S.L. Characterisation of thermally modified hard- and softwoods by 13C CPMAS NMR. *Carbohydr. Polym.* **2004**, *58*, 461–466. [CrossRef]
56. Čabalová, I.; Kačík, F.; Lagaňa, R.; Výbohová, E.; Bubeníková, T.; Čaňová, I.; Ďurkovič, J. Effect of thermal treatment on the chemical, physical, and mechanical properties of pedunculate oak (*Quercus robur* L.) wood. *BioResources* **2018**, *13*, 157–170. [CrossRef]
57. Čabalová, I.; Bélik, M.; Kučerová, V.; Jurczyková, T. Chemical and Morphological Composition of Norway Spruce Wood (*Picea abies*, L.) in the Dependence of its Storage. *Polymers* **2021**, *13*, 1619. [CrossRef]
58. Tureková, I.; Balog, K. Flame ignition parameters of polyethylene and activation energy of initiation of combustion process. *Res. Pap.* **2001**, *11*, 181–186.
59. Kačíková, D.; Makovická-Osvaldová, L. Wood burning rate of various tree parts from selected softwoods. *Acta Fac. Xylologiae Zvolen* **2009**, *51*, 27–32.

Article

Effect of Sunlight on the Change in Color of Unsteamed and Steamed Beech Wood with Water Steam

Michal Dudiak *, Ladislav Dzurenda and Viera Kučerová

Faculty of Wood Sciences and Technology, Technical University in Zvolen, T.G. Masaryka 24, 96001 Zvolen, Slovakia; dzurenda@tuzvo.sk (L.D.); viera.kucerova@tuzvo.sk (V.K.)
* Correspondence: xdudiak@tuzvo.sk; Tel.: +421-45-520-6367

Abstract: This paper presents the differences in the color changes of unsteamed and steamed beech wood (*Fagus sylvatica* L.) caused by long-term exposure to sunlight on the surface of wood in interiors for 36 months. The light white-gray color of the yellow tinge of native beech wood darkened under the influence of sunlight, and the wood took on a pale brown color of yellow tinge. The degree of darkening and browning is quantified by the value of the total color difference $\Delta E^* = 13.0$. The deep brown-red color of steamed beech under the influence of sunlight during the exposure brightened, and the surface of the wood took on a pale brown hue. The degree of lightening of the color of steamed beech wood in the color space CIE $L^*a^*b^*$ is quantified by the value of the total color difference $\Delta E^* = 7.1$. A comparison of the color changes of unsteamed and steamed beech wood through the total color difference ΔE^* due to daylight shows that the surface of steamed beech wood shows 52.2% smaller changes than unsteamed beech wood. The lower value of the total color difference of steamed beech wood indicates the fact that steaming of beech wood with saturated water steam has a positive effect on the color stability and partial resistance of steamed beech wood to the initiation of photochemical reactions induced by UV–VIS wavelengths of solar radiation. Spectra ATR-FTIR analyses declare the influence of UV–VIS components of solar radiation on unsteamed and steamed beech wood and confirm the higher color stability of steamed beech wood.

Keywords: beech wood; thermal treatment; saturated water steam; natural aging; wood color; ATR-FTIR spectroscopy

Citation: Dudiak, M.; Dzurenda, L.; Kučerová, V. Effect of Sunlight on the Change in Color of Unsteamed and Steamed Beech Wood with Water Steam. *Polymers* 2022, 14, 1697. https://doi.org/10.3390/polym14091697

Academic Editor: Jacques Lalevee

Received: 18 March 2022
Accepted: 18 April 2022
Published: 21 April 2022

Publisher's Note: MDPI stays neutral with regard to jurisdictional claims in published maps and institutional affiliations.

Copyright: © 2022 by the authors. Licensee MDPI, Basel, Switzerland. This article is an open access article distributed under the terms and conditions of the Creative Commons Attribution (CC BY) license (https://creativecommons.org/licenses/by/4.0/).

1. Introduction

The color of wood is a basic physical–optical property, which belongs to the group of macroscopic features on the basis of which the wood of individual woody plants differs visually. The color of the wood is formed by chromophores, i.e., functional groups of the type >C=O, -CH=CH-CH=CH-, -CH=CH-, aromatic nuclei found in the chemical components of wood (lignin and extractive substances such as dyes, tannins, resins and others), which absorb some components of the electromagnetic radiation of daylight and thus create the color of the wood surface perceived by human vision.

Using the coordinates of the color space CIE $L^*a^*b^*$ is one of the ways to quantify the given optical wood property objectively. Lab color space (according to CIE-Commission Internationale d'Eclairage) in accordance with ISO 11 664-4 is based on the measurement of three parameters: brightness L^* represents the darkest black at $L^* = 0$ and the brightest white at $L^* = 100$. The value of a^* is a measure of the red-green character of the color, with positive values for red shades ($+a^*$), and negative values for green ($-a^*$). The value of b^* gives the yellow-blue character of the color with positive values for yellow shades ($+b^*$) and negative for blue ($-b^*$).

Wood with long-term exposure to sunlight changes color on its surface. The surface of the wood darkens and mostly yellows and browns. This fact is also referred to in the professional literature as natural aging [1–3].

Solar radiation is electromagnetic radiation with wavelengths in the range from 100 to 3000 nm [4], which consists of ultraviolet radiation, visible radiation (light) and infrared radiation. Ultraviolet radiation (UV) with wavelengths of 100–380 nm makes up about 2% of the daylight spectrum. According to the effect of UV radiation on biological materials and their effects on these materials, UV radiation is divided into: UV-A radiation, with a wavelength of 320–380 nm; UV-B radiation, with a wavelength of 280–320 nm; and UV-C radiation, with a wavelength below 280 nm. The spectrum of UV radiation falls on the Earth's surface from solar radiation, which is made up of 90–99% UV-A radiation and 1–10% UV-B radiation. The most dangerous UV-C radiation is completely absorbed by the atmosphere. The visible light spectrum, referred to as VIS, with wavelengths from 380 to 780 nm, represents approximately 49% of the daylight spectrum. The rest consists of infrared IR radiation with wavelengths of 780–3000 nm. The wavelengths of visible and infrared radiation are absorbed or reflected by the wood surface. The reflected wavelengths of the visible spectrum allow a person to perceive its color when looking at a given object. The absorbed wavelengths of infrared solar radiation change to heat on the surface.

UV–VIS components of solar radiation (daylight) initiate wood photodegradation processes when impacting on the wood surface (photolytic and photo-oxidation reactions with lignin, polysaccharides and wood accessory substances), and carbohydrates absorb 5–20% and 2% of the accessory substance [5]. These reactions cleave both lignin macromolecules with the simultaneous formation of phenolic hydroperoxides, free radicals, carbonyl and carboxyl groups, as well as polysaccharides into polysaccharides, with a lower degree of polymerization to form carbonyl, carboxyl groups and gaseous products (CO, CO_2, H_2) [1,3,6–8].

The aim of this work is to compare the effect of solar radiation on the surface of thermally treated beech wood with saturated steam (steaming) and unsteamed beech wood. Through changes in the coordinates L^*, a^*, b^* in the color space CIE $L^*a^*b^*$ and the total color difference ΔE^*, color changes of native and steamed beech wood caused by UV–VIS components of sunlight (daylight) are evaluated.

2. Material and Methods

2.1. Material and Technology of Beech Wood Steaming

Blanks with dimensions 32 × 60 × 600 mm made of beech wood had a moisture content $w_p = 56.4 \pm 4.2\%$ and were divided into 2 groups. The blanks of the first group were not thermally steamed prior to drying. The blanks of the second group were steamed to modify the color of the beech wood. Steaming was performed in an APDZ 240 pressure autoclave (Himmasch AD, Haskovo, Bulgaria) installed at Sundermann Ltd. (Banská Štiavnica, Slovakia). The steaming mode of beech wood with saturated water steam is shown in Figure 1, and the technological parameters of the steaming mode are given in Table 1.

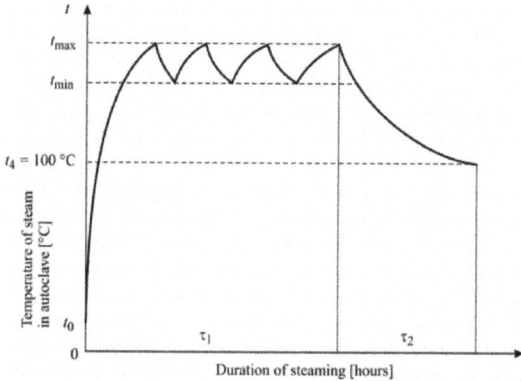

Figure 1. Mode of color modification of beech wood with saturated water steam.

Table 1. Mode of color modification of beech wood with saturated water steam.

Mode	Temperature of Saturated Water Steam (°C)			Time of Operation (h)		
	t_{min}	t_{max}	t_4	τ_1-Phase I	τ_2-Phase II	Total Time
Mode	132.5	137.5	100	6.0	1.0	7.5

2.2. Conditions of Beech Wood Exposure

Unsteamed and steamed beech blanks were dried by a low-temperature drying mode preserving the original wood color to a moisture content $w_k = 12 \pm 0.5\%$ in a convection hot air dryer: KC 1/50 (SUZAR Ltd., Považany, Slovakia) [9].

Samples of the following dimensions were produced from native and steamed beech wood blanks: 20 × 50 × 400 mm. The planed surface of unsteamed and steamed beech wood samples was exposed to daylight for a long time at an angle of 45° in the northern temperate zone (Slovakia, Central Europe) locality for 36 months. The temperature and relative humidity of the indoor air during the exposure were $t = 20 \pm 2.5\,°C$ and $\varphi = 50 \pm 10\%$.

The average density of incident solar radiation in Slovakia is 1100 kWh/m² per year. The intensity of the sun's rays changes throughout the year. The highest intensity of solar radiation is in the summer months of June and July when it reaches a value of 5.9 to 6.0 kWh/m² per day. During the autumn, the intensity of sunlight decreases and is lowest during the winter. In December, the intensity of solar radiation is the weakest, with an approximate value of 1.7 kWh/m² per day.

2.3. Color Measurement of Beech Wood

The surface color of beech samples before and during the exposure was evaluated in the color space CIE L*a*b* at monthly intervals using the Color Reader CR-10 (Konica Minolta, Osaka, Japan) colorimeter was measured. A D65 light source was used, and the diameter of the optical scanning aperture was 8 mm [10].

The total color difference ΔE^* of the beech wood surface change during the 36-month exposure to sunlight was determined according to the following ISO 11 664-4 equation:

$$\Delta E^* = \sqrt{\left(L_2^* - L_1^*\right)^2 + \left(a_2^* - a_1^*\right)^2 + \left(b_2^* - b_1^*\right)^2} \quad (1)$$

where L_1^*, a_1^*, b_1^* are the coordinates of the color space CIE L*a*b* on the surface of the dried, milled beech wood before exposure, and L_2^*, a_2^*, b_2^* are the coordinates of the color space CIE L*a*b* on the surface of the dried, milled beech wood during exposure.

2.4. Mathematical-Statistical Evaluation of Measured Data

The measured values on the brightness coordinate L^* and the chromaticity coordinates a^*, b^*, as well as the calculated values of the total color differences ΔE^* during the observed exposure periods, were statistically and graphically evaluated using Excel and Statistica v.12 programs (V12.0 SP2, Palo Alto, CA, USA).

2.5. Analysis of Changes in Lignin-Cellulose Matrix of Wood ATR-FTIR Spectroscopy

Fourier-transform infrared spectroscopy (FTIR) was used to follow chemical changes in beech wood after radiation in unsteamed and steamed beech wood. The measurements were carried out using a Nicolet iS10 spectrometer (Thermo Fisher Scientific, Madison, WI, USA) equipped with the Smart iTR ATR accessory.

The spectra were collected in an absorbance mode between 4000 and 650 cm^{-1} by accumulating 32 scans at a resolution of 4 cm^{-1} using diamond crystal. All analyses were performed in four replicates. The spectra were evaluated using the OMNIC 8.0 software.

3. Results and Discussion

Beech wood, according to [11–13], has a light white-brown-yellow color. In the steaming process with a steam–air mixture at atmospheric pressure, or thermal treatment of beech wood with saturated water steam, as reported by [13–16], depending on the temperature and length of the thermal treatment process, the wood darkens and acquires shades from a pale pink-yellow color shade to brown-red color. Table 2 shows the coordinates of the color space CIE $L^*a^*b^*$ of unsteamed and steamed beech wood at a moisture content of $w = 12\%$ on the planed surface before and after 36 months of dazzling. The values for the brightness coordinates L^* and the chromatic coordinates a^* and yellow b^* of the color space CIE $L^*a^*b^*$ of unsteamed beech wood given in Table 1 are similar to those reported by [13,17,18].

Table 2. Coordinate values of color space CIE $L^*a^*b^*$ of unsteamed and steamed beech wood.

Beech Wood	Unsteamed Beech Wood			Steamed Beech Wood		
	Coordinates of Color Space CIE $L^*a^*b^*$					
	L^*	a^*	b^*	L^*	a^*	b^*
Before exposure	76.6 ± 1.7	7.8 ± 1.5	19.8 ± 1.7	55.3 ± 1.6	14.5 ± 1.0	23.5 ± 0.8
After exposure	71.3 ± 0.9	12.2 ± 0.6	26.8 ± 0.9	57.6 ± 0.9	14.1 ± 0.8	27.5 ± 0.6

The color of native and steamed beech wood before and after exposure to daylight glare is shown in Figure 2.

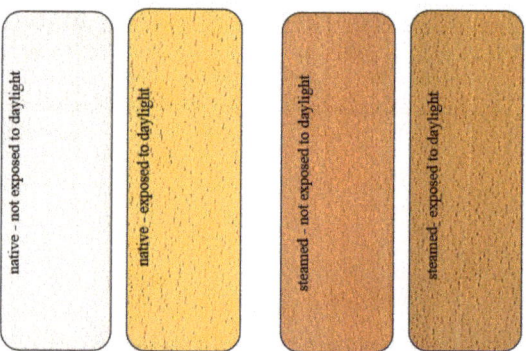

Figure 2. View of beech wood: native before and after exposure (**left**), steamed before and after exposure (**right**).

The courses of the measured values of beech wood color on the coordinates: L^*, a^*, b^* of the color space CIE $L^*a^*b^*$ in individual months, during 36 months of dazzling by the sunlight of daylight are shown in Figures 3 and 4.

The course of the measured values on the light coordinate L^*, the chromatic coordinates of color a^*, and the yellow color b^* in Figures 3 and 4 during 36 months of dazzling is not fluent. The fluctuations are attributed to the influence of the intensity of solar radiation during the individual seasons, causing photolytic and photo-oxidative reactions of daylight radiation with wood. Figures 5 and 6 show the magnitudes of changes in the average values of ΔL^*, Δa^*, Δb^* in individual seasons during the exposure.

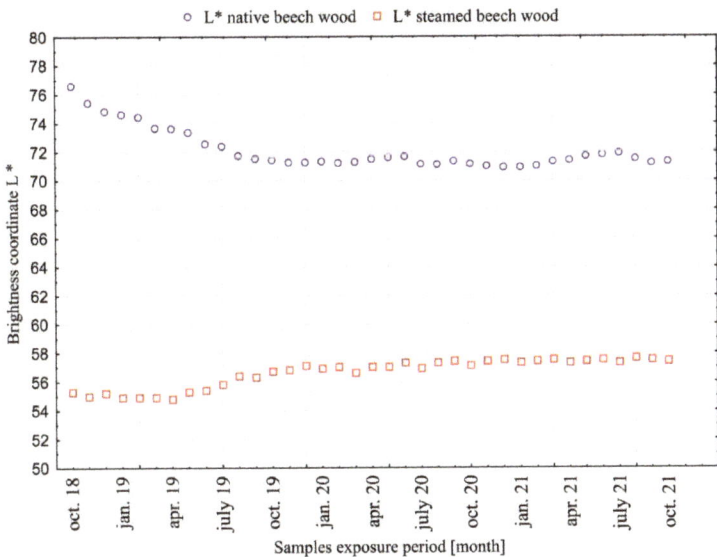

Figure 3. Values on the L^* coordinate of dazzled native and steamed beech wood over a period of 36 months (October 2019 to October 2021).

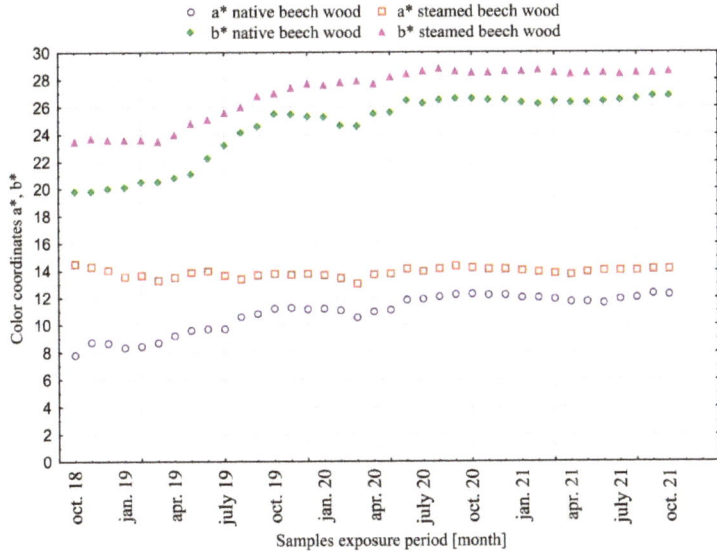

Figure 4. Values on chromatic coordinates of red color a^* and yellow color b^* of dazzling native and steamed beech wood over a period of 36 months (October 2019 to October 2021).

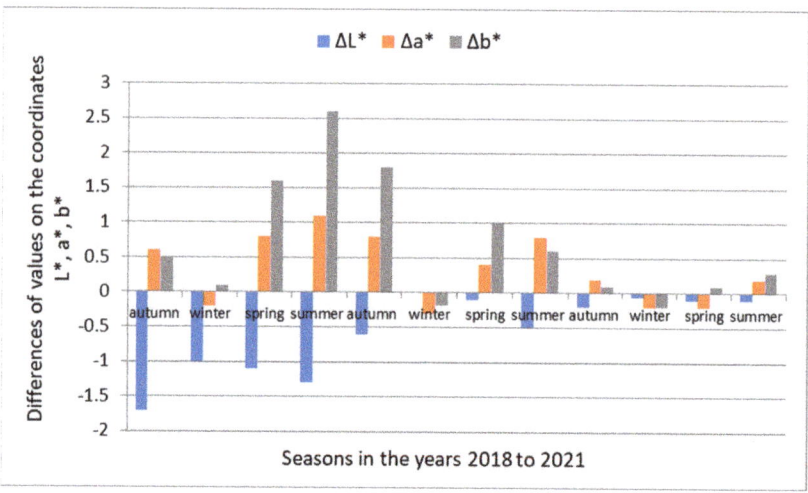

Figure 5. Magnitudes of changes in the ΔL^*, Δa^*, Δb^* values in the color space CIE $L^*a^*b^*$ native beech wood during the 36-month exposure to sunlight, depending on the season.

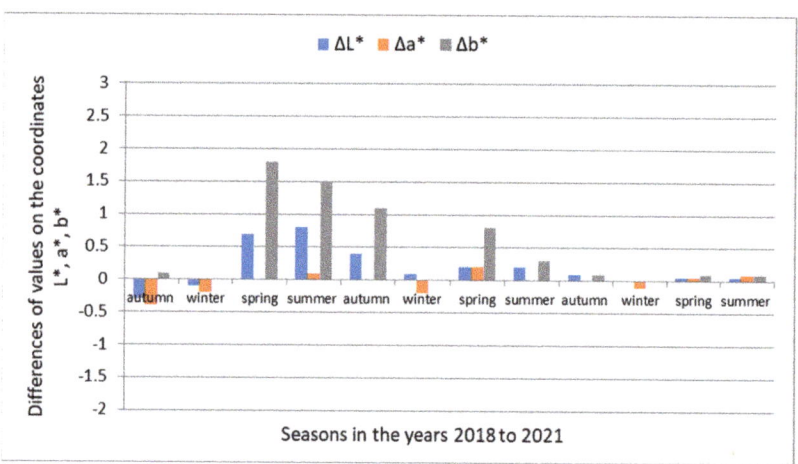

Figure 6. The magnitudes of changes in the values of ΔL^*, Δa^*, Δb^* in the color space CIE $L^*a^*b^*$ of steamed beech wood during the 36-month exposure to sunlight, depending on the season.

In Figure 7, the degree of the color change of the surface of native and steamed beech wood caused by solar radiation during 36 months is documented by the total color difference ΔE^*.

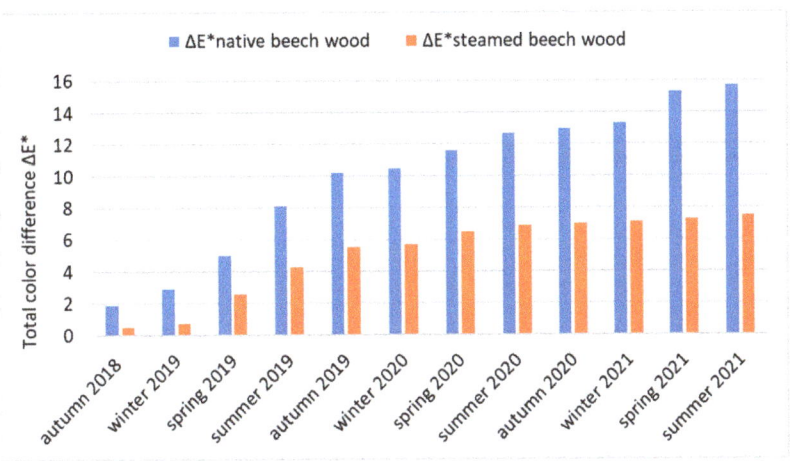

Figure 7. Values of the total color difference ΔE^* of native and steamed beech wood during 36 months of dazzling (October 2019 to October 2021).

From the comparison of wood colors in Figure 2 and the values presented at the coordinates L^*, a^*, b^* of unsteamed and steamed beech wood during the exposure, Figures 3 and 4 show that while the surface of the unsteamed beech wood darkened and browned, the brown-red color on the steamed wood lightened.

The darkening and browning of unsteamed beech wood numerically documents the shift of the brightness coordinate L^* from the value $L_0^* = 76.6$ to $L_{36}^* = 71.3$, i.e., by the value $\Delta L^* = -5.3$, and changes in chromatic coordinates: red color a^* from $a_0^* = 7.8$ to $a_{36}^* = 12.2$, i.e., by the value $\Delta a^* = +4.4$, and the yellow color b^* from the value $b_0^* = 19.8$ to $b_{36}^* = 26.8$, i.e., by the value $\Delta b^* = +10.0$. The largest darkening of unsteamed beech wood occurred during the first year of dazzling, when changes in the brightness coordinate ΔL^* reached 76.9% of the total change in brightness of beech wood caused by daylight; in the second year, it reached 16.8 and in the third year 6.3%. The browning of unsteamed beech wood is described by changes in the chromatic coordinates: red a^* and yellow b^*. The change in the red coordinate in the first year of exposure was 57.5% of the total change Δa^*, in the second year of dazzling 42.5% and in the third year oscillated around $a^* = 12$. In the yellow coordinate, the change Δb^* in the first year of dazzling was 57.7% of the total value of the change in Δb^* beech wood, 38.7% in the second year and 3.6% in the third year. The changes in the red a^* and yellow b^* coordinates of the color space CIE $L^*a^*b^*$ in the third year, as the measurements show, are small, and, in addition, the different seasons are contradictory, while in winter and spring, they show a decrease in values, so in summer, in times of more intense sunlight, they grow. The darkening of wood due to solar radiation is in line with the views of experts dealing with changes in the properties of wood due to long-term exposure to sunlight, who state that the wood surface darkens and mostly yellows and browns [2,3,8,19,20].

Steamed beech wood under the influence of sunlight for 36 months compared to unsteamed wood showed the opposite character of the color change, where the surface of the wood faded. Visually, this is documented in Figure 2, as well as the shift of the brightness coordinate L^* from the value $L_0^* = 55.3$ to $L_{36}^* = 57.2$, i.e., by the value $\Delta L^* = +1.9$, on the red coordinate a^* offset from $a_0^* = 14.5$ to $a_{36}^* = 14.1$, i.e., by the value $\Delta a^* = -0.4$, and on the chromatic coordinate of yellow color b^* from the value $b_0^* = 19.8$ to $b_{36}^* = 25.5$, i.e., by the value $\Delta b^* = +5.7$. On the basis of comparison of individual changes ΔL^*, Δa^*, Δb^* on the coordinates of the color space CIE $L^*a^*b^*$ of steamed beech wood caused by the action of sunlight with changes ΔL^*, Δa^*, Δb^* on the coordinates of the unmatched beech wood caused by daylight, it can be stated that the values expressing the magnitude of changes in steamed beech wood are smaller. The magnitude of changes in the brightness coordinates L^* and yellow b^*, similar to unsteamed beech wood, is largest in the first year of exposure. The red coordinate changes of a^* oscillated around the value of $a^* = 14.0$. In winter, at low sunlight intensity, the values on the red coordinate a^* decreased, and from spring to autumn, they increased at higher sunlight intensity. The rate of decline or the increase in red coordinate values decreases over the years. Based on the above findings, it can be stated that the functional groups of chromophores in beech wood with absorption of the electromagnetic radiation spectrum with a 630–750 nm red wavelength causing reddening of steamed beech wood were steamed and strongly eliminated for photochemical reactions of wood caused by daylight.

The authors of [21,22] point out the effect of lightening the surface of steamed wood under the action of UV radiation. In the work, a team of authors [21] report the lightening of the surface color of steamed maple wood after its irradiation in Xenotest with a 450 xenon lamp emitting UV radiation with a wavelength of 340 nm, 42 ± 2 W/m^2 intensity, for 7 days. The lightening of the red-brown color of steamed maple wood is declared by the increase in the values on the brightness coordinate from $L_1^* = 65.3$ to the value of $L_2^* = 70.7$, i.e., by the value $\Delta L^* = +5.4$, the increase in the value on the chromatic coordinate of the yellow color from $b_1^* = 19.4$ to the value $b_2^* = 28.9$, i.e., by the value $\Delta b^* = +9.3$, with a slight change in the red coordinate value from $a_1^* = 10.8$ to $a_2^* = 10.3$, i.e., by the value $\Delta a^* = -0.5$.

The influence of UV radiation on steamed acacia wood in [22] states that while the surface of steamed acacia wood darkened slightly at the steaming temperature $t = 100\ °C$, the surface of acacia wood lightened at the steaming temperature $t = 120\ °C$.

The authors [23] describe the positive effect of the steaming process on the decomposition of functional groups of maple wood chromophores manifested by the darkening and browning of maple wood and the elimination of photochemical reactions caused by daylight. They point to the fact that the greater the darkening of the maple wood in the steaming process, the smaller the color changes on the surface of the irradiated steamed maple wood by UV radiation. This is declared by the decrease in the total color difference from the value $\Delta E^* = 18.5$ for unsteamed maple wood to $\Delta E^* = 7.2$ and steamed maple wood with saturated water steam with temperature $t = 135\ °C$, as well as the results of FTIR analyses.

The contribution of the influence of beech wood steaming on the color fastness and resistance to the effects of sunlight declares a decrease in the value of the total color difference ΔE^* in Figure 7. While the change in the color of unsteamed beech wood caused by solar radiation expressed by the value of the total color difference over 3 years is $\Delta E^* = 15.7$, the change in the total color difference of steamed beech wood in the same period is $\Delta E^* = 7.5$, which is a decrease in color change by about 52.2%. This points to the fact that beech wood steaming has a positive effect on changes in the chromophore system of steamed beech wood and the partial resistance of steamed beech wood to the initiation of photochemical reactions induced by daylight wavelengths.

Chemical changes on the surface of unsteamed and steamed beech wood before and after solar irradiation were also monitored by ATR-FTIR spectroscopy. The FTIR spectra of the examined samples in Figure 8 show the whole range of wavenumbers 4000 to 650 cm^{-1}. In Figure 9, only spectra in the range 1800 to 800 cm^{-1} are presented, where most of the specific vibrations occurred.

During various thermal treatments of wood, the chemical composition changes in the wood, which depends on the experimental treatment conditions, such as the temperature, time and atmosphere used. However, many competitive reactions take place simultaneously, depending on experimental conditions. For these reasons, our results may differ from those of other authors.

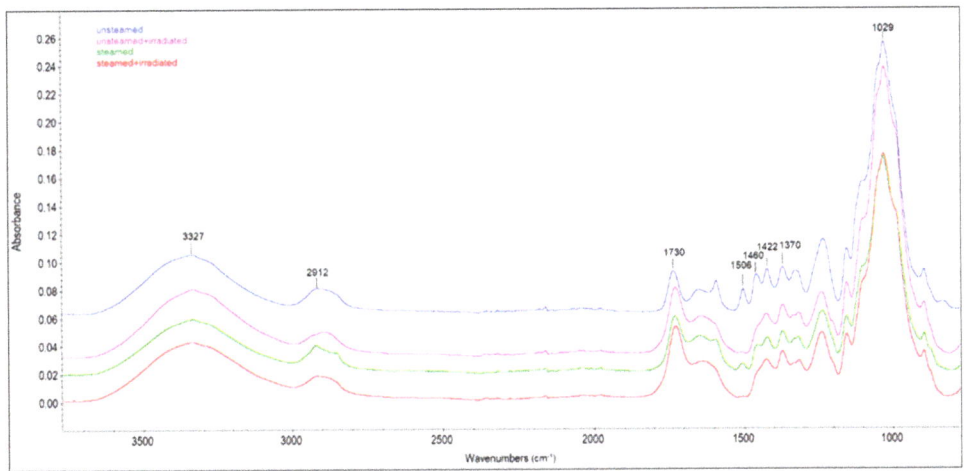

Figure 8. FTIR spectra of beech wood samples.

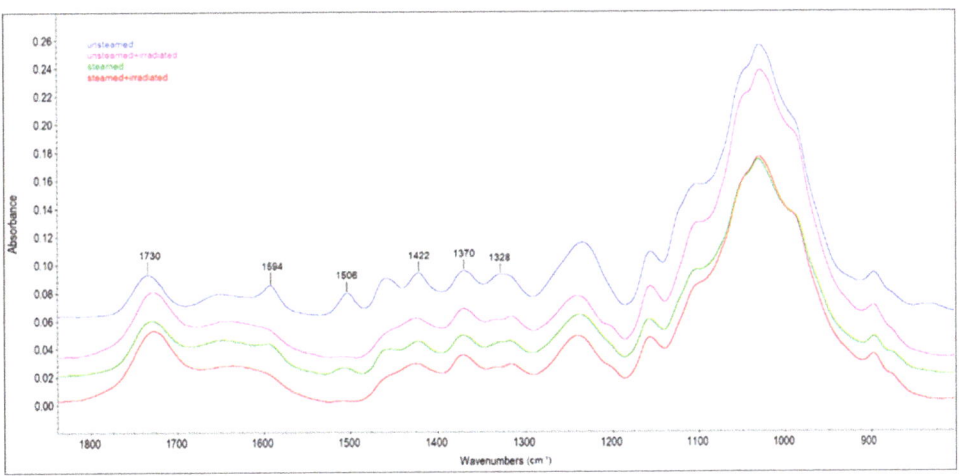

Figure 9. FTIR spectra of beech wood samples in the range 1800 to 800 cm^{-1}.

The bands in the range of 3800 to 2750 cm^{-1} are assigned to hydroxyl and methyl/methylene stretching vibrations [24]. In the FTIR spectra at 1730 cm^{-1} (assigned to unconjugated carbonyl groups), the increase in absorbance, between the original beech wood before and after irradiation and the steamed beech wood before and after irradiation, was due to an increase in carbonyl or carboxyl groups in lignin or carbohydrates (Figure 9). At this observed band, the highest increase in absorbance (by 34%) was recorded in the spectrum of the irradiated steamed beech wood sample. The results of the FTIR analysis also suggest that the lignin macromolecule changes after sun irradiation of the samples by reducing the absorption bands characteristic of lignin: 1594 cm^{-1} (belonging to aromatic skeletal vibration in lignin, and -C=O stretching [25]), 1506 cm^{-1} (C=C stretching of the aromatic skeletal vibrations in lignin [26]), 1422 cm^{-1} (assigned to aromatic skeletal vibrations combined with C-H deformation in carbohydrates [27]). The absorption band at 1506 cm^{-1} almost disappeared with the irradiated steamed sample, and the reduction of the band intensity was by more than 93% compared to the steamed sample. The reduction of band intensity for lignin in wood samples after UV irradiation was also observed by the authors [28].

In our research, we also recorded a decrease in absorbances at bands 1421 cm^{-1} (aromatic ring vibration in lignin combined with C–H deformation in carbohydrates) and 1328 cm^{-1} (C-O vibration in syringyl plus guaiacyl derivatives is characteristic for condensed structures in lignin) when comparing unsteamed beech wood samples before and after irradiation, as well as steamed wood and irradiated steamed beech wood. This decrease indicates cleavage of the methoxyl groups during sunlight, leading to gradual demethoxylation. The reduction in the intensity of the bands at 1124 cm^{-1} (C-H vibration of syringyl units in lignin) irradiated wood samples indicates cleavage of the ether bond in the lignin structure.

The intensity of the band at 1370 cm^{-1} (C-H deformations in polysaccharides) is not significantly affected by the treatment conditions of the wood samples, so this band was used as a reference to determine lignin degradation. To compare the changes in chemical composition on the surface of the wood samples used for the experiment, the ratios of intensities 1506/1370, 1730/1370 and 1730/1506 were calculated (Table 3).

Table 3. Ratios of absorption bands for the beech wood.

Assignment	Ratio of Intensities	Unsteamed	Unsteamed + Irradiated	Steamed	Steamed + Irradiated
Nonconjugated carbonyl/carbohydrates	1730/1370	2.04	2.72	2.49	3.19
Lignin/carbohydrates	1506/1370	1.03	0.08	0.43	0.03
Nonconjugated carbonyl/lignin	1730/1506	1.98	33.40	5.78	119.00

The decrease in the 1506/1370 ratio of intensities in the wood samples after solar irradiation demonstrates the decay of lignin, which occurs due to the interaction of radiation with wood. The increase in unconjugated carbonyl groups (carbonyl-containing chromophores) due to photo-oxidation is reflected in the increasing ratio of 1730/1370 (carbonyl/carbohydrates). These results are consistent with published research by other authors [29–31]. The relative increase in carbonyl groups in lignin or carbohydrates with the relative decrease in lignin in parallel is reflected in an increase in the 1730/1506 ratio. The authors of [32] explained the increase in this ratio by primarily lignin degradation and oxidation (photo-oxidation and thermo-oxidation resulting in an increase in carbonyl groups).

The authors of [33] used FTIR analysis of chemical changes in beech and oak wood after steaming and compression. Using FTIR, they observed changes in the hydroxyl groups, as well as in the C-O and C-H functional groups in the polysaccharides and in lignin on the surface of the samples. Other authors [34–38] state that the lightfastness of various woody plants to a large extent mainly affects lignin and extractives.

4. Conclusions

This paper presents the results of the color change of the surface of unsteamed and steamed beech wood with saturated water steam in a pressure autoclave, which was exposed to daylight interior for 36 months. The results of analyses of the effect of solar radiation on unsteamed and steamed beech wood showed:

1. The color surface of unsteamed beech wood changed under the influence of daylight during the exposure. The wood darkened and browned into a brown-yellow color shade. The opposite tendency, i.e., lightening, occurred during exposure on steamed beech wood samples.
2. The measured changes in the values at the coordinates of the color space CIE $L^*a^*b^*$ caused by solar radiation in unsteamed beech wood are: $\Delta L^* = -5.3$; $\Delta a^* = +4.4$; $\Delta b^* = +10.0$. Changes were recorded also for steamed beech wood: $\Delta L^* = +1.9$; $\Delta a^* = -0.4$; $\Delta b^* = 5.7$.
3. The evaluation of the color change of beech wood in the form of the total color difference ΔE^* shows that the surface of steamed beech wood shows approximately 52.2% less color change than the total color difference of unsteamed beech wood.

4. The decrease in the values of ΔL^*, Δa^*, and Δb^* and the total color difference ΔE^* of steamed beech wood indicate a positive effect of steaming of wood on the partial resistance of steamed beech wood to the initiation of photochemical reactions induced by wavelengths daylight.
5. The lightfastness of wood is greatly affected by lignin. The main change in the ATR-FTIR spectra is the decrease in lignin absorption bands in parallel with the increase in unconjugated carbonyl absorption after irradiation of unsteamed and steamed beech wood with daylight components. Photodegradation of irradiated unsteamed and steamed wood was associated with a reduction in the ratio of lignin to carbohydrates while at the same time increasing the ratio of unconjugated carbonyl to carbohydrates, which may affect the partial resistance of steamed beech wood to daylight.

Author Contributions: Conceptualization, M.D.; methodology, M.D., L.D. and V.K.; software, M.D. and V.K.; formal analysis, M.D., L.D. and V.K.; resources, M.D., L.D. and V.K.; data curation, M.D., L.D. and V.K.; writing—original draft preparation, M.D.; writing—review and editing, M.D., L.D. and V.K.; visualization, M.D.; project administration, L.D. All authors have read and agreed to the published version of the manuscript.

Funding: This research was funded by project APVV-17-0456 "Thermal modification of wood with saturated water steam for purposeful and stable change of wood color".

Institutional Review Board Statement: Not applicable.

Informed Consent Statement: Not applicable.

Acknowledgments: This experimental research was carried out under the grant project APVV-17-0456 "Thermal modification of wood with saturated water steam for purposeful and stable change of wood color" as the result of the work of the author and the considerable assistance of the APVV agency. This publication is the result of the project implementation: Progressive research of performance properties of wood-based materials and products (LignoPro), ITMS: 313011T720 (25%) supported by the Operational Programme Integrated Infrastructure (OPII) funded by the ERDF.

Conflicts of Interest: The authors declare no conflict of interest.

References

1. Hon, D.S.N. Weathering and Photochemistry in Wood. In *Wood and Cellulosic Chemistry*, 2nd ed.; Hon, D.S.N., Shiraishi, N., Eds.; Marcel Dekker: New York, NY, USA, 2001; pp. 513–546.
2. Reinprecht, L. *Wood Protection*; Technical University in Zvolen: Zvolen, Slovakia, 2008; 450p.
3. Baar, J.; Gryc, V. The analysis of tropical wood discoloration caused by simulated sunlight. *Eur. J. Wood Wood Prod.* **2011**, *70*, 263–269. [CrossRef]
4. Hrvoľ, J.; Tomlain, J. *Radiation in the Atmosphere*, 1st ed.; Comenius University in Bratislava: Bratislava, Slovakia, 1997; 136p.
5. Gandelová, L.; Horáček, P.; Šlezingerová, J. *The Science of Wood*; Mendel University of Agriculture and Forestry in Brno: Brno, Czech Republic, 2009; 176p.
6. Persze, L.; Tolvaj, L. Photodegradation of wood at elevated temperature: Colour change. *J. Photochem. Photobiol. B Biol.* **2012**, *108*, 44–47. [CrossRef] [PubMed]
7. Denes, L.; Lang, E.M. Photodegradation of heat-treated hardwood veneers. *J. Photochem. Photobiol. B Biol.* **2013**, *118*, 9–15. [CrossRef] [PubMed]
8. Geffertová, J.; Geffert, A.; Vybohová, E. The effect of UV irradiation on the colour change of the spruce wood. *Acta Fac. Xylologiae Zvolen* **2018**, *60*, 41–50.
9. Dzurenda, L. Mode for hot air drying of alder blanks that retain the color acquired during the steaming process. *Ann. Wars. Univ. Life Sci. For. Wood Technol.* **2021**, *114*, 86–92.
10. Makovíny, I. *Useful Properties and Use of Different Types of Wood*; Technical University in Zvolen: Zvolen, Slovakia, 2010; 104p.
11. Klement, I.; Réh, R.; Detvaj, J. *Basic Characteristics of Forest Trees—Wood Raw Material Processing in the Wood Processing Industry*; National Forestry Center: Zvolen, Slovakia, 2010; 82p.
12. Trebula, P. *Wood Drying and Hydrothermal Treatment*; Technical University in Zvolen: Zvolen, Slovakia, 1986; 255p.
13. Tolvaj, L.; Nemeth, R.; Varga, D.; Molnar, S. Colour homogenisation of beech wood by steam treatment. *Drewno* **2009**, *52*, 5–17.
14. Dzurenda, L. Colouring of Beech Wood during Thermal Treatment using Saturated Water Steams. *Acta Fac. Xylologiae Zvolen* **2014**, *56*, 13–22.
15. Banski, A.; Dudiak, M. Dependence of Color on the Time and Temperature of Saturated Water Steam in the Process of Thermal Modification of Beech Wood. *AIP Conf. Proc.* **2019**, *2118*, 030003.

16. Dzurenda, L.; Dudiak, M. Cross-correlation of color and acidity of wet beech wood in the process of thermal treatment with saturated steam. *Wood Res.* **2021**, *66*, 105–116. [CrossRef]
17. Babiak, M.; Kubovský, I.; Mamoňová, M. *Color space of selected local woods. Interaction of Wood with Various Forms of Energy*; Technical University in Zvolen: Zvolen, Slovakia, 2004; pp. 113–117.
18. Meints, T.; Teischinger, A.; Stingl, R.; Hansmann, C. Wood colour of central European wood species: CIELAB characterisation and colour intensification. *Eur. J. Wood Prod.* **2017**, *75*, 499–509. [CrossRef]
19. Chang, T.-C.; Chang, H.-T.; Wu, C.-L.; Chang, S.-T. Influences of extractives on the photodegradation of wood. *Polym. Degrad. Stab.* **2010**, *95*, 516–521. [CrossRef]
20. Kúdela, J.; Kubovský, I. Accelerated-ageing-induced photo-degradation of beech wood surface treated with selected coating materials. *Acta Fac. Xylologiae Zvolen* **2016**, *2*, 27–36.
21. Dzurenda, L.; Dudiak, M.; Banski, A. Influence of UV radiation on color stability of natural and thermally treated maple wood with saturated water steam. *Innov. Woodwork. Eng. Des. Int. Sci. J.* **2020**, 36–41.
22. Varga, D.; Tolvaj, L.; Molnar, Z.; Pasztory, Z. Leaching efect of water on photodegraded hardwood species monitored by IR spectroscopy. *Wood Sci. Technol.* **2020**, *54*, 1407–1421. [CrossRef]
23. Dzurenda, L.; Dudiak, M.; Výbohová, E. Influence of UV Radiation on the Color Change of the Surface of Steamed Maple Wood with Saturated Water Steam. *Polymers* **2022**, *14*, 217. [CrossRef] [PubMed]
24. Li, M.-Y.; Cheng, S.-C.; Li, D.; Wang, S.-N.; Huang, A.-M.; Sun, S.-Q. Structural characterization of steam-heat treated Tectona grandis wood analyzed by FT-IR and 2D-IR correlation spectroscopy. *Chin. Chem. Lett.* **2015**, *26*, 221–225.
25. Müller, G.; Schöpper, C.; Vos, H.; Kharazipour, A.P.; Polle, A. FTIR-ATR spectroscopic analyses of changes in wood properties during particle and fibreboard production of hard and softwood trees. *BioResources* **2009**, *4*, 49–71. [CrossRef]
26. Esteves, B.; Velez Marques, A.; Domingos, I.; Pereira, H. Chemical changes of heat treated Pine and Eucalypt wood monitored by FTIR. *Maderas-Cienc. Technol.* **2013**, *15*, 245–258. [CrossRef]
27. Mattos, B.D.; Lourençon, T.V.; Serrano, L.; Labidi, J.; Gatto, D.A. Chemical modification of fast-growing eucalyptus wood. *Wood Sci. Technol.* **2015**, *2*, 273–288. [CrossRef]
28. Zborowska, M.; Stachowiak-Wencek, A.; Waliszewska, B.; Prądzyński, W. Colorimetric and FTIR ATR spectroscopy studies of degradative effects of ultraviolet light on the surface of exotic ipe (*tabebuia* sp.) wood. *Cellul. Chem. Technol.* **2014**, *50*, 71–76.
29. Pandey, K.K.; Vuorinen, T. Comparative study of photodegradation of wood by a UV laser and axenon light source. *Polym. Degrad. Stab.* **2008**, *93*, 2138–2146. [CrossRef]
30. Srinivas, K.; Pandey, K.K. Photodegradation of thermally modified wood. *J. Photochem. Photobiol. B* **2012**, *117*, 140–145. [CrossRef] [PubMed]
31. Agresti, G.; Bonifazi, G.; Calienno, L.; Capobianco, G.; Lo Monaco, A.; Pelosi, C.; Picchio, R.; Serranti, S. Surface investigation of photo-degraded wood by color monitoring, infrared spectroscopy, and hyperspectral imaging. *J. Spectrosc.* **2013**, *2013*, 380536. [CrossRef]
32. Timar, M.C.; Varodi, A.M.; Gurău, L. Comparative study of photodegradation of six wood species after short-time UV exposure. *Wood Sci. Technol.* **2016**, *50*, 135–163. [CrossRef]
33. Báder, M.; Németh, R.; Sandak, J.; Sandak, A. FTIR analysis of chemical changes in wood induced by steaming and longitudinal compression. *Cellulose* **2020**, *27*, 6811–6829. [CrossRef]
34. Chang, T.C.; Chang, H.T.; Wu, C.L.; Lin, H.Y.; Chang, S.T. Stabilizing effect of extractives on the photo-oxidation of Acacia confusa wood. *Polym. Degrad. Stab.* **2010**, *95*, 1518–1522. [CrossRef]
35. Chang, T.C.; Lin, H.Y.; Wang, S.Y.; Chang, S.T. Study on inhibition mechanisms of light-induced woodradicals by Acacia confusa heartwood extracts. *Polym. Degrad. Stab.* **2014**, *105*, 42–47. [CrossRef]
36. Tolvaj, L.; Varga, D. Photodegradation of timber of three hardwood species caused by different lightsources. *Acta Silv. Lign. Hung.* **2012**, *8*, 145–155. [CrossRef]
37. Tolvaj, L.; Molnar, Z.; Nemeth, R. Photodegradation of wood at elevated temperature: Infrared spectroscopic study. *J. Photochem. Photobiol. B* **2013**, *12*, 132–136. [CrossRef]
38. *ISO 11664-4; Colorimetry—Part 4: CIE 1976 L*a*b* Color space*. Internation Organization for Standardization: Geneve, Switzerland, 2008.

Article

Influence of UV Radiation on the Color Change of the Surface of Steamed Maple Wood with Saturated Water Steam

Ladislav Dzurenda *, Michal Dudiak and Eva Výbohová

Faculty of Wood Sciences and Technology, Technical University in Zvolen, T.G. Masaryka 24, 96001 Zvolen, Slovakia; xdudiak@tuzvo.sk (M.D.); vybohova@tuzvo.sk (E.V.)
* Correspondence: dzurenda@tuzvo.sk; Tel.: +421-45-520-6365

Abstract: The wood of maple (*Acer Pseudopatanus* L.) was steamed with a saturated steam-air mixture at a temperature of $t = 95$ °C or saturated steam at $t = 115$ °C and $t = 135$ °C, in order to give a pale pink-brown, pale brown, and brown-red color. Subsequently, samples of unsteamed and steamed maple wood were irradiated with a UV lamp in a Xenotest Q-SUN Xe-3-H after drying, in order to test the color stability of steamed maple wood. The color change of the wood surface was evaluated by means of measured values on the coordinates of the color space CIE $L^* a^* b^*$. The results show that the surface of unsteamed maple wood changes color markedly under the influence of UV radiation than the surface of steamed maple wood. The greater the darkening and browning color of the maple wood by steaming, the smaller the changes in the values at the coordinates L^*, a^*, b^* of the steamed maple wood caused by UV radiation. The positive effect of steaming on UV resistance is evidenced by the decrease in the overall color difference ΔE^*. While the value of the total color diffusion of unsteamed maple wood induced by UV radiation is $\Delta E^* = 18.5$, for maple wood steamed with a saturated steam-air mixture at temperature $t = 95$ °C the ΔE^* decreases to 12.6, for steamed maple wood with saturated water steam with temperature $t = 115$ °C the ΔE^* decreases to 10.4, and for saturated water steam with temperature $t = 135$ °C the ΔE^* decreases to 7.2. Differential ATR-FTIR spectra declare the effect of UV radiation on unsteamed and steamed maple wood and confirm the higher color stability of steamed maple wood.

Keywords: maple wood; color difference; ATR-FTIR spectroscopy; steaming; saturated water steam

Citation: Dzurenda, L.; Dudiak, M.; Výbohová, E. Influence of UV Radiation on the Color Change of the Surface of Steamed Maple Wood with Saturated Water Steam. *Polymers* **2022**, *14*, 217. https://doi.org/10.3390/polym14010217

Academic Editor: Antonios N. Papadopoulos

Received: 14 December 2021
Accepted: 31 December 2021
Published: 5 January 2022

Publisher's Note: MDPI stays neutral with regard to jurisdictional claims in published maps and institutional affiliations.

Copyright: © 2022 by the authors. Licensee MDPI, Basel, Switzerland. This article is an open access article distributed under the terms and conditions of the Creative Commons Attribution (CC BY) license (https://creativecommons.org/licenses/by/4.0/).

1. Introduction

The color of wood is a basic physical-optical property, which belongs to the group of macroscopic features on the basis of which the wood of individual woody plants differs visually. The color of the wood is formed by chromophores, i.e., functional groups of the type: >C=O, –CH=CH–CH=CH–, –CH=CH–, aromatic nuclei found in the chemical components of wood (lignin and extractive substances, such as dyes, tannins, resins, etc.), which absorb some components of the electromagnetic radiation of daylight and thus create the color of the wood surface perceived by human vision.

The color of wood changes in thermal processes, such as wood drying, wood steaming, and thermo-wood production technologies. The wood darkens more or less and, depending on the wood, acquires color shades of pink, red, and brown to dark brown-gray color [1–11].

Wood steaming is a physico-chemical process, in which wood placed in an environment of hot water, saturated water steam or saturated humid air is heated and changes its physical, mechanical, and chemical properties. The action of heat initiates the chemical reactions in wet wood, such as the extraction of water-soluble substances, degradation of polysaccharides, cleavage of free radicals, and phenolic hydroxyl groups in lignin, resulting in the formation of new chromophoric groups causing a change in the color of the wood. These facts are used for the full-volume modification of wood color into non-traditional color shades of wood of individual trees. Beech wood, depending on the length of the steaming time, acquires a pale pink to red-brown color shade [5,7,9,12–15]. Oak wood, as reported by the works [11,16,17] and depending on the steaming conditions, achieves color shades from a pale brown-yellow color to a dark dark-gray color. The light white-yellow color of maple wood in the process of steaming the wood with saturated water steam acquires shades of pale pink-brown to brown-red color [18,19].

The color of the wood also changes due to the long-term effects of sunlight on its surface. The surface of the wood darkens and is mostly yellow and brown. This fact is also referred to in the professional literature as natural aging [20–22].

Solar radiation falling on the wood surface is partly absorbed and partly reflected from the surface. The absorbed spectrum of infrared electromagnetic radiation is converted into heat. In addition, the photon flux of ultraviolet and part of the visible radiation of wavelengths λ = 200–400 nm are the source of initiation of photolytic and photooxidation reactions with lignin, polysaccharides, and accessory substances of wood. Of the chemical components of wood, lignin is the most subject to photodegradation, which captures 80–85% of UV radiation, while carbohydrates absorb 13–18% and 2% of accessory substances [23]. These reactions cleave the lignin macromolecule with the simultaneous formation of phenolic hydroperoxides, free radicals, carbonyl and carboxyl groups, and to a lesser extent depolymerize polysaccharides to polysaccharides with a lower degree of polymerization to form carbonyl, carboxyl groups, and gaseous products (CO, CO_2, H_2). Although the photodegradation of natural wood is a widely studied phenomenon [20,22,24–29], less attention has been paid to the issue of photodegradation and color stability of steamed wood.

The aim of the work is to investigate the color fastness of maple wood obtained by the process of steaming with a saturated steam-air mixture or with saturated water steam through a simulated aging process-UV radiation in Xenotest Q-SUN Xe-3-H. The color fastness of the wood is evaluated by changes in the coordinates L^*, a^*, b^* of the color space CIE $L^* a^* b^*$, the total color difference ΔE^*, and changes in the values of differential absorbance A_d of selected bands in FTIR spectra.

2. Material and Methods

2.1. Material

The wet wood of maple blanks with the following dimensions: Thickness: h = 40 mm, width w = 100 mm, length d = 750 mm, and moisture content w_p = 57.8 ± 4.8%, was steamed with a saturated steam-air mixture at a temperature of t = 95 °C or saturated steam at t = 115 °C and t = 135 °C for τ = 9 h, in order to obtain a pale pink-brown, pale brown, and brown-red color in a pressure autoclave: APDZ 240 in Sundermann s.r.o. (Banská Štiavnica, Slovakia). The steamed and unsteamed maple wood blanks were subsequently dried to a moisture content of w = 10 ± 0.5%. Samples measures with the following dimensions: Thickness: h = 15 mm, width w = 50 mm, and length d = 100 mm, were made to test the color fastness of the wood.

2.2. Color Measurement of Maple Wood

The color of steamed and unsteamed maple wood before and after irradiation was measured in the color space CIE $L^*a^*b^*$. To measure the color of maple wood, in the color space CIE $L^*a^*b^*$, the color reader CR-10 (Konica Minolta, Osaka, Japan) was measured. A D65 light source was used and the diameter of the optical scanning aperture was 8 mm.

The color measurement was performed on a radial surface machined by planning. The color coordinates of maple wood samples in the color space CIE $L^*a^*b^*$ before irradiation are given in Table 1.

Table 1. Coordinate values of color space CIE L*a*b* of native and thermally treated maple wood.

Maple Wood	Wood Color	Color Coordinates in the Color Space CIE L*a*b*		
		L*	a*	b*
unsteamed maple wood	pale white-yellow	84.2 ± 2.4	7.1 ± 1.5	15.6 ± 1.9
steamed at $t = 95 \pm 2.5\ °C$	pale pink-brown	75.0 ± 1.8	10.3 ± 1.4	16.0 ± 1.6
steamed at $t = 115 \pm 2.5\ °C$	light brown	69.7 ± 1.6	11.9 ± 1.4	16.4 ± 1.4
steamed at $t = 135 \pm 2.5\ °C$	brown-red	61.1 ± 1.4	12.6 ± 1.3	16.9 ± 1.6

The total color difference ΔE^* of the color change of the surface of the maple wood samples under the influence of UV radiation is determined according to the following equation ISO 11 664-4:

$$\Delta E^* = \sqrt{\left(L^*_{298} - L^*_0\right)^2 + \left(a^*_{298} - a^*_0\right)^2 + \left(b^*_{298} - b^*_0\right)^2} \quad (1)$$

where L^*_0, a^*_0, b^*_0 are values on the coordinates of the color space of the surface of the dried milled native and thermally treated maple wood before exposure.

$L^*_{298}, a^*_{298}, b^*_{298}$ are values on the surface color coordinates of the dried milled native and thermal treated maple wood during UV exposure.

2.3. Irradiation of Maple Wood in Xenon Test Chamber

In the Q-SUN Xe-3-H Xenon test chamber, Q-Lab Corporation, Westlake, OH, USA (1800 W Xenon arc lamp-full spectrum, irradiation 0.35 W/m^2-340 nm, black panel temperature 63 °C), the samples were irradiated for τ = 298 h. During the exposure, the color of the irradiated surface was measured regularly at τ = 24 h intervals.

2.4. Analysis of Changes in Lignin-Cellulose Matrix of Wood ATR-FTIR Spectroscopy

Infrared spectroscopy was used to monitor changes in maple wood components induced by UV radiation in unsteamed and steamed wood. The FTIR surface analysis of wood samples was performed on a Nicolet iS 10 FTIR spectrometer (Thermo Fisher Scientific, Madison, WI, USA) using the attenuated total reflectance (ATR-FTIR) technique. The measurements were performed on a diamond crystal in the range of 4000–650 cm^{-1}. For each sample, 64 scans were performed at a resolution of 4 cm^{-1}. The obtained spectral records were evaluated by the spectroscopic software OMNIC 8. The calculation of the values of the differential absorbance A_d describes the relation:

$$A_d = \frac{\left(A_i - A_{ref}\right)}{A_{ref}} \cdot 100 \quad (2)$$

where A_d is the differential absorbance, A_i is the absorbance at a given wavelength in the spectrum of the irradiated sample, and A_{ref} is the absorbance at a given wavelength in the spectrum of the unirradiated sample.

3. Results and Discussion

3.1. Color Analysis

The color of unsteamed and steamed maple wood before and after UV irradiation in the Q-SUN Xe-3-H test chamber (Q-Lab Corporation, Westlake, OH, USA) is shown in Figure 1.

According to the visual evaluation of the color of maple wood before and after the UV radiation, it can be stated that while the light white-yellow color of untreated maple wood darkens and acquires a yellow-red-brown color shade due to photodegradation reactions induced by UV radiation, which is the pale pink-brown of court wood treated with a steam-air mixture with a temperature of t = 95 °C, it darkened slightly with UV radiation and took on a pale brown-yellow color shade. The brown-red color of steamed maple wood obtained by steaming with saturated steam with a temperature of t = 135 °C lightened.

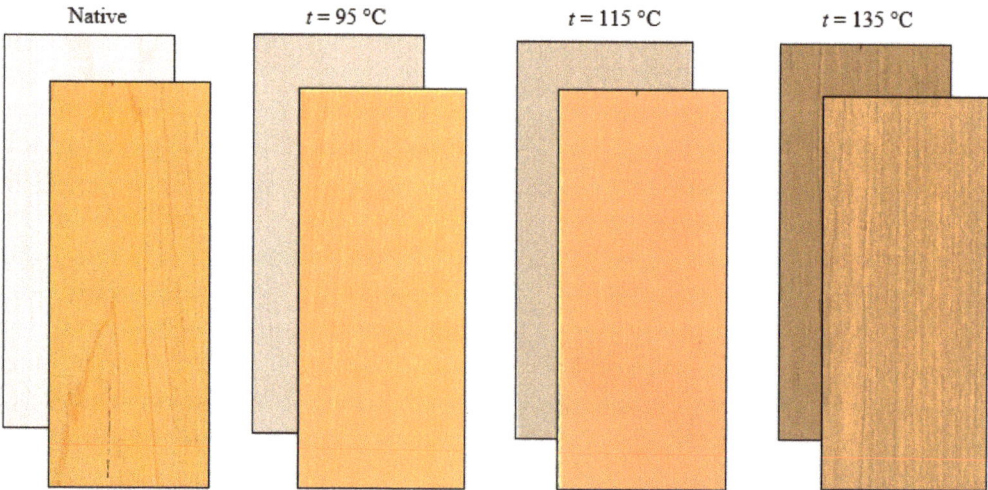

Figure 1. View of maple wood before and after UV irradiation: Native; steamed at t = 95 °C; steamed at t = 115 °C; steamed at t = 135 °C.

The course of color changes of unsteamed and steamed maple wood in the color space CIE $L^*a^*b^*$ under the influence of UV radiation in Xenotest Q-SUN Xe-3-H for 298 h, are shown in Figures 2–4.

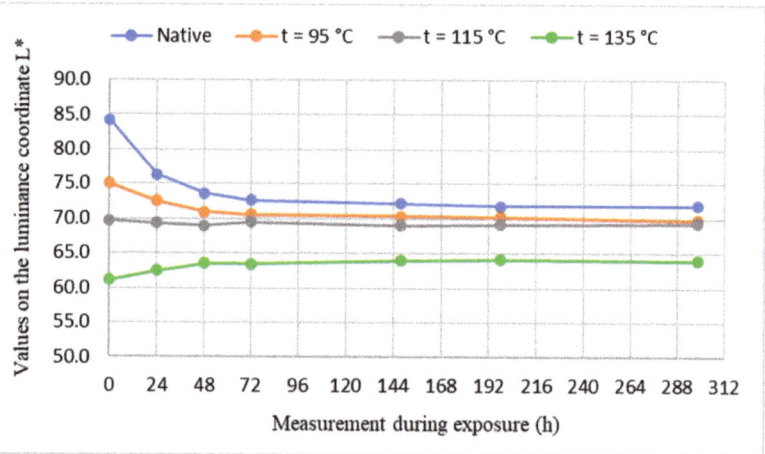

Figure 2. The course of changes of values on the light coordinate L^* in the process of UV irradiation of samples of unsteamed and steamed maple wood.

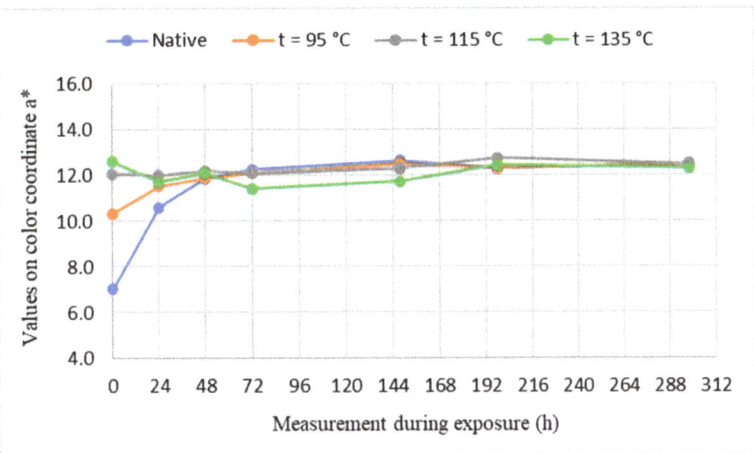

Figure 3. The course of changes of values on the coordinate of red color and a^* in the process of UV irradiation of samples of unsteamed and steamed maple wood.

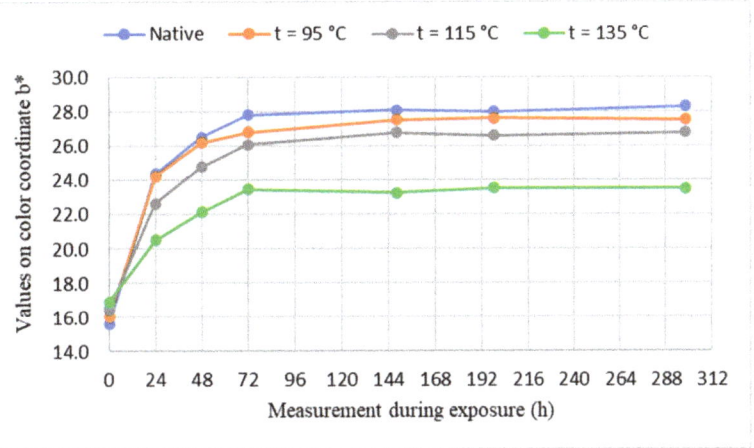

Figure 4. The course of changes of values on the coordinate of yellow color b^* in the process of UV irradiation of samples.

Based on the experimentally determined values of color changes on the luminance coordinate L^*, which are the chromaticity coordinates red a^* and yellow color b^* of maple wood samples induced by photodegradation reactions of individual maple wood components with UV radiation in Xenotest Q-SUN Xe-3-H, it can be stated that significant changes in the color of the wood occur in the first 72 h of UV radiation. The greater the darkening and browning color of the maple wood by steaming, the smaller the changes in the values at the coordinates L^*, a^*, b^* of the steamed maple wood caused by UV radiation. The greater the darkening and browning color of the maple wood by steaming, the smaller the changes in the values at the coordinates L^*, a^*, b^* of the steamed maple wood caused by UV radiation. The change of color of unsteamed maple wood is greater than the steamed maple wood. Numerically, this is documented by the shifts on the individual coordinates of the color space CIE $L^*a^*b^*$ of the analyzed maple wood samples before and after UV irradiation, as shown in Table 2.

Table 2. Sizes of changes in ΔL^*, Δa^*, Δb^* values in the CIE $L^*a^*b^*$ color space of unsteamed and steamed maple wood before and after UV irradiation in the Q-SUN Xe-3-H test chamber.

Maple Wood	Color Coordinates of Samples in the CIE $L^*a^*b^*$ Sample Area before and after UV Irradiation in the Q-SUN Xe-3-H Test Chamber								
	L_0^*	L_{298}^*	ΔL^*	a_0^*	a_{298}^*	Δa^*	b_0^*	b_{289}^*	Δb^*
unsteamed maple wood	84.2	71.9	−12.3	7.0	12.5	+5.8	15.6	28.3	+12.7
steamed at $t = 95 \pm 2.5$ °C	75.0	69.8	−5.2	10.3	12.5	+2.1	16.0	27.5	+11.5
steamed at $t = 115 \pm 2.5$ °C	69.7	69.3	−0.4	11.9	12.5	+0.6	16.4	26.8	+10.4
steamed at $t = 135 \pm 2.5$ °C	61.1	63.9	+2.8	12.6	12.3	−0.3	16.9	23.5	+6.6

The degree of darkening and browning of the unmapped maple during 298 h of UV irradiation in the CIE $L^*a^*b^*$ color space is declared by a decrease in the luminance coordinate by $\Delta L^* = -12.3$ and an increase in points in the red chromatic coordinate by $\Delta a^* = +5.8$ a on the yellow coordinate b^* by the value $\Delta b^* = +12.7$. The above findings on darkening of wood due to UV radiation are in accordance with the opinions of experts dealing with changes in the properties of wood due to solar radiation, respectively UV radiation is shown in [20,22,24,25,29–31].

The differences in the color of steamed maple wood before and after UV irradiation on the light coordinate L^* and the chromaticity coordinate of red a^* and yellow b^* are smaller compared to the changes of unsteamed maple wood.

While the darkness of the thermal treated maple wood with the steam-air mixture with the temperature $t = 95$ °C due to UV radiation increased, the darkness of the maple wood treated with the saturated steam increased by decreasing the values from $L_0^* = 75$ to $L_{298}^* = 69.8$, i.e., $\Delta L^* = -5.2$ temperature $t = 115$ °C did not change due to UV radiation and the brightness of steamed maple wood with saturated water steam with temperature $t = 135$ °C increased from $L_0^* = 61.1$ to $L_{298}^* = 63.9$, i.e., the value of $\Delta L^* = +2.8$. The decrease in the darkening of maple wood steamed at higher steaming temperatures due to UV radiation or to achieving the opposite effect—lightening the surface of steamed maple wood with saturated steam with temperature $t = 135$ °C indicates changes in the chromatic system caused by steaming, which affects the photochemical reactions of UV radiation with functional groups of the chromophore system of steamed maple wood. The work of [32,33] also points that steamed wood, unlike unsteamed wood, is more or less resistant to UV radiation.

The effect of fading of the red-brown color of the beech wood surface is achieved by steaming with a saturated steam with temperature $t = 120 \pm 2$ °C after UV irradiation in the Xenotest 450 Xenon lamp. This emits UV radiation with a wavelength of 340 nm and an intensity of 42 ± 2 W/m^2 for 7 days, as stated in the work of [32].

In [33], the effect of UV radiation on steamed agate wood states that while the surface of steamed agate wood darkened slightly at a steaming temperature $t = 100$ °C, the surface of agate wood brightened at a steaming temperature $t = 120$ °C.

The positive effect of maple wood steaming on the resistance to the effects of UV radiation is declared by the decrease in values Δa^* and Δb^* on the chromatic coordinates given in Table 2. The pale pink-brown color obtained by steaming with a steam-air mixture with temperature $t = 95$ °C by absorbing UV radiation increased on the red coordinate by $a^* = +2.1$. In addition, the pale-brown color of steamed maple wood formed by steaming with saturated steam with temperature $t = 115$ °C due to UV radiation increased by the value $\Delta a^* = +0.6$. Moreover, the effect of UV radiation on the surface of steamed maple wood steamed with saturated steam with a temperature of $t = 135$ °C was manifested by a decrease in the value on the red coordinate from $a_0^* = 12.6$ to $a_{298}^* = 12.3$, i.e., value $\Delta a^* = -0.3$. Similarly, at the yellow coordinate there are decreases in Δb^* values caused by UV radiation on steamed maple wood. The value of the change Δb^* on the yellow coordinate induced by UV radiation on the surface of maple wood steamed with a steam mixture with temperature $t = 95$ °C is $\Delta b^* = +11.5$. In addition, maple wood steamed with saturated water steam with temperature $t = 115\ °C$ is $\Delta b^* = +10.4$ and saturated water steam with temperature $t = 135\ °C$ is $\Delta b^* = +6.6$. Based on the above findings, it can be stated that the functional groups of the maple wood chromophoric system absorbing electromagnetic radiation spectra with a wavelength of red 630–750 nm and a wavelength of 570–590 nm of yellow color were significantly eliminated for photochemical reactions of wood induced by UV radiation for the red color and to a lesser extent for the yellow color.

The changes in the values at the individual coordinates of the color space CIE $L^*a^*b^*$ induced on the surface of unsteamed and steamed maple wood by UV radiation in Xenotest 450 are reflected in the quantification of the color change of the maple wood surface expressed by the total color difference ΔE^*. The influence of UV radiation on the magnitude of color changes of the analyzed maple wood samples in the form of the total color difference ΔE^* is shown in Figure 5.

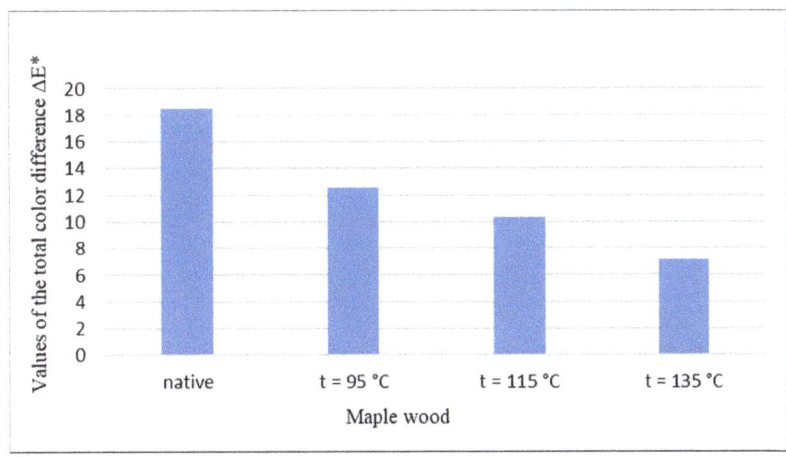

Figure 5. Influence of UV radiation on the size of the total color difference ΔE^* of unsteamed and steamed maple wood.

The lower values of the total color difference ΔE^* of steamed maple wood indicate the benefit of steaming on the resistance of steamed maple wood to UV radiation causing the color change in the process of natural aging. While the color change of unsteamed maple wood caused by UV radiation reaches the value $\Delta E^* = 18.5$, for steamed maple wood steamed with temperature $t = 95\ °C$ it is $\Delta E^* = 12.6$, which is a decrease of 31.8% compared to the total color difference of unsteamed maple wood, for steamed maple wood with saturated steam with temperature $t = 115\ °C$ is $\Delta E^* = 10.4$, which is a decrease of 43.8%, and for steamed maple wood steamed with saturated steam with temperature $t = 135\ °C$ is $\Delta E^* = 7.2$, which is a decrease of 61.1%.

3.2. ATR-FTIR Spectroscopy Analysis

The color changes of maple wood samples caused by photodegradation reactions initiated by UV radiation are also documented by FTIR analyses of the surface of unsteamed and steamed maple wood after UV radiation in Figure 6 and Table 3.

The results of FTIR analysis indicate the formation of new carbonyl C=O groups in the spectra of the samples manifested by an increase in the intensity of the absorption band at a maximum of 1720 cm^{-1}. In this section, an overlap of several absorption bands can be observed, which is a manifestation of vibrations of conjugated and unconjugated C=O bonds, as well as carboxyl groups. These can come not only from the main constituents of wood (lignin, cellulose, hemicelluloses), but also from extractives [33]. We recorded the most significant increase in the intensity of the C=O group bands in the spectrum of the irradiated native sample, by more than 57%. In the case of irradiated steamed wood samples, this increase ranges from 16 to 21%.

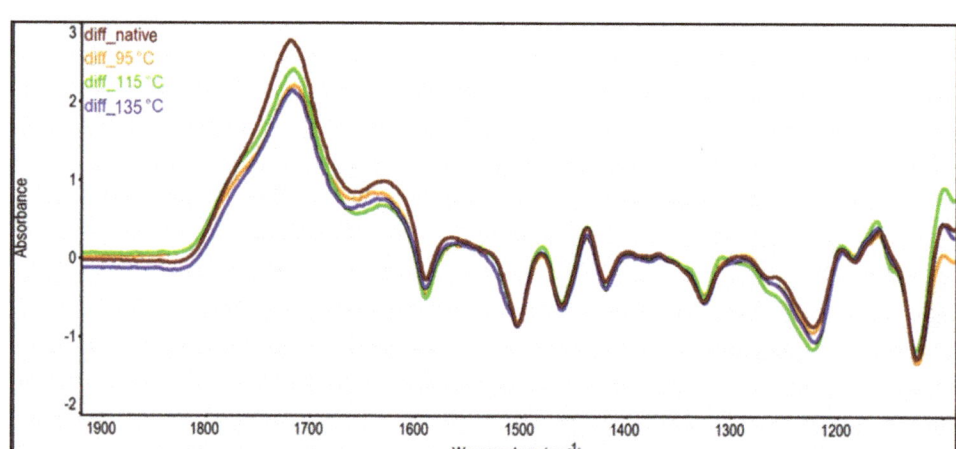

Figure 6. Differential ATR-FTIR spectra expressing the effect of UV radiation on unsteamed and steamed maple wood.

Table 3. Values of differential absorbances of selected absorption bands expressing the influence of UV radiation on unsteamed and steamed maple wood.

Maple Wood Samples	$A_{d\ (1720)}$ (%)	$A_{d\ (1504)}$ (%)	$A_{d\ (1235)}$ (%)
unsteamed maple wood	157.23	−96.02	−25.30
steamed at $t = 95 \pm 2.5\ °C$	116.80	−96.24	−31.56
steamed at $t = 115 \pm 2.5\ °C$	121.43	−97.37	−35.04
steamed at $t = 135 \pm 2.5\ °C$	118.23	−100.00	−31.69

The performed FTIR analyses also indicate a significant degradation of the lignin macromolecule due to UV radiation. After irradiating the steamed wood sample at 135 °C, we recorded a complete loss of the absorption band at wavenumber 1504 cm^{-1}, in order to be able to speak of complete degradation of lignin in the surface layer of the samples. After irradiation of the samples steamed at 95 and 115 °C, the intensity of the said characteristic lignin absorption band decreased by 96.24% or 97.37%. The reduction can also be observed by comparing the intensities of other absorption bands characteristic for lignin, at wavenumber 1593, 1462, 1422, 1225, and 830 cm^{-1}. However, it should be noted that at 1462 and 1225 cm^{-1}, not only vibrations of lignin, but also hemicelluloses, occur.

Several studies have confirmed that lignin is the most sensitive to UV radiation among all of the wood components [24,29,34,35]. By absorbing energy, the bonds are cleaved and new functional groups (carbonyl and carboxyl) are formed, as well as radicals, which further induce lignin depolymerization and condensation reactions. Aromatic phenoxyl radicals react with oxygen to form unsaturated carbonyl compounds (quinones), which contribute to the color changes of wood [25,36].

Based on the decrease in the intensities of the absorption bands at the wavenumber 1328 and 1126 cm^{-1}, we can state that in addition to lignin, polysaccharide degradation also occurs. While the first band corresponds to the vibrations of cellulose macromolecule, the second band belongs to the symmetric valence vibrations of the ether bond and the glucose ring [33,37].

Since the formation of new C=O bonds is considered to be the main cause of wood color changes during its exposure to UV radiation, the results of FTIR analysis confirm the positive effect of thermal steaming of wood on its color stability.

4. Conclusions

The surface color of unsteamed maple wood changes more markedly than the surface color of steamed maple wood due to UV radiation. The more pronounced the darkening and browning color of the steamed maple wood, the smaller the UV-induced changes in the color of the steamed maple wood. This is evidenced by the degree of darkening of the surface of unsteamed and steamed maple wood at $t = 95\ °C$ and $t = 125\ °C$ after UV irradiation by decreasing values on the luminance coordinate L^*, as well as the rate of decrease of Δa^*, Δb^* values on chromatic coordinates. The decrease in Δa^* and Δb^* values on the chromatic coordinates indicates that the functional groups of the maple wood chromophore system absorbing electromagnetic radiation spectra with a wavelength of red of 630–750 nm and a wavelength of 570–590 nm were eliminated to a lesser extent by steaming for photochemical reactions of wood caused by UV radiation.

The positive effect of maple wood steaming on the limiting effect of initiating photo degradation reactions induced by UV radiation on the surface of steamed maple wood is evidenced by the decrease in the overall color difference ΔE^*. While the change in color of unsteamed maple wood caused by UV radiation expressed by the total color difference is $\Delta E^* = 18.5$, for steamed maple wood the stated changes in color difference values depending on the steaming temperature decrease from $\Delta E^* = 12.6$ to $\Delta E^* = 7.2$, which is a decrease from 31.8 to 61.1%.

This is confirmed by the results of FTIR analyses. While in the case of unsteamed wood we recorded an increase in the intensity of absorption bands of chromophoric C=O groups due to UV radiation by more than 57%, in the case of steamed wood samples this increase is lower and ranges from 16 to 21%.

Author Contributions: L.D. designed the whole study; M.D., L.D., and E.V. conducted data collection, modeling, and results analysis; L.D. wrote the original draft paper; M.D. and E.V. revised and edited the paper. All authors have read and agreed to the published version of the manuscript.

Funding: This research was funded by project APVV-17-0456 "Thermal modification of wood with saturated water steam for purposeful and stable change of wood color".

Institutional Review Board Statement: Not applicable.

Informed Consent Statement: Not applicable.

Data Availability Statement: The data presented in this study are available on request from the corresponding author.

Conflicts of Interest: The authors declare no conflict of interest.

References

1. Barański, J.; Klement, I.; Vilkovská, T.; Konopka, A. High temperature drying process of beech wood (*Fagus sylvatica* L.) with different zones of sapwood and red false heartwood. *BioResources* **2017**, *12*, 1861–1870. [CrossRef]
2. Barcik, Š.; Gašparík, M.; Razumov, E.Y. Effect of thermal modification on the colour changes of oak wood. *Wood Res.* **2015**, *60*, 385–396.
3. Bekhta, P.; Niemz, P. Effect of high temperature on the change in colour, dimensional stability and mechanical properties of spruce wood. *Holzforschung* **2003**, *57*, 539–546. [CrossRef]
4. Cividini, R.; Travan, L.; Allegretti, O. White beech: A tricky problem in drying process. In Proceedings of the International Scientific Conference on Hardwood Processing, Quebec City, QC, Canada, 24–26 September 2007.
5. Deliiski, N. Metod dlja ocenki stepeni oblagoraživanija bukovych pilomaterialov vo vremja ich proparki. In *Current Problems and Perspectives of Beech Lumber Drying*; ES-VŠLD: Zvolen, Slovakia, 1991; pp. 37–44.
6. Dudiak, M.; Dzurenda, L. Changes in the physical and chemical properties of alder wood in the process of thermal treatment with saturated water steam. *Coatings* **2021**, *11*, 898. [CrossRef]
7. Dzurenda, L. Colouring of Beech Wood during Thermal Treatment using Saturated Water Steams. *Acta Fac. Xylologiae Zvolen* **2014**, *56*, 13–22.
8. Gonzalez-Pena, M.M.; Hale, M.D.C. Colour in thermally modified wood of beech, Norway spruce and Scots pine. Part 1: Colour evolution and colour changes. *Holzforschung* **2009**, *63*, 385–393. [CrossRef]
9. Milić, G.; Todorović, N.; Popadić, R. Influence of steaming on drying quality and colour of beech timber. *Glasnik Šumarskog Fakulteta* **2015**, 83–96. [CrossRef]

10. Molnar, S.; Tolvaj, L. Colour homogenisation of different wood species by steaming. In *Interaction of Wood with Various Forms of Energy*; Technical University in Zvolen: Zvolen, Slovakia, 2002; pp. 119–122.
11. Todaro, L.; Zuccaro, L.; Marra, M.; Basso, B.; Scopa, A. Steaming effects on selected wood properties of Turkey oak by spectral analysis. *Wood Sci. Technol.* **2012**, *46*, 89–100. [CrossRef]
12. Tolvaj, L.; Nemeth, R.; Varga, D.; Molnar, S. Colour homogenisation of beech wood by steam treatment. *Drewno* **2009**, *52*, 5–17.
13. Geffert, A.; Vybohová, E.; Geffertová, J. Characterization of the changes of colour and some wood components on the surface of steamed beech wood. *Acta Fac. Xylologiae Zvolen* **2017**, *59*, 49–57.
14. Dzurenda, L.; Dudiak, M. Cross-correlation of color and acidity of wet beech wood in the process of thermal treatment with saturated steam. *Wood Res.* **2021**, *66*, 105–116. [CrossRef]
15. Banski, A.; Dudiak, M. Dependence of color on the time and temperature of saturated water steam in the process of thermal modification of beech wood. *AIP Conferen. Proc.* **2019**, *2118*, 30003.
16. Tolvaj, L.; Molnár, S. Colour homogenisation of hardwood species by steaming. *Acta Silv. Lignaria Hung.* **2006**, *2*, 105–112.
17. Dzurenda, L. The shades of color of *Quercus robur* L. wood obtained through the processes of thermal treatment with saturated water vapour. *BioResources* **2018**, *13*, 1525–1533. [CrossRef]
18. Dzurenda, L. Colour modification of *Robinia pseudoacacia* L. during the processes of heat treatment with saturated water steam. *Acta Fac. Xylologiae Zvolen* **2018**, *60*, 61–70.
19. Dudiak, M. Modification of maple wood colour during the process of thermal treatment with saturated water steam. *Acta Fac. Xylologiae Zvolen* **2021**, *63*, 25–34.
20. Hon, D.S.N. Weathering and photochemistry in wood. In *Wood and Cellulosic Chemistry*, 2nd ed.; Hon, D.S.N., Shiraishi, N., Eds.; MarcelDekker: New York, NY, USA, 2001; pp. 513–546.
21. Reinprecht, L. *Wood Protection*; Technical University in Zvolen: Zvolen, Slovakia, 2008; p. 450.
22. Baar, J.; Gryc, V. The analysis of tropical wood discoloration caused by simulated sunlight. *Eur. J. Wood Wood Prod.* **2011**, *70*, 263–269. [CrossRef]
23. Gandelová, L.; Horáček, P.; Šlezingerová, J. *The Science of Wood*; Mendel University of Agriculture and Forestry in Brno: Brno, Czech Republic, 2009; p. 176.
24. Müller, U.; Rätzsch, M.; Schwanninger, M.; Steiner, M.; Zöbl, H. Yellowing and IR-changes of spruce wood as result of UV-irradiation. *J. Photochem. Photobiol. B Biol.* **2003**, *69*, 97–105. [CrossRef]
25. Pandey, K.K. Study of the effect of photo-irradiation on the surface chemistry of wood. *Polym. Degrad. Stab.* **2005**, *90*, 9–20. [CrossRef]
26. Persze, L.; Tolvaj, L. Photodegradation of wood at elevated temperature: Colour change. *J. Photochem. Photobiol. B Biol.* **2012**, *108*, 44–47. [CrossRef]
27. Denes, L.; Lang, E.M. Photodegradation of heat-treated hardwood veneers. *J. Photochem. Photobiol. B Biol.* **2013**, *118*, 9–15. [CrossRef] [PubMed]
28. Zivkovic, V.; Arnold, M.; Radmanovic, K.; Richter, K.; Turkulin, H. Spectral sensitivity in the photodegradation of fir wood (*Abies alba* Mill.) surfaces: Colour changes in natural weathering. *Wood Sci. Technol.* **2014**, *48*, 239–252. [CrossRef]
29. Geffertová, J.; Geffert, A.; Vybohová, E. The effect of UV irradiation on the colour change of the spruce wood. *Acta Fac. Xylologiae Zvolen* **2018**, *60*, 41–50.
30. Chang, T.C.; Chang, H.T.; Chang, S.T. Influences of extractives on the photodegradation of wood. *Polym. Degrad. Stab.* **2010**, *95*, 516–521. [CrossRef]
31. Kúdela, J.; Kubovský, I. Accelerated-ageing-induced photo-degradation of beech wood surface treated with selected coating materials. *Acta Fac. Xylologiae Zvolen* **2016**, *2*, 27–36.
32. Dzurenda, L. The effect of UV radiation in Xenotest 450 on the colour of steamed beech wood during the process of simulated ageing. *Ann. Wars. Univ. Life Sci. For. Wood Technol.* **2019**, *106*, 114–119. [CrossRef]
33. Varga, D.; Tolvaj, L.; Molnar, Z.; Pasztory, Z. Leaching efect of water on photodegraded hardwood species monitored by IR spectroscopy. *Wood Sci. Technol.* **2020**, *54*, 1407–1421. [CrossRef]
34. Cogulet, A.; Blanchet, P.; Landry, V. Wood degradation under UV irradiation: A lignin characterization. *J. Photochem. Photobiol. B Biol.* **2016**, *158*, 184–191. [CrossRef]
35. Zborowska, M.; Stachowiak-Wencek, A.; Waliszewska, B.; Prądzyński, W. Colorimetric and FTIR ATR spectroscopy studies of degradative effects of ultraviolet light on the surface of exotic ipe (*Tabebuia* sp.) wood. *Cellulose Chem. Technol.* **2016**, *50*, 71–76.
36. Teaca, C.-A.; Rosu, D.; Bodirlau, R.; Rosu, L. Structural changes in wood under arteficial UV light irradiation determined by FTIR spectroscopy and color measurement—A brief review. *BioResources* **2014**, *8*, 1478–1507.
37. Colom, X.; Carrillo, F.; Nogués, F.; Garriga, P. Structural analysis of photodegraded wood by means of FTIR spectroscopy. *Polym. Degrad. Stab.* **2003**, *80*, 543–549. [CrossRef]

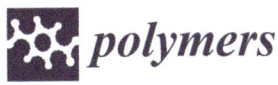

Article

Enzymatic Functionalization of Wood as an Antifouling Strategy against the Marine Bacterium *Cobetia marina*

Daniel Filgueira [1,2], Cristian Bolaño [1], Susana Gouveia [1] and Diego Moldes [1,3,*]

1. CINTECX, Department of Chemical Engineering, Campus Universitario as Lagoas-Marcosende, Universidade de Vigo, 36310 Vigo, Spain; daniel.martinez@tecnalia.com (D.F.); cbolano@uvigo.es (C.B.); gouveia@uvigo.es (S.G.)
2. TECNALIA, Basque Research and Technology Alliance (BRTA), Area Anardi 5, 20730 Azpeitia, Spain
3. Research Group of Bioengineering and Sustainable Processes, Department of Chemical Engineering, Edificio Fundición, Lagoas Marcosende s/n, University of Vigo, 36310 Vigo, Spain
* Correspondence: diego@uvigo.es

Citation: Filgueira, D.; Bolaño, C.; Gouveia, S.; Moldes, D. Enzymatic Functionalization of Wood as an Antifouling Strategy against the Marine Bacterium *Cobetia marina*. *Polymers* **2021**, *13*, 3795. https://doi.org/10.3390/polym13213795

Academic Editors: Ľuboš Krišťák, Roman Réh and Ivan Kubovský

Received: 5 October 2021
Accepted: 29 October 2021
Published: 2 November 2021

Publisher's Note: MDPI stays neutral with regard to jurisdictional claims in published maps and institutional affiliations.

Copyright: © 2021 by the authors. Licensee MDPI, Basel, Switzerland. This article is an open access article distributed under the terms and conditions of the Creative Commons Attribution (CC BY) license (https://creativecommons.org/licenses/by/4.0/).

Abstract: The protection of wood in marine environments is a major challenge due to the high sensitivity of wood to both water and marine microorganisms. Besides, the environmental regulations are pushing the industry to develop novel effective and environmentally friendly treatments to protect wood in marine environments. The present study focused on the development of a new green methodology based on the laccase-assisted grafting of lauryl gallate (LG) onto wood to improve its marine antifouling properties. Initially, the enzymatic treatment conditions (laccase dose, time of reaction, LG concentration) and the effect of the wood specie (beech, pine, and eucalyptus) were assessed by water contact angle (WCA) measurements. The surface properties of the enzymatically modified wood veneers were assessed by X-ray photoelectron spectroscopy (XPS), Fourier transform-infrared spectroscopy (FTIR). Antifouling properties of the functionalized wood veneers against marine bacterium *Cobetia marina* were studied by scanning electron microscopy (SEM) and protein measurements. XPS and FTIR analysis suggested the stable grafting of LG onto the surface of wood veneers after laccase-assisted treatment. WCA measurements showed that the hydrophobicity of the wood veneers significantly increased after the enzymatic treatment. Protein measurements and SEM pictures showed that enzymatically-hydrophobized wood veneers modified the pattern of bacterial attachment and remarkably reduced the bacterium colonization. Thus, the results observed in the present study confirmed the potential efficiency of laccase-assisted treatments to improve the marine antifouling properties of wood.

Keywords: laccase; lauryl gallate; wood; *Cobetia marina*; antifouling

1. Introduction

Wood is a renewable, cheap, and biodegradable bioresource with interesting mechanical properties which make it a competitive material in construction applications. In marine environments, wood is used as a raw material for the construction of waterfront structures, e.g., groynes, jetties, dolphins [1] and classic boats. Besides, wood is used in marine platforms for the aquaculture of mollusks and crustaceous species. However, wood is highly hydrophilic and very sensitive to the attack of biological organisms [2]. These characteristics are a major challenge in marine environments, due to the regular wet conditions and the great number of living organisms in seawater. Under these conditions, all the immersed surfaces are rapidly colonized by different microorganisms leading to the formation of a biofilm, but also larger organisms could colonize the substrata after the biofilm formation. This phenomenon is commonly known as marine biofouling [3].

The biofilm formation starts with the adsorption of free organic material onto the surface of the substrata [3]. Then, bacteria and other microorganisms (e.g., diatoms, protozoa) are the first living organisms in colonizing the substrata [4]. Bacteria have the capacity to secrete extracellular polymeric substances (EPS) which are mainly composed of carbohydrates and proteins [5], forming a 3D structure of bacterial aggregates which facilitates their adhesion onto the substrata [6]. Such biofilm may change the physicochemical properties of the substrata, influencing the future adhesion of larger fouling organisms (e.g., barnacles, mollusks) [7]. A wide variety of paints and coatings have been used to limit the marine biofouling. Most of them involved the use of heavy metals (e.g., copper, zinc) or organic compounds that are toxic for fouling organisms, but also for non-target organisms such as algae, mollusks, crustaceans and fishes and they could be introduced in the food chain [8]. One of the most extended antifouling chemicals used in paints (tributyltin) was banned years ago and other hazardous treatments should be avoided to protect the marine ecosystems. Environmental factors and physicochemical properties of surfaces are the two main parameters which may favor or hinder the biofilm formation [9]. Since environmental factors are not controllable, the modification of the physicochemical properties of the substrata may disturb the pattern of bacterial colonization, minimizing the subsequent adhesion of marine fouling organisms. Based on such a mechanism, different non-biocide antifouling treatments have been proposed in the last decade [7,10]; most of them aim to interfere with the adhesion of marine microorganisms adjusting the surface free energy of the substrata between a specific range [11], which is usually controlled with hydrophobic compounds e.g., fluorinated polymers or silicones [12].

Another important issue regarding the protection of wood in marine environments is the leaching of the antifouling coatings and paints, since this release results in water pollution but also in the requirement of a new antifouling treatment. Covalent grafting of antifouling compounds onto surfaces could significantly reduce or avoid their leaching, providing stable protection against fouling organisms. Regarding wood as a material, a sustainable alternative to conventional chemical and thermal treatments is the use of laccase, which enable the stable grafting of several chemical compounds. Such lignolytic enzyme can oxidize phenolic and amine groups of a target compound [13,14], leading to the formation of radicals that may link to the aromatic structure of woody lignin by radical polymerization. By this mechanism, laccase could be used for functionalization of lignocellulosic materials or their components, mainly lignin. It is worth noting that laccase-assisted reactions are performed in mild conditions, minimizing the energy requirements. Thus, the physicochemical properties of wood can be tailored by means of an environmentally-friendly pathway. Enzymatic hydrophobization of wood veneers was successfully performed with different chemical compounds such as alkylamines [15] and fluorophenols [16]. Alkyl gallates are laccase-specific substrates with an aliphatic chain at the *para*-position that could provide stable hydrophobic properties if grafted onto wood. In fact, these compounds have been enzymatically grafted on different lignocellulosic substrates for such purpose [17,18], but also for providing antibacterial properties [19] since the aliphatic tail also presents this characteristic.

Therefore, the reaction conditions namely lauryl gallate (LG) concentration, laccase dose and time of reaction for the enzymatic hydrophobization of wood veneers were assessed in the present study. Water contact angle (WCA) measurements were carried out for the assessment of the wettability of the enzymatically-hydrophobized wood veneers. In addition, the surface of the hydrophobized wood samples was characterized by X-ray photoelectron spectroscopy (XPS) and Fourier transform infrared spectroscopy (FTIR). Finally, the potential antifouling properties of the enzymatically-hydrophobized wood veneers were studied against the marine bacterium *Cobetia marina* by means of protein measurements and scanning electron microscopy (SEM), in order to assess the capability of hydrophobic functionalization of wood as a new antifouling strategy for such material.

2. Materials and Methods

2.1. Materials

Beech (*Fagus sylvatica*), eucalyptus (*Eucalyptus globulus*) and pine (*Pinus pinaster*) wood veneers were supplied by Foresa (Caldas de Reis, Spain). Beech and eucalyptus veneers were vaporized before supplying. These wood species were selected because of their importance regarding production and commercial interest in Galicia (NW of Spain).

Commercial laccase from *Myceliophthora thermophila* (NS51003) was provided by Novozymes (Bagsværd, Denmark). The activity of the enzyme was calculated by the 2,2′-azino-bis(3-ethylbenzothiazoline-6-sulphonic acid) (ABTS) oxidation assay. One unit of activity is defined as the amount of enzyme that oxidizes 1 μmol of ABTS per minute at 25 °C and pH 7 in a 0.1 M phosphate buffer.

Cobetia marina (CECT4278) bacterium was purchased from the Colección Española de Cultivos Tipo—CECT (Valencia, Spain).

Bradford reactive was provided by Bio-Rad (Hercules, CA, USA).

All the other chemical reagents were acquired from Sigma-Aldrich (St. Louis, MO, USA) at reagent grade and used without any kind of purification.

2.2. Enzymatic Hydrophobization of Wood Veneers

Wood veneers from beech, pine or eucalyptus were cut in square plugs (50 × 50 mm), washed with distilled water at 50 °C for 30 min and oven-dried at 50 °C for 4 h. Each wood sample was then immersed in 58 mL of a phosphate buffer (pH 7, 0.1 M): acetone solution (60:40, v/v) at 50 °C. An experimental design was performed for assessing the treatment conditions which could provide the highest hydrophobic properties to wood veneers. Three levels of each of the selected parameters were chosen: laccase dose (3.36, 5.00 and 6.72 U/cm^2 of wood), LG concentration (2.5, 5 and 10 mM) and treatment time (1, 2 and 4 h). All the possible combinations of such parameters were assayed (Table S1). The enzymatically treated samples were then dried at a room temperature (18–20 °C) for 12 h and washed with an aqueous solution of acetone (50:50, v/v) at 50 °C for 1 h. Finally, the samples were washed with distilled water three times and oven-dried at 50 °C for 4 h. Control treatments without adding laccase and/or LG were performed in parallel. All the experiments were performed in triplicate.

2.3. Water Contact Angle Measurements

The hydrophobization of wood veneers was assessed by water contact angle (WCA) measurements. A goniometer MobileDrop GH11 (Krüss GmbH, Hamburg, Germany) was used to measure the contact angle of one drop of distilled water on the surface of wood veneers. DSA2 software (Krüss GmbH) was utilized to analyze the drop shape. WCA measurements were performed after water drop deposition with intervals of 30 s for 5 min. A minimum of two points per each side of the wood veneers were measured.

2.4. X-ray Photoelectron Spectroscopy (XPS)

The surface of wood veneers was analyzed by X-ray photoelectron spectroscopy (XPS). The analyses were performed in a Thermo Scientific K-Alpha device using a monochromized Al-K-α-X-ray source (1486.6 eV). As the samples were not conductive, an electron flood gun was used to minimize surface changing. A constant analyzed energy mode (CAE) with 100 eV pass energy for survey spectra was used to perform the measurements. Neutralization of the surface charge was implemented with a low-energy flood gun (electrons in the range 0–14 eV) and a low-energy Ar ions gun. Binding energy was adjusted using C1 (BE 284.6 eV). The elemental composition of the surface of wood samples was defined plotting on the standard Scofield photoemission cross-sections. The O/C ratio was calculated by means of the survey spectra collected (from 0 to 1350 eV) which showed the components C1s and O1s.

2.5. Fourier Transform Infrared Spectroscopy (FTIR)

The surface of wood pieces was also studied with a 4100 Fourier transform infrared spectrometer (Jasco), equipped with an attenuated total reflectance (ATR) device. A total of 32 scans were picked between 600 and 4000 cm^{-1} with a resolution of 4 cm^{-1}. The spectra were scale normalized and analyzed with OMNIC Suite software v 7.3 (Thermo Scientific, Waltham, MA, USA).

2.6. Stock Bacteria and Culture Conditions

Marine broth (5 g of bacteriological peptone, 3 g of yeast extract, 750 mL filtered marine water, 250 mL distilled water and pH adjusted to 7.4) was used as culture media in all assays.

Cobetia marina CECT4278 was provided lyophilized. It was resuspended in marine broth, which was then used for the inoculation of two flasks with 50 mL of fresh marine broth. Such culture media was incubated in an orbital shaker at 100 rpm and 27 °C for 24 h. Optical density (OD) was measured with Unicam Helios Beta spectrophotometer (Thermo Fisher Scientific) at 600 nm. When the value of OD was 1.0, 1.5 mL of bacterial culture medium was used to inoculate fresh culture media (3% of inoculum, v/v), in duplicate. After 24 h of incubation in an orbital shaker at 100 rpm and 27 °C, the culture medium was centrifuged (10,000 rpm) and stored as stock culture in aliquots of 3 mL with glycerol (30%) at −20 °C.

Working cultures were prepared in the same way, using stock cultures as inoculum but discarding the supernatant, suspending the pellet with 1.5 mL of culture media; then 750 µL of this suspension was used for inoculating 50 mL of culture media (3% of inoculum, v/v).

2.7. Biofilm Assay

Pine wood samples were hydrophobized according to the results obtained from the study of the treatment conditions. After the enzymatic treatment, the veneers were cut in square plugs (25 × 25 mm), dried at 50 °C for 24 h. Finally, the pine veneers were autoclaved at 121 °C for 20 min.

For the biofilm assay, each hydrophobized pine veneer was submerged in working cultures (50 mL of marine broth inoculated with *C. marina* 3% (v/v)). The flask was incubated in an agitator with orbital shaker at 100 rpm and 27 °C for 5 days. The colonized pine veneers were gently washed with distilled water to remove unattached bacteria from their surface. Control tests using untreated pine veneers and, also control test without *C. marina* were performed. Six replicas of biofilm colonization were carried out for each experiment.

2.8. Bacterial Surface Hydrophobicity

Surface hydrophobicity of *C. marina* was measured by means of the microbial adhesion to hydrocarbons (MATH) method. The test was performed according to Warne Zoueki et al. [20], who adapted the method previously proposed by Rosenberg et al. [21]. Firstly, 50 mL of marine broth were inoculated (3% (v/v) *C. marina*) and incubated for 24 h. Such culture was then centrifuged (8000 rpm) at 4 °C for 20 min into a Falcon tube. The supernatant was discarded, and the recovered biomass was suspended with KCl solution (150 mM, pH 7). Centrifugation and supernatant removal steps were performed twice. Then, the biomass was suspended with KCl solution (150 mM, pH 7) and diluted with the same solution in order to obtain an absorbance of 1.0 at 600 nm (ABS_i). After that, in our experiment, 5 mL of the diluted bacterial solution were mixed with 300 µL of dodecane in an acid washed Pyrex glass tube. Subsequently, the mixture was stirred for 10 min at 27 °C and left for 15 min to separate the aqueous phase from the organic phase. The aqueous phase was recovered, and its absorbance was measured at 600 nm (ABS_f). The partition coefficient (P) was calculated as follows:

$$P = 1 - (ABS_f / ABS_i) \quad (1)$$

2.9. Protein Measurement

Each colonized pine veneer was immersed in 15 mL of distilled water in a 50 mL centrifuge tube. The tube content was then sonicated with a HD 2070 Sonoplus sonicator

(Bandelin GmbH, Berlin, Germany) for 10 min (power 65%, cycle 70%) divided in two equal intervals of 5 min, with a pause of 2 min. After sonication, the wood samples were removed, and the aqueous solution was centrifuged at 7500 rpm for 40 min. Finally, the protein amount of the supernatant was measured using the Bradford's method, with slight modifications: 800 µL of the supernatant were mixed with 200 µL of Bradford's concentrated reagent and stirred with a vortex device; the mixture was left for 20 min and its absorbance was measured at 595 nm. A standard curve (5–25 mg protein/L) for protein quantification was performed using bovine serum albumin. Protein measurements of bacteria biomass on the wood veneers were performed after one, three and five days of incubation of *C. marina*.

2.10. SEM Analysis

Colonized pine veneers were introduced into plastic tubes with 2% of glutaraldehyde in cacodylate buffer (0.1 M, pH 7.4) and left at 4 °C for 2 h. Part of the solution was discarded avoiding the contact of the samples with atmospheric air and sodium phosphate buffer was added and the samples left for 30 min. Such a step was performed twice. Dehydration was performed with graded ethanol series, graded ethanol:amyl acetate substitution series and CO_2 critical point drying (73 atm, 31.3 °C). Then, the samples were coated with a layer of gold (thickness: 10–20 nm) (K550X Sputter Coater, Emitech, Ashford, UK). A Philips XL30 (SEMTech Solutions, Billerica, MA, USA) scanning electron microscope (SEM) was used to perform the SEM analysis. The applied acceleration voltage was 12kW and the magnifications were $100\times$, $200\times$, $500\times$, $1000\times$, $1500\times$ and $2500\times$.

3. Results and Discussion

3.1. Enzymatic Hydrophobization

Nowadays, acetylation is the only processes available at industrial scale for the manufacturing of hydrophobic wood. However, acetylation is a high-cost process with a relatively high environmental impact. Hence, the wood industry needs to develop new sustainable routes for the manufacturing of hydrophobic wood. In this sense, laccase-assisted hydrophobization is a promising pathway to achieve an efficient hydrophobization of wood with a low environmental impact but also a competitive production cost.

Several parameters of laccase-mediated hydrophobization of beech veneers, namely laccase dose, LG concentration ([LG]) and time of reaction, were assessed to optimize the enzymatic hydrophobization process on beech veneers. The hydrophobicity of the treated beech veneers is showed as a function of the water contact angle (WCA).

It was expected that long reaction time combined with high [LG] improved the hydrophobicity of the beech veneers. Nevertheless, long reaction times (higher than 3 h) did not improve the surface hydrophobicity of the wood veneers. In fact, the time of reaction necessary to achieve the highest WCA on the surface of beech veneers was between 2–3 h for the different [LG] and laccase doses studied (Figure 1). It is worth to mention that besides the grafting of LG onto the surface of beech veneers, self-polymerization of LG monomers could be expected leading to the formation of oligomers [22]. In fact, several authors have already reported the laccase-catalyzed polymerization of phenolic compounds such as epigallocatechin [23], lignin model compounds [24], or condensed tannins [25]. Hence, stable LG adsorption without correct orientation and/or grafting/adsorption of LG oligomers without proper orientation after the grafting of LG monomers could explain the lower hydrophobicity obtained after 3 h of treatment. These results allow us to set 2 h as optimum time of treatment to achieve the highest enzymatic hydrophobization of beech veneers. The [LG] that yielded the highest WCA was in the range of 5–6 mM. Moreover, [LG] higher than 6 mM reduced the hydrophobicity of the wood samples. It is likely that a [LG] higher than 5 mM favored the grafting of LG oligomers which apparently did not provide high a hydrophobization of the wood veneers. Therefore, the optimum [LG] was set at 5 mM.

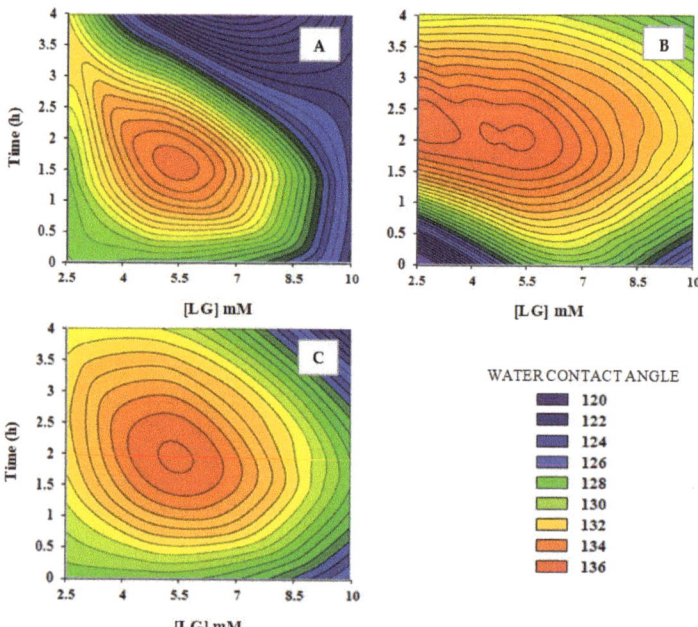

Figure 1. Water contact angle values after one second of drop deposition of the hydrophobized beech veneers as a function of time of reaction and lauryl gallate concentration. Laccase dose: 3.36 U/cm² wood (**A**); 5.00 U/cm² wood (**B**); 6.72 U/cm² wood (**C**).

Regarding the laccase dose, the results observed suggest that the higher the laccase dose the higher the WCA of the enzymatically treated beech veneers. The oxidation of the phenolic moieties of lignin present in the beech veneers enables the stable grafting of LG. Hence, the higher the laccase dose the higher the lignin oxidation and, therefore, a higher grafting of LG could be achieved. A medium laccase dose (5 U/cm² of wood) was set as the optimum value considering the small differences in the WCA in the tested range. Thus, the experimental design permitted to identify the conditions to improve the hydrophobicity of treated beech veneers while minimizing the requirements of energy and both chemical and enzymatic reagents.

The WCA of the beech veneers treated with laccase and LG was compared with those samples treated only with laccase or LG. As noted in Figure 2, veneers treated with both laccase and LG showed a stable WCA and always higher than 125°, after 5 min of water drop deposition. In addition, samples treated only with laccase showed a remarkable hydrophobicity, but much lower than those samples treated with both laccase and LG. Probably, laccase oxidized the hydroxyl groups of lignin's phenolic structures [26], leading to the formation of oxidized functionalities which, apparently, increased the hydrophobicity of lignin. Samples treated with LG alone showed a WCA similar to the observed on the untreated beech veneers. Thus, the activity of laccase was necessary to achieve a stable grafting of LG onto wood veneers surface.

Regarding the different wood species tested (beech, pine, and eucalyptus), enzymatically treated beech veneers showed a WCA slightly higher than pine and a much higher hydrophobicity than eucalyptus veneers (Figure 3). Nonetheless, the highest change in hydrophobicity was achieved in the enzymatically hydrophobized pine veneers, since their WCA, after 5 min of drop deposition, was about 120° whereas untreated pine veneers had already absorbed the water droplet. Enzymatically hydrophobized beech veneers showed a WCA around 70% higher than untreated samples and such gap was much lower (20%) for eucalyptus wood veneers. It is worth mentioning that LG was expected to be enzymatically grafted on lignin moieties and both eucalyptus and beech are hardwood species with a similar lignin content (≈25%). Lignin biopolymer is composed of phenylpropane units (C_6–C_3) which differ one each other in their methoxy substitutions on the aromatic ring, e.g., guaiacyl (G), syringyl (S) and *p*-hydroxyphenyl (H). The S/G ratio in eucalyptus wood is proximately to 6.25 which is much higher than the S/G ratio in beech wood, 0.71 [27,28]. In addition, the lignin of pine has a clear predomination of G units [29,30]. Thus, these results obtained with the WCA measurements suggest that the laccase-assisted grafting of LG was more efficient in wood species in which there is a relative low amount of S units. G units present higher tendency to addition reaction in comparison with S units, due to the lower content of methoxy groups in *ortho*-position. Otherwise, the radicals produced by laccase in S units should be more stable than those produced in G units, since S units have two electro-donating groups (methoxy) in *ortho*-position instead one methoxy group which possess G units [31]. This extra methoxy group of S units seems to have a remarkable effect in the laccase-mediated polymerization of technical lignins [32]. However, our results suggest that the free *ortho*-position in G units apparently favored the laccase-mediated grafting of LG.

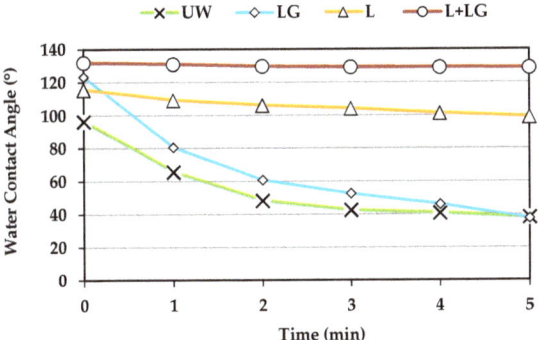

Figure 2. Water contact angle of beech wood veneers. Untreated (UW); treated with lauryl gallate (LG); laccase (L); and laccase and lauryl gallate (L + LG).

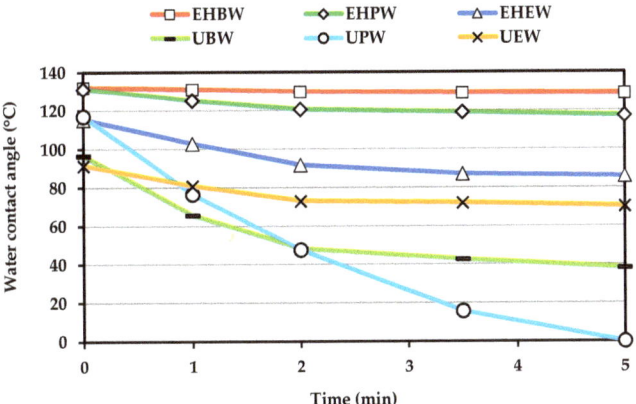

Figure 3. Water contact angle of wood veneers. Enzymatically hydrophobized beech wood (EHBW); pine wood (EHPW); and eucalyptus wood (EHEW). Untreated beech wood (UBW); pine wood (UPW); and eucalyptus wood (UEW).

3.2. XPS Study

X-ray photoelectron spectroscopy (XPS) analysis was performed to obtain a better knowledge of the surface chemistry of the enzymatically hydrophobized beech veneers. Such technique has been previously used to study the surface chemistry of several wood species [33,34], pulp and paper [35,36], cellulose nanocrystals [37,38] and biocomposites [39,40]. Nevertheless, XPS analysis of some lignocellulosic species, e.g., softwood must be carefully conducted due to the migration of lipophilic extractives to the surface of the material which could be induced by the vacuum conditions necessary to perform the analysis [41]. XPS analysis of beech wood samples provided the elemental composition e.g., carbon (C1s), oxygen (O1s) and nitrogen (N1s), but also the functional groups which are present in the surface of the wood veneers. Thus, the results obtained were a powerful tool to study the changes induced by the laccase-assisted grafting of LG onto beech veneers.

The elemental composition results confirmed that beech samples treated enzymatically with LG showed a clear increase of carbon atoms (4.8%) and a decrease of oxygen (5.9%) which were related to the presence of the aliphatic tail of LG onto the veneers surface (Table 1). In addition, wood samples treated with laccase showed a higher nitrogen content than untreated wood samples which suggests that laccase remained partially adsorbed on the veneers surface even after the washing process. Importantly, adsorption of laccase seems to be reduced when LG is present in the reaction [33].

Table 1. Elemental quantification (%) and functional groups relative percent (%) of beech veneers. Untreated beech wood veneers (UW); beech veneers treated with lauryl gallate (LG); treated with laccase (L); treated with laccase and lauryl gallate (L + LG).

Sample	Elements			C1s Components				O1s Components		O/C Ratio	C1/C2 Ratio
	C1s	O1s	N1s	C1%	C2%	C3%	C4%	O1%	O2%		
UW	65.29	30.85	1.29	47.25	38.98	11.30	2.47	16.82	83.19	0.48	1.21
LG	66.60	29.06	1.72	53.79	30.63	11.05	4.54	17.30	82.71	0.44	1.76
L	65.39	26.75	6.42	38.70	42.46	16.60	2.25	30.40	69.61	0.41	0.91
L + LG	70.09	24.95	3.24	53.68	31.99	11.77	2.56	26.68	73.32	0.36	1.68

Four different components were obtained after deconvolution of C1s spectra. C1 was related to C-C or C-H bonds; C2 comprised a carbon bonded to single non-carbonyl oxygen atom (C-O); C3 corresponded to a carbonyl group (C=O) or a carbon bonded to two non-carboxyl oxygen atoms (O-C-O); C4 was assigned to a carbon bonded to a carbonyl group and non-carbonyl oxygen (O-C=O). Beech veneers treated with laccase showed a remarkable increase of C3 component which was related with the oxidation of hydroxyl groups in the phenolic moieties of lignin to form carbonyl groups. The presence of hydrophobic compounds on wood surface could be related with both a low O/C ratio and a high C1/C2 ratio [42]. Samples treated only with laccase showed an important reduction of C-C or C-H bonds and a higher proportion of both ether (C2) and carbonyl (C3) groups which could explain the increase in the hydrophobicity of the laccase-treated wood samples (Figure 2). Samples treated with LG without laccase addition showed a high C1/C2 ratio but also a relative high O/C ratio which means that LG could be adsorbed on the wood surface but not stably bonded. Nevertheless, the LG-treated wood samples did not improve their hydrophobicity (Figure 2) which suggests that the adsorption of LG was not meaningful, or the adsorbed LG was not properly oriented. On the contrary, beech veneers enzymatically treated with LG showed the lowest O/C ratio and a much higher C1/C2 ratio than untreated samples. Therefore, the increase of the C1/C2 ratio (39%) and the decrease of the O/C ratio (25%) suggest that LG was stably bonded onto the wood samples by means of the laccase-mediated treatment.

3.3. FT-IR Analysis

Several works have shown that FT-IR spectra could be a useful technique to assess the grafting of hydrophobic compounds onto the surface of biobased materials after their laccase-mediated hydrophobization [18,19,43]. Generally, the main differences between both hydrophobized and unmodified lignocellulosic materials are due to the vibration of the chemical groups which are present in the aliphatic chain of the grafted compound. Thus, FT-IR spectra analysis of the enzymatically hydrophobized beech veneers showed two small peaks at 2921 and 2854 cm^{-1} which were related with the stretching of methyl (-CH$_3$) and methylene (-CH$_2$-) groups of the 12-carbon aliphatic chain of LG (Figure 4). However, such small peaks were not detected in the spectra of unmodified samples. Therefore, FT-IR spectra analysis evidenced that there was a stable link between LG and beech veneers surface after the laccase-assisted treatment.

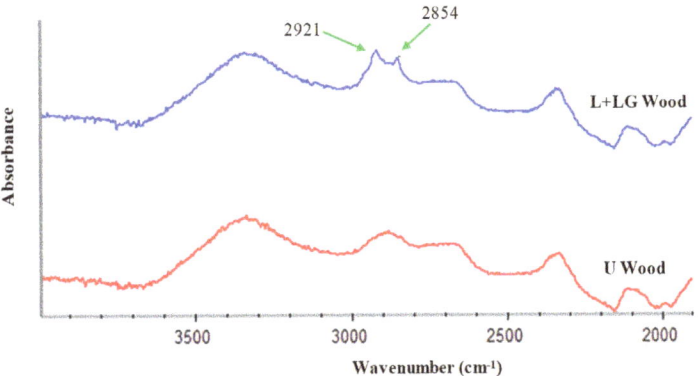

Figure 4. FT-IR spectra between 2000 and 4000 cm^{-1} of untreated beech wood veneers (U Wood) and beech wood treated with laccase and lauryl gallate (L + LG Wood).

3.4. Biofilm Assay

The biofilm assay was performed for the assessment of the potential antifouling properties provided by the laccase-mediated grafting of LG onto the surface of pine veneers surface. Pine wood was the specie used for the antifouling assay due to its high hydrophobicity (Figure 3) but also its economic importance in the Atlantic area of Europe. The conditions for the enzymatic grafting of LG were those found in the factorial design (5 mM LG, 5 U of laccase/cm^2 of wood and 2 h of treatment).

Regarding the antimicrobial properties of LG, [44] found that the antimicrobial activity of alkylphenols was related with the hydrophobicity of the *para*-substituent. In addition, it was proved that LG has antibacterial properties, specifically against Gram-positive bacteria due to inhibition of their membrane respiratory chain [45]. However, the bacterium (*C. marina*) in this study is a Gram-negative bacterium which suggest that the potential antifouling properties of the enzymatically-hydrophobized pine veneers were mostly related with their hydrophobicity. It was showed that the antifouling properties of a substrate are directly linked to its surface energy [46]. It was observed that the lowest surface retention of fouling organisms was achieved when the surface energy of the substrata was between 20–30 mN/m. Since surface energy is inversely proportional to WCA, hydrophobic substrata present low surface energy. In fact, they showed that the atomic groups that performed best antifouling properties were hydrophobic domains such as methyl groups. Thus, it was expected that the long aliphatic chain of LG provided a hydrophobicity high enough to hinder the *C. marina* adhesion onto pine veneers.

At the same time, bacterial surface chemistry is also an important parameter since its hydrophilicity/hydrophobicity character will determine its chemical compatibility and the strength of its initial attachment with the hydrophobized pine veneers. Therefore, the surface chemistry of *C. marina* was studied to assess its hydrophilicity degree. Hence, *C. marina* was suspended in a solution containing both aqueous and organic phases and the migration of the bacteria to the organic phase was measured. According to Karunakaran and Biggs [47], a bacterial migration higher than 50% would mean that bacteria possess a hydrophobic surface. Nevertheless, the partition coefficient was 19.65 ± 2.86% which means that more than 80% of *C. marina* remained in the aqueous phase. Therefore, the surface chemistry of the *C. marina* strain (CECT4278) was mostly hydrophilic.

3.5. Protein Measurement

The potential antifouling properties of the enzymatically-hydrophobized pine veneers were studied by incubating the wood samples in a marine broth inoculated at 3% of *C. marina* (*v/v*) for 1, 3 and 5 days. The amount of bacteria onto the veneers surface was indirectly quantified by measuring the protein content on the surface of the wood samples. Such protein measurement of the colonized wood samples showed that the laccase-assisted grafting of LG reduced substantially the *C. marina* adhesion (Figure 5). After one day, the hydrophobized pine veneers showed a protein content of 44.83% lower than untreated veneers. Such trend was also observed after 3 and 5 days of incubation of the wood veneers in the marine broth inoculated with *C. marina*. Apparently, the hydrophobicity induced by the laccase-assisted grafting of LG onto the surface of pine veneers modified the pattern of *C. marina* adhesion, restricting the bacterial attachment and/or the secretion of EPS.

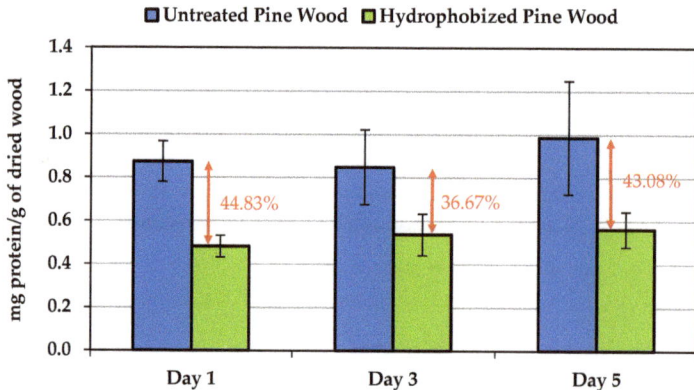

Figure 5. Protein measurement of the hydrophobized pine veneers submerged for 5 days in a marine broth containing a *C. marina* culture.

It is worth noting that LG was grafted onto the aromatic moieties of lignin, which means that the hydroxyl groups of cellulose and hemicelluloses were not significantly modified [22]. Therefore, the surface of the enzymatically treated pine veneers resembles to an amphiphilic surface, since both hydrophilic (hydroxyl groups from cellulose and hemicellulose) and hydrophobic (alkyl groups from LG) domains are present on the veneers surface. Such particular chemical composition could hinder the adhesion of a wide range of microorganisms [48,49], but also reduce the impact of the secreted proteins which has an important role in the microorganisms attachment [50,51]. The results proved the efficiency of the laccase-assisted functionalization to provide antifouling properties to wood.

3.6. SEM Analyses

The surface of the colonized pine veneers was studied by SEM to confirm the results obtained through the protein measurements. It was expected to detect the EPS produced by *C. marina* after their attachment onto veneers surface [52]. Such EPS are a real problem since they may protect bacteria against biocides and antibiotics [53]. In addition, EPS have chelating properties which are used by bacteria to enhance the nutrients availability [54].

Therefore, by hindering the production of EPS it could be easier to attack the bacteria and reduce their biofouling activity.

Untreated pine veneers showed a relatively high density of EPS after one day of in-cubation in the marine broth and, such EPS density was even higher after the fifth day of test (Figure 6A,B). These pictures suggest that C. marina can adhere easily to the surface of untreated pine veneers. On the contrary, SEM pictures of the enzymatically hydrophobized pine veneers showed a much lower density of EPS compared with untreated veneers. Moreover, there were not significant differences in the apparent EPS density of the hydrophobized pine veneers between the first and the fifth day of incubation in the marine broth. These results agreed with the data obtained in the measurements of the amount of protein which was present on the veneers surface. There are two plausible mechanisms that could explain the big differences observed between the SEM pictures of the untreated and the enzymatically hydrophobized pine veneers. One the one hand, the 12-carbons aliphatic tails of the LG that was enzymatically grafted onto the surface of pine veneers could disrupt notably the normal secretion of EPS. On the other hand, the hydrophobization of the pine veneers could delay or directly hinder the surface colonization of *C. marina* and, therefore the EPS could not be secreted yet.

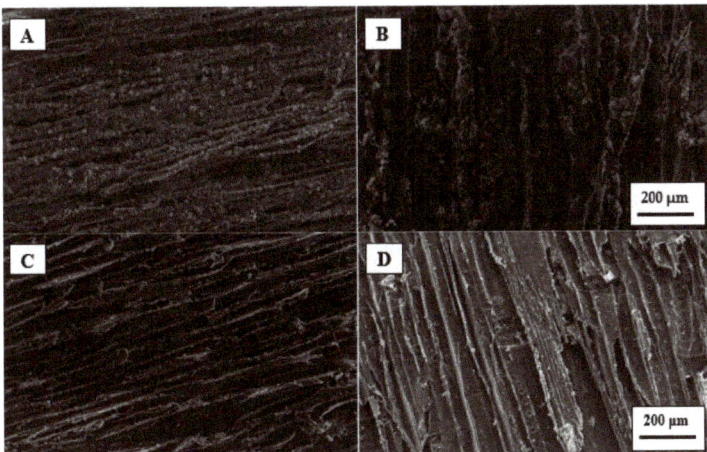

Figure 6. SEM pictures of pine wood veneers colonized by *C. marina*. Untreated pine wood veneers after one day of anti-biofilm assay (**A**); and after five days (**B**). Enzymatically hydrophobized pine wood veneers after one day of anti-biofilm assay (**C**); and after five days (**D**). All the pictures were acquired at 100× magnification.

It is worth noting that both untreated and hydrophobized pine veneers showed an important surface roughness which apparently affected to the bacterial adhesion. In fact, the higher density of EPS was observed on the angled edges of the veneers surface, which means that such angled zones likely favored the attachment of *C. marina* (Figure 7). Nonetheless, hydrophobized pine veneers showed a much lower density of EPS onto such angled edges than untreated veneers which, confirm the effectiveness of the enzymatic grafting of LG. These results confirm that the laccase-assisted grafting of LG modified the normal pattern of *C. marina* adhesion and/or the secretion of EPS. Thus, the marine antifouling properties of the pine veneers were remarkably improved by means of the laccase-assisted grafting of LG onto the pine veneers surface. These results could lead to the development of a new environmentally friendly treatment to protect wood in marine environments.

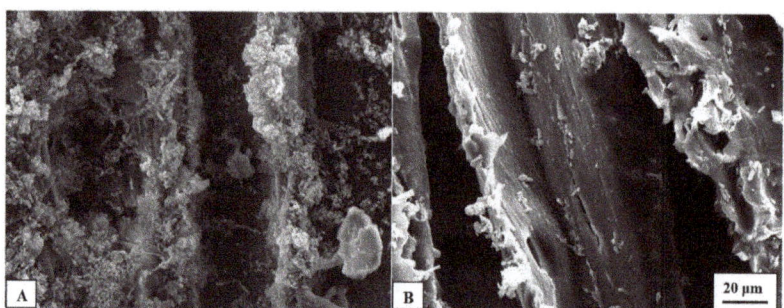

Figure 7. SEM pictures of colonized pine wood veneers after five days of antibiofilm assay. Untreated pine wood veneers (**A**); and enzymatically hydrophobized pine wood veneers (**B**). The pictures were acquired at 1000× magnification.

4. Conclusions

A new environmentally friendly strategy to improve the marine antifouling properties of wood has been proposed in the present study. Such new green strategy is based on the laccase-assisted grafting of LG onto wood veneers. Different wood species (beech, pine, and eucalyptus) were effectively hydrophobized through the enzymatic treatment. It was observed that the reaction conditions played an important role on the extent of hydrophobization, but the treated wood species were also a major factor. Based on these results, pine wood was selected to study the impact of the laccase-mediated hydrophobization on the marine antifouling properties of wood. SEM pictures and protein measurements confirmed that the hydrophobized wood veneers modified the colonization pattern of *Cobetia marina*, revealing that the proposed enzymatic methodology could act as a new marine antifouling treatment for lignocellulosic materials. Future studies should analyze the antifouling properties of the enzymatically treated wood against other marine microorganisms and the "in situ" response of the hydrophobized wood under marine environments. The combination of the proposed treatment with other conventional ones is also of interest for future research.

Supplementary Materials: The following are available online at https://www.mdpi.com/article/10.3390/polym13213795/s1, Table S1. Treatment conditions for the enzymatic grafting of lauryl gallate (LG) on beech veneers tested in the experimental design.

Author Contributions: D.F. conducted the experiments of wood modification and wrote the manuscript. C.B. designed and conducted the biofouling experiments and edited the manuscript. S.G. conducted the experiments of wood modification and edited the manuscript. D.M. supervised the work, designed the experiments, and edited the manuscript. All authors have read and agreed to the published version of the manuscript.

Funding: This research was funded by ERDF and Xunta de Galicia (Grant Numbers 09TMT012E, EM2014/041 and Eq3-GRC2017-I751).

Institutional Review Board Statement: Not applicable.

Informed Consent Statement: Not applicable.

Data Availability Statement: The data presented in this study are available on request from the corresponding author.

Acknowledgments: The support of CACTI from the University of Vigo for XPS analysis and acquisition of microscopy images is appreciated. The authors want to thank to Foresa (Caldas de Reis, Spain) for providing wood samples and to Novozymes (Bagsværd, Denmark) for supplying the enzymes.

Conflicts of Interest: The authors declare no conflict of interest.

References

1. Crossman, M.; Simm, J. Sustainable Coastal Defences—The Use of Timber and Other Materials. *Proc. Inst. Civ. Eng.-Munic. Eng.* **2002**, *151*, 207–211. [CrossRef]
2. Nilsson, T.; Rowell, R. Historical Wood—Structure and Properties. *J. Cult. Herit.* **2012**, *13* (Suppl. 3), 5–9. [CrossRef]
3. Jain, A.; Bhosle, N.B. Biochemical Composition of the Marine Conditioning Film: Implications for Bacterial Adhesion. *Biofouling* **2009**, *25*, 13–19. [CrossRef]
4. Cooksey, K.E.; Wigglesworth-Cooksey, B. Adhesion of Bacteria and Diatoms to Surfaces in the Sea: A Review. *Aquat. Microb. Ecol.* **1995**, *9*, 87–96. [CrossRef]
5. More, T.T.; Yadav, J.S.S.; Yan, S.; Tyagi, R.D.; Surampalli, R.Y. Extracellular Polymeric Substances of Bacteria and Their Potential Environmental Applications. *J. Environ. Manag.* **2014**, *144*, 1–25. [CrossRef]
6. Mieszkin, S.; Callow, M.E.; Callow, J.A. Interactions between Microbial Biofilms and Marine Fouling Algae: A Mini Review. *Biofouling* **2013**, *29*, 1097–1113. [CrossRef]
7. Gittens, J.E.; Smith, T.J.; Suleiman, R.; Akid, R. Current and Emerging Environmentally-Friendly Systems for Fouling Control in the Marine Environment. *Biotechnol. Adv.* **2013**, *31*, 1738–1753. [CrossRef]
8. Callow, M.E.; Callow, J.A. Marine Biofouling: A Sticky Problem. *Biologist* **2002**, *49*, 10–14.
9. Krsmanovic, M.; Biswas, D.; Ali, H.; Kumar, A.; Ghosh, R.; Dickerson, A.K. Hydrodynamics and Surface Properties Influence Biofilm Proliferation. *Adv. Colloid Interface Sci.* **2021**, *288*, 102336. [CrossRef]
10. Callow, M.E.; Callow, J.A. Trends in the Development of Environmentally Friendly Fouling-Resistant Marine Coatings. *Nat. Commun.* **2011**, *2*, 210–244. [CrossRef]
11. Baier, R.E. Surface Behaviour of Biomaterials: The Theta Surface for Biocompatibility. *J. Mater. Sci. Mater. Med.* **2006**, *17*, 1057–1062. [CrossRef]
12. Brady, R.F. Properties Which Influence Marine Fouling Resistance in Polymers Containing Silicon and Fluorine. *Prog. Org. Coat.* **1999**, *35*, 31–35. [CrossRef]
13. Riva, S. Laccases: Blue Enzymes for Green Chemistry. *Trends Biotechnol.* **2006**, *24*, 219–226. [CrossRef]
14. Kudanga, T.; Nyanhongo, G.S.; Guebitz, G.M.; Burton, S. Potential Applications of Laccase-Mediated Coupling and Grafting Reactions: A Review. *Enzym. Microb. Technol.* **2011**, *48*, 195–208. [CrossRef]
15. Kudanga, T.; Prasetyo, E.N.; Sipilä, J.; Guebitz, G.M.; Nyanhongo, G.S. Reactivity of Long Chain Alkylamines to Lignin Moieties: Implications on Hydrophobicity of Lignocellulose Materials. *J. Biotechnol.* **2010**, *149*, 81–87. [CrossRef]
16. Kudanga, T.; Prasetyo, E.N.; Widsten, P.; Kandelbauer, A.; Jury, S.; Heathcote, C.; Sipilä, J.; Weber, H.; Nyanhongo, G.S.; Guebitz, G.M. Laccase Catalyzed Covalent Coupling of Fluorophenols Increases Lignocellulose Surface Hydrophobicity. *Bioresour. Technol.* **2010**, *101*, 2793–2799. [CrossRef]
17. Filgueira, D.; Holmen, S.; Melbø, J.K.; Moldes, D.; Echtermeyer, A.; Chinga-Carrasco, G. Enzymatic-Assisted Modification of Thermomechanical Pulp Fibres To Improve the Interfacial Adhesion with Poly (*Lactic acid*) for 3D Printing. *ACS Sustain. Chem. Eng.* **2017**, *5*, 9338–9346. [CrossRef]
18. Garcia-Ubasart, J.; Vidal, T.; Torres, A.L.; Rojas, O.J. Laccase-Mediated Coupling of Nonpolar Chains for the Hydrophobization of Lignocellulose. *Biomacromolecules* **2013**, *14*, 1637–1644. [CrossRef]
19. Hossain, K.M.G.; González, M.D.; Lozano, G.R.; Tzanov, T. Multifunctional Modification of Wool Using an Enzymatic Process in Aqueous-Organic Media. *J. Biotechnol.* **2009**, *141*, 58–63. [CrossRef]
20. Warne Zoueki, C.; Ghoshal, S.; Tufenkji, N. Bacterial Adhesion to Hydrocarbons: Role of Asphaltenes and Resins. *Colloids Surf. B Biointerfaces* **2010**, *79*, 219–226. [CrossRef]
21. Rosenberg, M.; Gutnick, D.; Rosenberg, E. Adherence of Bacteria to Hydrocarbons: A Simple Method for Measuring Cell-Surface Hydrophobicity. *FEMS Microbiol. Lett.* **1980**, *9*, 29–33. [CrossRef]
22. Saastamoinen, P.; Mattinen, M.L.; Hippi, U.; Nousiainen, P.; Sipilä, J.; Lille, M.; Suurnäkki, A.; Pere, J. Laccase Aided Modification of Nanofibrillated Cellulose with Dodecyl Gallate. *BioResources* **2012**, *7*, 5749–5770. [CrossRef]
23. Itoh, N.; Katsube, Y.; Yamamoto, K.; Nakajima, N.; Yoshida, K. Laccase-Catalyzed Conversion of Green Tea Catechins in the Presence of Gallic Acid to Epitheaflagallin and Epitheaflagallin 3-O-Gallate. *Tetrahedron* **2007**, *63*, 9488–9492. [CrossRef]
24. Areskogh, D.; Li, J.; Nousiainen, P.; Gellerstedt, G.; Sipilä, J.; Henriksson, G. Oxidative Polymerisation of Models for Phenolic Lignin End-Groups by Laccase. *Holzforschung* **2010**, *64*, 21–34. [CrossRef]
25. Filgueira, D.; Moldes, D.; Fuentealba, C.; García, D.E. Condensed Tannins from Pine Bark: A Novel Wood Surface Modifier Assisted by Laccase. *Ind. Crops Prod.* **2017**, *103*, 185–194. [CrossRef]
26. Grönqvist, S.; Viikari, L.; Niku-Paavola, M.L.; Orlandi, M.; Canevali, C.; Buchert, J. Oxidation of Milled Wood Lignin with Laccase, Tyrosinase and Horseradish Peroxidase. *Appl. Microbiol. Biotechnol.* **2005**, *67*, 489–494. [CrossRef] [PubMed]
27. Pinto, P.C.; Evtuguin, D.V.; Neto, C.P. Effect of Structural Features of Wood Biopolymers on Hardwood Pulping and Bleaching Performance. *Ind. Eng. Chem. Res.* **2005**, *44*, 9777–9784. [CrossRef]
28. Choi, J.W.; Faix, O.; Meier, D. Characterization of Residual Lignins from Chemical Pulps of Spruce (*Picea abies* L.) and Beech (*Fagus sylvatica* L.) by Analytical Pyrolysis-Gas Chromatography/Mass Spectrometry. *Holzforschung* **2001**, *55*, 185–192. [CrossRef]
29. Saito, K.; Kato, T.; Tsuji, Y.; Fukushima, K. Identifying the Characteristic Secondary Ions of Lignin Polymer Using ToF-SIMS. *Biomacromolecules* **2005**, *6*, 678–683. [CrossRef]

30. Alves, A.; Schwanninger, M.; Pereira, H.; Rodrigues, J. Calibration of NIR to Assess Lignin Composition (H/G Ratio) in Maritime Pine Wood Using Analytical Pyrolysis as the Reference Method. *Holzforschung* **2006**, *60*, 29–31. [CrossRef]
31. Cañas, A.I.; Camarero, S. Laccases and Their Natural Mediators: Biotechnological Tools for Sustainable Eco-Friendly Processes. *Biotechnol. Adv.* **2010**, *28*, 694–705. [CrossRef] [PubMed]
32. Gouveia, S.; Fernández-Costas, C.; Sanromán, M.A.; Moldes, D. Polymerisation of Kraft Lignin from Black Liquors by Laccase from Myceliophthora Thermophila: Effect of Operational Conditions and Black Liquor Origin. *Bioresour. Technol.* **2013**, *131*, 288–294. [CrossRef]
33. Fernández-Fernández, M.; Sanromán, M.A.; Moldes, D. Potential of Laccase for Modification of Eucalyptus Globulus Wood: A XPS Study. *Wood Sci. Technol.* **2014**, *48*, 151–160. [CrossRef]
34. Vázquez, G.; Ríos, R.; Freire, M.S.; Antorrena, G.; González-Álvarez, J. Surface Characterization of Eucalyptus and Ash Wood Veneers by XPS, TOF-SIMS, Optic Profilometry and Contact Angle Measurements. *WIT Trans. Eng. Sci.* **2011**, *72*, 187–198. [CrossRef]
35. Jiang, L.; Tang, Z.; Clinton, R.M.; Hess, D.W.; Breedveld, V. Fabrication of Highly Amphiphobic Paper Using Pulp Debonder. *Cellulose* **2016**, *23*, 3885–3899. [CrossRef]
36. George, M.; Mussone, P.G.; Bressler, D.C. Surface and Bulk Transformation of Thermomechanical Pulp Using Fatty Acyl Chlorides: Influence of Reaction Parameters on Surface, Morphological, and Thermal Properties. *J. Wood Chem. Technol.* **2016**, *36*, 114–128. [CrossRef]
37. Vanhatalo, K.; Maximova, N.; Perander, A.; Johansson, L.; Haimi, E.; Dahl, O. Comparison of Conventional and Lignin-Rich. *BioResources* **2016**, *11*, 4037–4054. [CrossRef]
38. Javanbakht, T.; Raphael, W.; Tavares, J.R. Physicochemical Properties of Cellulose Nanocrystals Treated by Photo-Initiated Chemical Vapour Deposition (PICVD). *Can. J. Chem. Eng.* **2016**, *94*, 1135–1139. [CrossRef]
39. Granda, L.A.; Espinach, F.X.; Tarrés, Q.; Méndez, J.A.; Delgado-Aguilar, M.; Mutjé, P. Towards a Good Interphase between Bleached Kraft Softwood Fibers and Poly(Lactic) Acid. *Compos. Part B Eng.* **2016**, *99*, 514–520. [CrossRef]
40. González, D.; Santos, V.; Parajó, J.C. Silane-Treated Lignocellulosic Fibers as Reinforcement Material in Polylactic Acid Biocomposites. *J. Thermoplast. Compos. Mater.* **2012**, *25*, 1005–1022. [CrossRef]
41. Nguila Inari, G.; Pétrissans, M.; Dumarcay, S.; Lambert, J.; Ehrhardt, J.J.; Šernek, M.; Gérardin, P. Limitation of XPS for Analysis of Wood Species Containing High Amounts of Lipophilic Extractives. *Wood Sci. Technol.* **2011**, *45*, 369–382. [CrossRef]
42. Sernek, M.; Kamke, F.A.; Glasser, W.G. Comparative Analysis of Inactivated Wood Surfaces Comparative Analysis of Inactivated Wood Surfaces. *Holzforschung* **2004**, *58*, 22–31. [CrossRef]
43. Fernández-Fernández, M.; Sanromán, M.Á.; Moldes, D. Wood Hydrophobization by Laccase-Assisted Grafting of Lauryl Gallate. *J. Wood Chem. Technol.* **2015**, *35*, 156–165. [CrossRef]
44. Etoh, H.; Ban, N.; Fujiyoshi, J.; Murayama, N.; Sugiyama, K.; Watanabe, N.; Sakata, K.; Ina, K.; Miyoshi, H.; Iwamura, H. Quantitative Analysis of the Antimicrobial Activity and Membrane-Perturbation Potency of Antifouling Para-Substituted Alkylphenols. *Biosci. Biotechnol. Biochem.* **1994**, *58*, 467–469. [CrossRef]
45. Kubo, I.; Xiao, P.; Fujita, K. Anti-MRSA Activity of Alkyl Gallates. *Bioorg. Med. Chem. Lett.* **2002**, *12*, 113–116. [CrossRef]
46. Baier, R.E.; Shafrin, E.G.; Zisman, W.A. Adhesion: Mechanisms That Assist or Impede It. *Science* **1968**, *162*, 1360–1368. [CrossRef]
47. Karunakaran, E.; Biggs, C.A. Mechanisms of Bacillus Cereus Biofilm Formation: An Investigation of the Physicochemical Characteristics of Cell Surfaces and Extracellular Proteins. *Appl. Microbiol. Biotechnol.* **2011**, *89*, 1161–1175. [CrossRef]
48. Li, J.; Xie, Z.; Wang, G.; Ding, C.; Jiang, H.; Wang, P. Preparation and Evaluation of Amphiphilic Polymer as Fouling-Release Coating in Marine Environment. *J. Coat. Technol. Res.* **2017**, *14*, 1237–1245. [CrossRef]
49. Van Zoelen, W.; Buss, H.G.; Ellebracht, N.C.; Lynd, N.A.; Fischer, D.A.; Finlay, J.; Hill, S.; Callow, M.E.; Callow, J.; Kramer, E.J.; et al. Sequence of Hydrophobic and Hydrophilic Residues in Amphiphilic Polymer Coatings A Ff Ects Surface Structure and Marine Antifouling/Fouling Release Properties. *ACS Macro Lett.* **2014**, *3*, 364–368. [CrossRef]
50. Bauer, S.; Arpa-Sancet, M.P.; Finlay, J.A.; Callow, M.E.; Callow, J.A.; Rosenhahn, A. Adhesion of Marine Fouling Organisms on Hydrophilic and Amphiphilic Polysaccharides. *Langmuir* **2013**, *29*, 4039–4047. [CrossRef]
51. Dundua, A.; Franzka, S.; Ulbricht, M. Improved Antifouling Properties of Polydimethylsiloxane Films via Formation of Polysiloxane/Polyzwitterion Interpenetrating Networks. *Macromol. Rapid Commun.* **2016**, *37*, 2030–2036. [CrossRef] [PubMed]
52. Lelchat, F.; Cérantola, S.; Brandily, C.; Colliec-Jouault, S.; Baudoux, A.C.; Ojima, T.; Boisset, C. The Marine Bacteria Cobetia marina DSMZ 4741 Synthesizes an Unexpected K-Antigen-like Exopolysaccharide. *Carbohydr. Polym.* **2015**, *124*, 347–356. [CrossRef] [PubMed]
53. Stewart, P.S.; Costerton, J.W. Antibiotic Resistance of Bacteria in Biofilms. *Lancet* **2001**, *358*, 135–138. [CrossRef]
54. Ordax, M.; Marco-Noales, E.; López, M.M.; Biosca, E.G. Exopolysaccharides Favor the Survival of *Erwinia amylovora* under Copper Stress through Different Strategies. *Res. Microbiol.* **2010**, *161*, 549–555. [CrossRef]

Article

Effect of UV Radiation on Optical Properties and Hardness of Transparent Wood

Igor Wachter *, Tomáš Štefko, Peter Rantuch, Jozef Martinka and Alica Pastierová

Department of Integrated Safety, Faculty of Materials Science and Technology in Trnava, Slovak University of Technology in Bratislava, Botanická 49, 917 24 Trnava, Slovakia; tomas.stefko@stuba.sk (T.Š.); peter.rantuch@stuba.sk (P.R.); jozef.martinka@stuba.sk (J.M.); alica.pastierova@stuba.sk (A.P.)
* Correspondence: igor.wachter@stuba.sk; Tel.: +421-904-398-793

Abstract: Optically transparent wood is a type of composite material, combining wood as a renewable resource with the optical and mechanical properties of synthetic polymers. During this study, the effect of monochromatic UV-C (λ—250 nm) radiation on transparent wood was evaluated. Samples of basswood were treated using a lignin modification method, to preserve most of the lignin, and subsequently impregnated with refractive-index-matched types of acrylic polymers (methyl methacrylate, 2-hydroxyethyl methacrylate). Optical (transmittance, colour) and mechanical (shore D hardness) properties were measured to describe the degradation process over 35 days. The transmittance of the samples was significantly decreased during the first seven days (12% EMA, 15% MMA). The average lightness of both materials decreased by 10% (EMA) and 17% (MMA), and the colour shifted towards a red and yellow area of CIE $L^*a^*b^*$ space coordinates. The influence of UV-C radiation on the hardness of the samples was statistically insignificant (W+MMA 84.98 ± 2.05; W+EMA 84.89 ± 2.46), therefore the hardness mainly depends on the hardness of used acrylic polymer. The obtained results can be used to assess the effect of disinfection of transparent wood surfaces with UV-C radiation (e.g., due to inactivation of SARS-CoV-2 virus) on the change of its aesthetic and mechanical properties.

Keywords: transparent wood; UV-C radiation; optical properties; basswood; hardness; chromophores deactivation

Citation: Wachter, I.; Štefko, T.; Rantuch, P.; Martinka, J.; Pastierová, A. Effect of UV Radiation on Optical Properties and Hardness of Transparent Wood. *Polymers* **2021**, *13*, 2067. https://doi.org/10.3390/polym13132067

Academic Editor: Roman Réh

Received: 31 May 2021
Accepted: 21 June 2021
Published: 23 June 2021

Publisher's Note: MDPI stays neutral with regard to jurisdictional claims in published maps and institutional affiliations.

Copyright: © 2021 by the authors. Licensee MDPI, Basel, Switzerland. This article is an open access article distributed under the terms and conditions of the Creative Commons Attribution (CC BY) license (https://creativecommons.org/licenses/by/4.0/).

1. Introduction

Wood, as a renewable and earth-abundant resource, is a well-established material in many applications due to its good physical and chemical properties, including high strength, low thermal conductivity, non-toxicity, and biodegradability [1,2]. The future of sustainable development depends on how humans will transfer their dependability from finite fossil-based materials to sustainable and renewable materials to combat the climate change. Recently, there is an increasing number of articles dealing with eco-friendly composites [3,4].

Wood composite materials are engineered and produced with tailored physical and mechanical properties appropriate for a wide variety of applications, both known and not discovered yet [5].

Transparent wood is a composite material consisting of a modified wood component (deactivated chromophores or delignification) and an in situ polymerized, transparent component. Transparent wood has received much attention, owing to its great potential for applications in light-transmitting buildings, which can partially replace artificial light with sunlight and therefore save energy [6,7]. Transparent wood can be used to produce building [8,9], solar cells [10] and magnetic materials [11]. Additional functionalization has been demonstrated, such as lasing [12], heat shielding [13], thermal energy storage [14], electroluminescent devices [15], and combined with conducting polymers in electromechanical devices [16].

To fabricate transparent wood, two steps are typically involved: completely removing the light-absorbing lignin from the cell walls of natural wood by a solution-based immersion method and infiltrating a refractive index matching polymer into the delignified wood matrix to minimize light absorption and scattering, respectively [9,17–20].

Alternatively, many studies have focused on physically and/or chemically modifying lignin structures to reduce lignin colour and impart new functionalities, including fractionation [21], grind [22], acetylated lignin [23,24], fragmented lignin [25,26] and metal-decorated lignin [27,28].

Although considered critical in previous publications, delignification processes are time consuming and not necessarily environmentally friendly because of the production of odorous components and chlorinated compounds. Moreover, the removed lignin significantly weakens the wood structure so it can be challenging to work with such a fragile material, and this also lowers the number of suitable wood species for transparent wood preparation; pine and spruce, for example, breaks into pieces after the delignification step. The lignin modification method is superior to the delignification process in the following four aspects [29]:

(1) The lignin modification method is completed in a short time.
(2) Lignin is largely retained, and the wood structure is therefore better preserved.
(3) Lignin-modified wood templates show better mechanical properties.
(4) The lignin modification method is a green process since toxic effluents are minimized.

Also, it is important to note that wood consists of around 30 wt% of lignin which provides structural support and therefore the transparent wood fabricated by this process could be considered more environmentally friendly because less synthetic polymer is needed for its fabrication.

There is an increasing number of studies with various applications for the use of transparent wood where it needs to withstand outdoor weather conditions from which UV radiation may cause its degradation. Such applications include perovskite solar cells assembled directly on transparent wood substrates [30], anisotropic transparent paper with high efficiency as a light management coating layer for GaAs solar cell [10], smart photo-responsive windows with energy storage capabilities [31], radiative cooling structural materials [32], smart and energy-saving buildings applications [9,33], and structural elements in architectural construction. [34,35] In addition, due to the worldwide pandemic caused by COVID-19, the use of UV-C radiation for sanitation of surfaces and internal spaces has risen dramatically.

Based on the arguments described above, it is necessary to understand the influence of UV radiation on this type of material that has huge potential applications in the future. According to a review made by [36], there are questions which should be addressed in the future studies to allow industrialization of the technology, such as optical and mechanical stability and the desirability of increased cellulose content. The increased cellulose content was addressed by [29].

A study, conducted by [37], evaluated colour, chemical and optical (transmittance) changes of a transparent wood composite made from poplar wood and epoxy resin with a UV absorber when exposed to UV-A (340 nm) light. To the best of our knowledge, to this date, it is the only study dealing with this issue. Therefore, the aim of this study is to further examine the effect of UV radiation (UV-C, 250 nm) on optical (transmittance and colorimetry) and mechanical stability (hardness) of lignin retaining transparent wood obtained by lignin chromophores deactivation.

2. Materials and Methods

2.1. Materials and Chemicals

Radially cut basswood (*Tilia*) was purchased from JAF Holz Slovakia s. r. o. (density: 0.53–0.56 g cm^{-3}). Sodium silicate, sodium hydroxide, magnesium sulfate, DTPA and

H$_2$O$_2$ (35%), propanol, acetone were purchased from CentralChem s.r.o. Deionised water was prepared directly in the laboratory. MMA and 2,2'-azobis(2-methylpropionitrile) were purchased from Sigma-Aldrich and 2-hydroxyethyl-methacylate with activator were purchased from Epoxy s.r.o.

2.2. Lignin Modification

The lignin modification procedure was originally proposed by [29]. Basswood samples with dimensions of 100 × 50 × 1.2 mm (±0.1 mm) were submerged into a lignin modifying solution at 70 °C until the wood became white. The solution was prepared by mixing chemicals in the following order: deionized water, sodium silicate (3.0 wt%), sodium hydroxide solution (3.0 wt%), magnesium sulphate (0.1 wt%), DTPA (0.1 wt%), and then H$_2$O$_2$ (4.0 wt%). Gradually, H$_2$O$_2$ (35% vol) was added to the solution until the samples became completely white. The samples were then washed with hot deionized water to remove traces of residual chemicals. Finally, the samples were dehydrated with propanol and acetone, subsequently, and stored until polymer infiltration.

2.3. Transparent Wood Preparation

Before polymer infiltration, wood samples were dehydrated with ethanol and acetone sequentially. Each solvent-exchange step was repeated three times. MMA was pre-polymerized before infiltration to remove the dissolved oxygen. Pre-polymerization was carried out at 75 °C for 15 min with 0.3 wt% 2,2'-azobis(2-methylpropionitrile) as initiator and the solution was then cooled to room temperature. Subsequently, the bleached wood template was fully vacuum-infiltrated in a pre-polymerized PMMA solution. Finally, the infiltrated wood was sandwiched between two glass slides, packaged in aluminium foil, and the polymerization was performed in an oven at 75 °C for 4 h. Infiltration of 2-hydroxyethyl-methacylate was carried out without pre-polymerization. Activator (0.2 wt%) was mixed with 2-hydroxyethyl-methacrylate and it was allowed to dissolve for 1 h. After full vacuum-infiltration the samples were sandwiched between two glass slides, packaged in aluminium foil, and the polymerization was performed in an oven at 90 °C for 4 h. In total 20 pieces of samples were prepared (10 for each methacrylate). After fabrication of the samples, the resulting product can be considered to be around 40% renewable.

2.4. Characterization

According to the study by Li et al. [29], the transparent wood samples retained up to 80 wt% of lignin leading to a stronger wood template compared to the de-lignified alternative. In this study, the weight loss of the wood component due to the modification of lignin was 21.9 ± 0.9%. After polymer infiltration, a high-lignin content transparent wood with a transmittance of 83%, a haze of 75%, a thermal conductivity of 0.23 Wm K^{-1}, and work-to fracture of 1.2 MJ m^{-3} (a magnitude higher than glass) was obtained (MMA samples). Samples prepared for this study did not reach the values of previously mentioned research because of the use of different (more dense) wood. Figure 1 shows a boxplot of wood and acrylate polymer weight in the samples. The average proportion of wood component in the W+MMA and W+EMA samples was 27% and 29%, respectively.

The colorimetry of the samples was performed using by a Colorimeter NR200 Precision (Threenh Technology Co., Ltd.; Shenzhen, China) with the following characterizations: Measuring aperture Φ8 mm, Colour space CIE $L^*a^*b^*$ and Light Source D65. The colour change caused by UV radiation was monitored using the CIE $L^*a^*b^*$ colour space coordinates. In this way, the colour of the measured surface is expressed using three coordinates:

- L^*—coordinate on the axis indicating lightness
- a^*—coordinate on the axis between red and green
- b^*—coordinate on the axis between yellow and blue

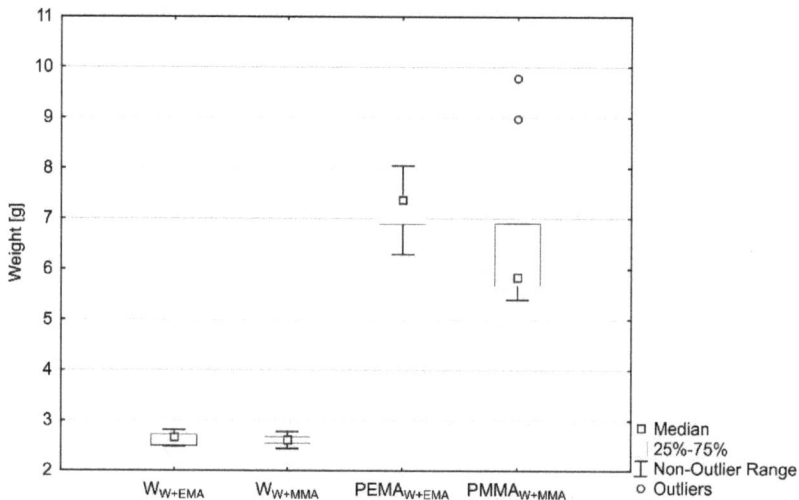

Figure 1. Boxplot of wood and acrylate polymer weight in the samples.

To describe the total shift in this colour space, the total colour difference is used, which can be expressed as follows:

$$dE_t = \sqrt{(L_t^* - L_0^*)^2 + (a_t^* - a_0^*)^2 + (b_t^* - b_0^*)^2} \qquad (1)$$

where dE_t is the total colour difference at time t, L_t^* is the value of L^* at time t, L_0^* is the value of L^* before exposure to UV radiation, a_t^* is the value of a^* at time t, a_0^* is the value a^* before exposure to UV radiation, b_t^* is the value of b^* at time t, b_0^* is the value of b^* before exposure to UV radiation.

The transmittance was measured using a modified photometer RMG2.1 (Heil Metalle GmbH, Mülheim, a. d. Ruhr, Germany). The measuring area was 20 mm × 20 mm.

For the measurement of the hardness Digital Shore D Hardness Tester—Sauter HD (Sauter GmbH, Balingen, Germany) was used.

The UV ageing (an accelerated weathering test) has been carried out in a UV chamber. The samples were irradiated for 35 days. All measurements have been done after 7 days of UV exposition. The ageing has been done under a temperature of 50 °C. As a source of UV-C radiation, 4 germicidal fluorescent lamps Philips TUV 15 W (Piła, Poland) were used. The efficiency of the fluorescent lamp was 32%. UV-C radiation (wavelength 250 nm) reached a power output of 4.9 W and the volume of the chamber was 50 L. The samples were placed 100 mm from the UV lamps in every direction. The irradiance flux density was 16.07 W m^{-2} and the inner surface of the chamber was made of stainless steel with a 50% reflectance factor.

For FTIR analysis, Varian FT-IR Spectrometer 660 (Agilent Technologies, Inc., Santa Clara, CA, USA) samples were directly applied to a diamond crystal of ATR GladiATR (PIKE Technology Inc., Madison, WI, USA) and the resulting spectra were corrected for background air absorbance. The spectra were recorded using a Varian Resolutions Pro and samples were measured in the region 4000–400 cm^{-1}; each spectrum was measured 146 times, at resolution 4.

All of the measurements were carried out using Stat Soft STATISTICA 10 (StatSoft s.r.o., Praha, Czechia) software. The impact of the exposure time of UV radiation on the total colour difference, transmittance and hardness were evaluated by the Duncan's test.

3. Results

3.1. Colorimetry

In terms of colour change due to UV radiation, the most significant changes occurred during the first 7 days. This fact is clearly visible in Figure 2. A significant change was observed in all three coordinates, which was subsequently reflected in the value of the overall colour difference.

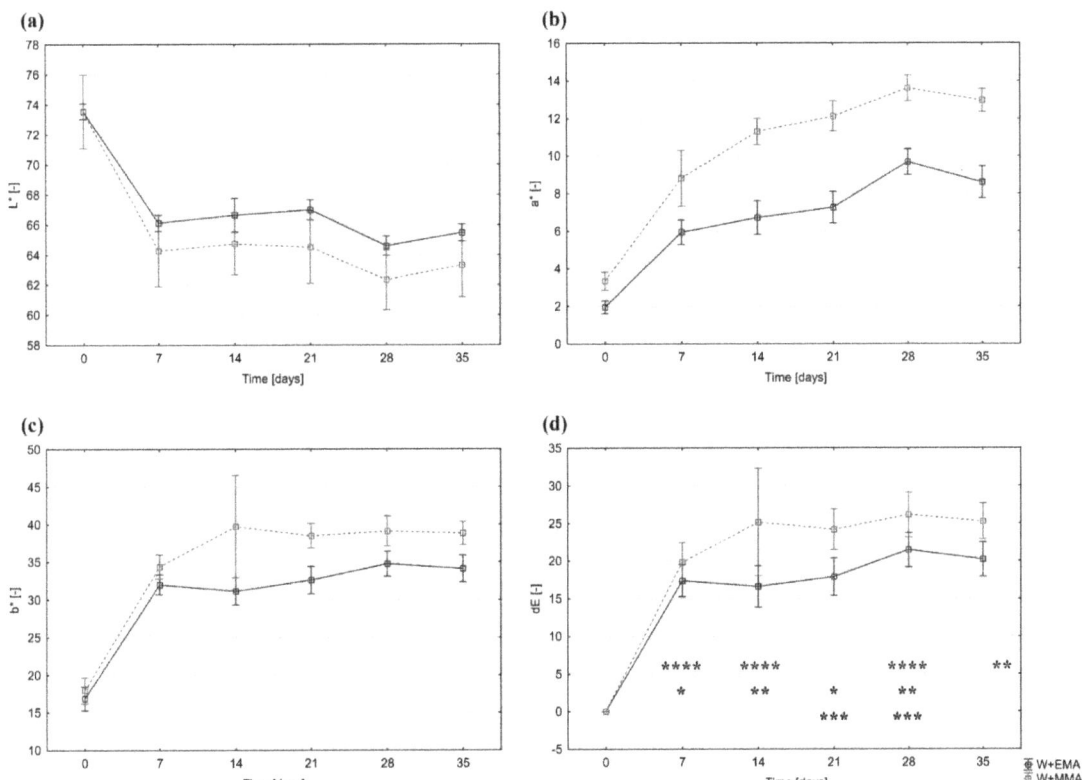

Figure 2. Time dependence of colour components in CIE $L^*a^*b^*$ space and total colour difference: (**a**) Lightness; (**b**) the coordinate a^*; (**c**) b^* coordinate; (**d**) total colour difference *, **, *** denoted exposure times of W+EMA and **** denoted exposure times of W+MMA at which the total colour differences were statistically significant based on the results of Duncan's test (difference between 0 days of exposure and other exposure times is obvious without statistical test).

The average lightness of both measured materials was approximately 73.5. However, after the first 7 days of UV exposure, it decreased to values of around 66 for W+EMA samples and to an average of 64 for W+MMA samples, which represents a reduction of 10% and 17%, respectively.

The a^* coordinate also changed most rapidly at the onset of UV exposure. In 7 days, its average value increased from 2 to almost 6 (W+EMA) and from 3.4 to 9 (W+MMA). A less pronounced increase subsequently continued until day 28 of the test. Subsequently, there was a very slight decrease. As with L^*, W+EMA samples proved to be less susceptible to changes due to UV radiation.

In the case of the b^* values, it is possible to see a similar course as in the case of a^*, but after the initial significant increase it changes only slightly over a period of more than 7 days. Although the b^* of both materials is very similar in the samples before exposure to UV radiation, the samples of W+MMA acquire higher values than W+EMA due to its influence.

The results of the overall colour difference reflect the changes described in the individual coordinates. A significant colour change occurs mainly during the first 7 days, followed by only a slight increase.

As already mentioned, the graphs corresponding to the values of a^* and b^* have a similar course. The following equations can be determined from the graphical representation of the measured values (Figure 3):

$$b^*_{W+EMA} = 9.96 + 26.24 \times \log a^*_{W+EMA} \qquad (2)$$

$$b^*_{W+MMA} = 1.83 + 33.91 \times \log a^*_{W+MMA} \qquad (3)$$

where b^*_{W+EMA} is the coordinate b^* for the sample W+EMA, a^*_{W+EMA} is the coordinate a^* for the sample W+EMA, b^*_{W+MMA} is the coordinate b^* for the sample W+MMA and a^*_{W+MMA} is the coordinate a^* for the sample W+MMA. The coefficients of determination in these cases are 0.9376 (W+EMA) and 0.8191 (W+MMA).

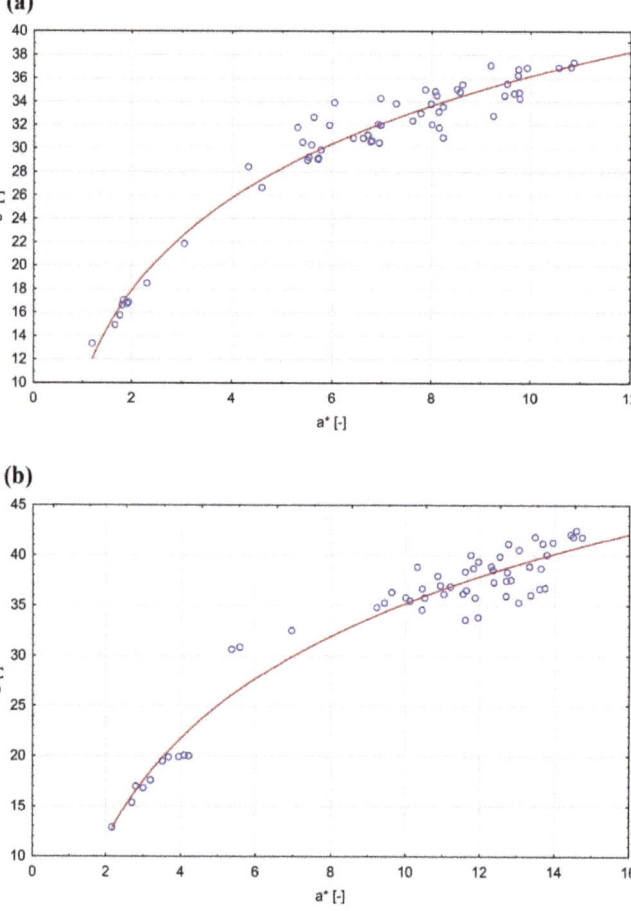

Figure 3. Interdependence of colour space coordinates a^* and b^*: (**a**) W+MMA; (**b**) W+EMA.

The comparison of colour changes in transparent wood infiltrated by 2-hydroxyethyl-methacrylate (a,b) and methyl methacrylate (c,d) before and after 840 h of UV-C irradiation is shown in Figure 4a,b, and in Figure 4c,d. The significant colour darkening (photo yellowing) was observed within the first few hours of exposure, which increases with further exposure.

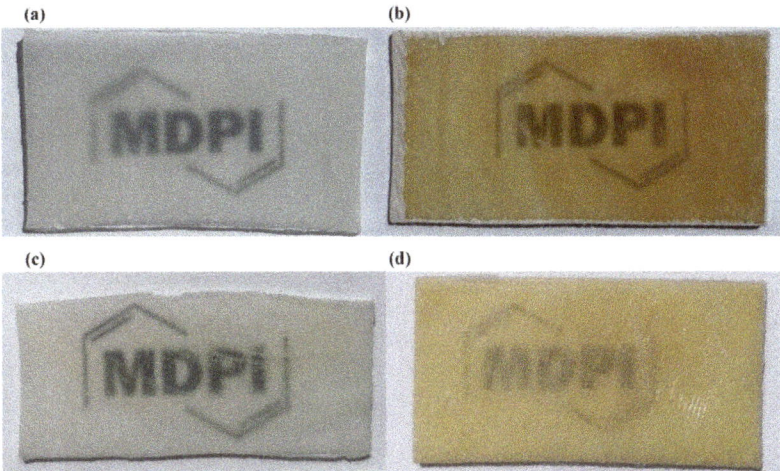

Figure 4. Change of transmittance and colour: W+EMA (**a**) before UV-C irradiation; (**b**) after 35 days of UV-C irradiation; W+MMA (**c**) before UV-C irradiation; (**d**) after 35 days of UV-C irradiation (Source of logo: https://www.mdpi.com/journal/polymers, accessed on 28 May 2021).

3.2. Transmittance

Depending on the acrylic polymer used, the transmittance values differ significantly, even for samples not exposed to UV-C radiation (Figure 5). While W+EMA transmits almost 69% of light, W+MMA is about 58%. After 7 days, these values decrease to 57% resp. 43% and consequently their change is almost negligible. Throughout the experiment, the W+EMA samples remained significantly more transparent, with the difference between them highlighted by the action of UV-C radiation, as is shown in Figure 5.

3.3. FTIR Analysis

Changes to the chemical structure (bond scission/forming) of W+EMA and W+MMA sample after UV-C irradiation are displayed in Figure 6 as infrared spectrum. There are two characteristic bands attributed to the stretching C–O and CH_3–O of methyl ester, peak at wavenumber 2916 cm^{-1} and 2848 cm^{-1} the –CH stretching aliphatic band of the ethylene segment. It is seen, a very strong peak is visible at 1720 cm^{-1} due to carbonyl (–C=O) stretching vibration of the acrylate ester group, in both samples. Two peaks at 1435 and 1381 cm^{-1} can be attributed to CH_3 symmetric and asymmetric deformation. At wavenumber 958 cm^{-1} can be seen C–O–C stretching vibration and at 746 cm^{-1} is band characteristic for C–H stretching [38–41]. However, after irradiation by UV-C, major changes were observed evidencing chemical changes in the polymer samples. The bands that undergo prominent changes are the functionalities of hydroxyl O–H, carbonyl C=O and ester (C–O–C) in region of wavenumber from 746 to 1435 cm^{-1}. Other photo products, e.g., carbonyl groups or double bonds may be weakened from the surfaces, leading to reduced absorption [42].

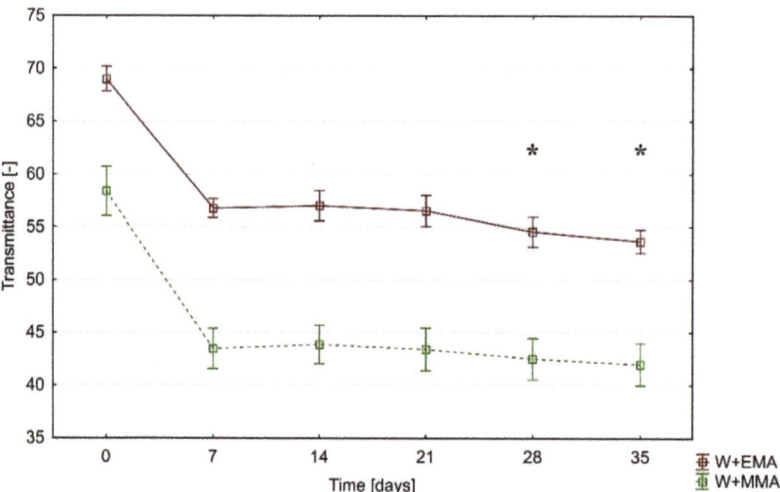

Figure 5. Change of transmittance during the experiment * denoted exposure time of W+EMA at which the total colour differences was statistically significant (in interval from 7 to 35 days) based on results of Duncan's test (difference between 0 days of exposure and other exposure times is obvious without statistical test).

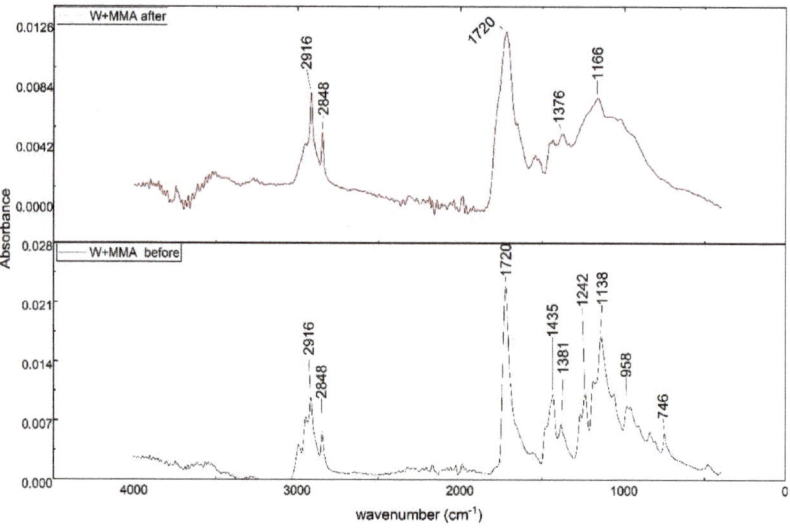

Figure 6. *Cont.*

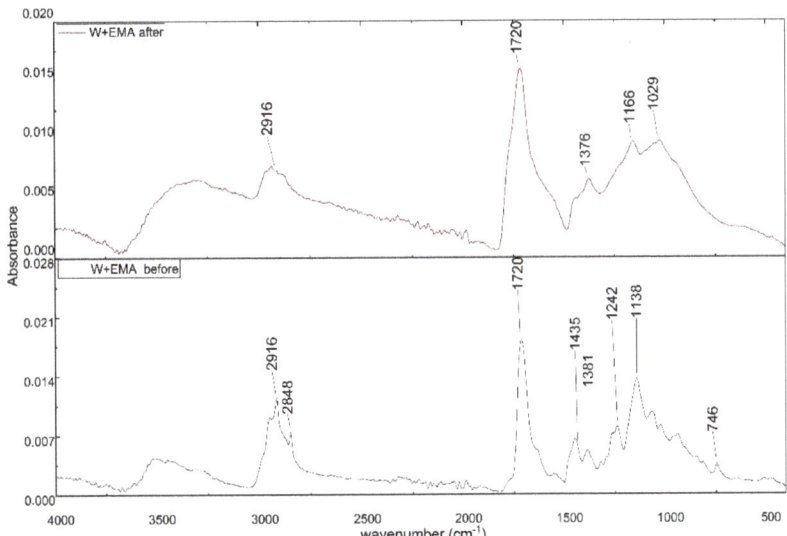

Figure 6. Changes to the chemical structure of W+EMA and W+MMA samples before and after the UV-C irradiation.

3.4. Hardness

The hardness of both types of materials was at a very similar level. Its values were initially around 86. In contrast to the optical properties, the hardness of the measured samples is significantly less affected by the UV-C radiation to which the transparent wood samples were exposed. Figure 7 shows its course as a function of the time of UV-C treatment. The hardness of both types of samples was practically the same during the experiment and no differences are apparent between them.

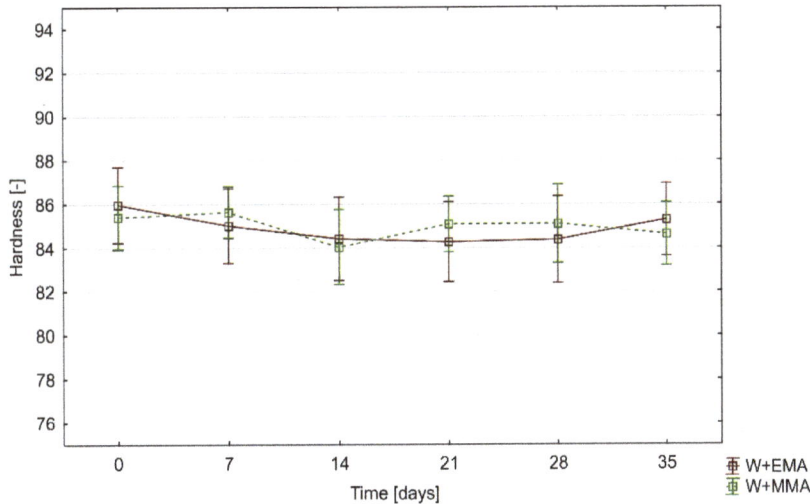

Figure 7. Change of hardness during the experiment (the Duncan's test proved no statistically significant differences between hardness at all investigated times for both investigated samples).

4. Discussion

Shore A hardness of PEMA is according to Hourston, Satgurunathan and Varma 78 [43]. A value higher than 95 can be deducted from the graph in the work of Hourston and Schäfer [44]. The W+EMA samples had a hardness of 86 ± 2.3 and were therefore among the data of the mentioned authors.

The hardness of PMMA in the Shore D scale can be found in a higher number of works than in the case of PEMA. According to Poomali, Suresha and Lee, its value is 90 ± 1 [45]. Seeger et al. state a value of 87.5 ± 0.4 [46] and Akinci, Sen and Sen 79 [47]. The data measured in this study for W+PMMA are 85.4 ± 1.9, which is in agreement with the reported values. Because of these similarities as well as the high proportion of acrylic polymer in the samples (approximately 35%), it can be stated that the hardness is much more significantly influenced by the properties of the resin as a component of delignified wood. It was also observed that the samples were more brittle during hardness testing. No cracks and fractures were observed directly after UV-C irradiation. This behaviour was confirmed by the study of [48] where ultraviolet radiation altered PMMA stiffness, resulting in changes in tensile properties, such as reduction in elongation at break and tensile strength.

Light exposure is a major cause of wood degradation, leading to colour change and loss in mechanical properties [49–51]. Significant changes were observed after 4 h of irradiation. All *m/z* signals of lignin were either absent or their intensity was considerably reduced, suggesting that lignin underwent an extensive degradation. The irradiation promoted a reduction in the transparency, due to the yellowing [47].

UV degradation of poly(methyl methacrylate) and its vinyltriethoxysilane containing copolymers, was tested using a mercury lamp with a wavelength of 259 nm, situated 10 cm away from the samples and found out that UV irradiation causes changes in the mechanical properties of PMMA [52].

Wochnowskia et al. [53] irradiated PMMA by UV-laser light with different wavelengths (193 nm, 248 nm and 308 nm) in order to investigate the photolytic degradation of the physico-chemical molecular structure and reported that, during the UV-irradiation (248 nm), there was the existence of methyl formate, a great amount of methanate, methanol and additionally the occurrence of methyl and other molecule fragments of the polymer side-chain even at a low irradiation dose. At this irradiation dose, side chain cleavage from the polymer main chain takes place yielding mechanical densification of the polymeric material due to Van-der-Waals forces with a subsequent increase in the refractive index.

From the above-mentioned arguments we can conclude that the change of favourable optical properties of transparent wood (transmittance and colour) was caused by the degradation of both components, the acrylic polymer as well as the wood itself.

5. Conclusions

Transparent wood, combining many advantageous properties, is an emerging new material for light-transmitting and environmentally friendly applications. There is an increasing number of research teams who introduce new methods of fabrication and new ways to use transparent wood. Therefore, it is crucial to know how this material behaves under various conditions.

Exposure to UV-C sources has a significant effect on the colour of transparent wood. It was mostly pronounced from the beginning of the test (during the first 7 days). Samples became darker with increasing exposure time and their colour shifts towards shades of red and yellow which can be possibly explained by the reactivation of chromophores. The values of the coordinates a^* and b^* show an interdependence that appears to be logarithmic. W+MMA samples are more prone to discolouration due to UV-C radiation than W+EMA samples.

The transmittance of light through the measured samples of transparent wood was significantly affected by the action of UV-C radiation. As in the case of colour changes, the UV-C effect was most pronounced at the beginning and had only a minimal effect in the later stages. W+EMA had higher light transmission and its reduction due to UV-C was less pronounced than in the case of W+MMA.

The influence of UV-C on shore D hardness of W+EMA and W+MMA is significantly lower than in the case of optical properties. The differences between these materials are not statistically significant. The measured values show that the resulting hardness of transparent wood depends mainly on the hardness of the acrylic polymer used.

In a previously mentioned study, the impact of UV-B radiation on the optical and mechanical properties of transparent wood has been investigated. The UV-B radiation was used for ageing acceleration. Due to the SARS-CoV-2 virus pandemic, the UV-C radiation (for virus deactivation purpose) began to be used massively. However, there were no data concerning the impact of UV-C radiation on transparent wood key properties before this study. This is the first study revealing the impact of UV-C radiation on key optical and mechanical parameters of transparent wood. Obtained results also proved that UV-C radiation (at irradiance flux of 16 $W \cdot m^{-2}$ during 35 days) has virtually no effect on the transparent wood (W+EMA and W+MMA) shore D hardness. Obtained results also proven that the impact of UV-C radiation on the optical characteristics of transparent wood (at stated irradiance flux) is significant only for the first 7 days (in the following days the impact was only negligible).

In future research, it is necessary to evaluate the effect of different wavelengths on the properties of transparent wood and also to describe the time period during which the highest degradation occurs.

Author Contributions: Conceptualization, I.W. and T.Š.; methodology, I.W.; software, P.R.; validation, I.W., T.Š. and P.R.; formal analysis, I.W., A.P. and T.Š.; investigation, I.W.; resources, I.W. and J.M.; data curation, I.W., T.Š., J.M. and P.R.; writing—original draft preparation, I.W.; writing—review and editing, I.W. and A.P.; visualization, I.W. and P.R; supervision, I.W.; project administration, I.W.; funding acquisition, I.W. and J.M. All authors have read and agreed to the published version of the manuscript.

Funding: This work was supported by the Slovak Research and Development Agency under the contract No. APVV-16-0223. This work was also supported by the KEGA agency under the contracts No. 016STU-4/2021 and No. 001TU Z-4/2020. This work was also supported by the Institutional project—FiTraW of MTF STU No. 1617.

Data Availability Statement: Not applicable.

Conflicts of Interest: The authors declare no conflict of interest. The funders had no role in the design of the study; in the collection, analyses, or interpretation of data; in the writing of the manuscript, or in the decision to publish the results.

References

1. Cabane, E.; Keplinger, T.; Merk, V.; Hass, P.; Burgert, I. Renewable and Functional Wood Materials by Grafting Polymerization Within Cell Walls. *ChemSusChem* **2014**, *7*, 1020–1025. [CrossRef] [PubMed]
2. Vay, O.; De Borst, K.; Hansmann, C.; Teischinger, A.; Müller, U. Thermal conductivity of wood at angles to the principal anatomical directions. *Wood Sci. Technol.* **2015**, *49*, 577–589. [CrossRef]
3. Antov, P.; Jivkov, V.; Savov, V.; Simeonova, R.; Yavorov, N. Structural Application of Eco-Friendly Composites from Recycled Wood Fibres Bonded with Magnesium Lignosulfonate. *Appl. Sci.* **2020**, *10*, 7526. [CrossRef]
4. Antov, P.; Krišťák, L.; Réh, R.; Savov, V.; Papadopoulos, A.N. Eco-Friendly Fiberboard Panels from Recycled Fibers Bonded with Calcium Lignosulfonate. *Polymers* **2021**, *13*, 639. [CrossRef]
5. Krišťák, L.; Réh, R. Application of Wood Composites. *Appl. Sci.* **2021**, *11*, 3479. [CrossRef]
6. Li, Y.; Fu, Q.; Yu, S.; Yan, M.; Berglund, L. Optically Transparent Wood from a Nanoporous Cellulosic Template: Combining Functional and Structural Performance. *Biomacromolecules* **2016**, *17*, 1358–1364. [CrossRef]
7. Zhu, M.; Song, J.; Li, T.; Gong, A.; Wang, Y.; Dai, J.; Yao, Y.; Luo, W.; Henderson, D.; Hu, L. Highly Anisotropic, Highly Transparent Wood Composites. *Adv. Mater.* **2016**, *28*, 5181–5187. [CrossRef]
8. Li, T.; Zhu, M.; Yang, Z.; Song, J.; Dai, J.; Yao, Y.; Luo, W.; Pastel, G.; Yang, B.; Hu, L. Wood Composite as an Energy Efficient Building Material: Guided Sunlight Transmittance and Effective Thermal Insulation. *Adv. Energy Mater.* **2016**, *6*, 1601122. [CrossRef]
9. Yaddanapudi, H.S.; Hickerson, N.; Saini, S.; Tiwari, A. Fabrication and characterization of transparent wood for next generation smart building applications. *Vacuum* **2017**, *146*, 649–654. [CrossRef]

10. Jia, C.; Li, T.; Chen, C.; Dai, J.; Kierzewski, I.M.; Song, J.; Li, Y.; Yang, C.-P.; Wang, C.; Hu, L. Scalable, anisotropic transparent paper directly from wood for light management in solar cells. *Nano Energy* **2017**, *36*, 366–373. [CrossRef]
11. Gan, W.; Gao, L.; Xiao, S.; Zhang, W.; Zhan, X.; Li, J. Transparent magnetic wood composites based on immobilizing Fe3O4 nanoparticles into a delignified wood template. *J. Mater. Sci.* **2017**, *52*, 3321–3329. [CrossRef]
12. Vasileva, E.; Li, Y.; Sychugov, I.; Mensi, M.; Berglund, L.; Popov, S. Lasing from Organic Dye Molecules Embedded in Transparent Wood. *Adv. Opt. Mater.* **2017**, *5*, 1700057. [CrossRef]
13. Yu, Z.; Yao, Y.; Yao, J.; Zhang, L.; Chen, Z.; Gao, Y.; Luo, H. Transparent wood containing CsxWO3 nanoparticles for heat-shielding window applications. *J. Mater. Chem. A* **2017**, *5*, 6019–6024. [CrossRef]
14. Montanari, C.; Li, Y.; Chen, H.; Yan, M.; Berglund, L.A. Transparent Wood for Thermal Energy Storage and Reversible Optical Transmittance. *ACS Appl. Mater. Interfaces* **2019**, *11*, 20465–20472. [CrossRef] [PubMed]
15. Zhang, T.; Yang, P.; Chen, M.; Yang, K.; Cao, Y.; Li, X.; Tang, M.; Chen, W.; Zhou, X. Constructing a Novel Electroluminescent Device with High-Temperature and High-Humidity Resistance based on a Flexible Transparent Wood Film. *ACS Appl. Mater. Interfaces* **2019**, *11*, 36010–36019. [CrossRef] [PubMed]
16. Wang, M.; Li, R.; Chen, G.; Zhou, S.; Feng, X.; Chen, Y.; He, M.; Liu, D.; Haisong, Q.; Qi, H. Highly Stretchable, Transparent, and Conductive Wood Fabricated by in Situ Photopolymerization with Polymerizable Deep Eutectic Solvents. *ACS Appl. Mater. Interfaces* **2019**, *11*, 14313–14321. [CrossRef] [PubMed]
17. Zhu, M.; Li, T.; Davis, C.S.; Yao, Y.; Dai, J.; Wang, Y.; AlQatari, F.; Gilman, J.W.; Hu, L. Transparent and haze wood composites for highly efficient broadband light management in solar cells. *Nano Energy* **2016**, *26*, 332–339. [CrossRef]
18. Li, Y.; Liu, Y.; Chen, W.; Wang, Q.; Liu, Y.; Li, J.; Yu, H. Facile extraction of cellulose nanocrystals from wood using ethanol and peroxide solvothermal pretreatment followed by ultrasonic nanofibrillation. *Green Chem.* **2016**, *18*, 1010–1018. [CrossRef]
19. Vasileva, E.; Chen, H.; Li, Y.; Sychugov, I.; Yan, M.; Berglund, L.; Popov, S. Light Scattering by Structurally Anisotropic Media: A Benchmark with Transparent Wood. *Adv. Opt. Mater.* **2018**, *6*, 1800999. [CrossRef]
20. Frey, M.; Widner, D.; Segmehl, J.S.; Casdorff, K.; Keplinger, T.; Burgert, I. Delignified and Densified Cellulose Bulk Materials with Excellent Tensile Properties for Sustainable Engineering. *ACS Appl. Mater. Interfaces* **2018**, *10*, 5030–5037. [CrossRef] [PubMed]
21. Zhang, H.; Bai, Y.; Yu, B.; Liu, X.; Chen, F. A practicable process for lignin color reduction: Fractionation of lignin using methanol/water as a solvent. *Green Chem.* **2017**, *19*, 5152–5162. [CrossRef]
22. Zhang, H.; Chen, F.; Liu, X.; Fu, S. Micromorphology Influence on the Color Performance of Lignin and Its Application in Guiding the Preparation of Light-colored Lignin Sunscreen. *ACS Sustain. Chem. Eng.* **2018**, *6*, 12532–12540. [CrossRef]
23. Qian, Y.; Deng, Y.; Li, H.; Qiu, X. Reaction-Free Lignin Whitening via a Self-Assembly of Acetylated Lignin. *Ind. Eng. Chem. Res.* **2014**, *53*, 10024–10028. [CrossRef]
24. Zhang, H.; Liu, X.; Fu, S.; Chen, Y. High-value utilization of kraft lignin: Color reduction and evaluation as sunscreen ingredient. *Int. J. Biol. Macromol.* **2019**, *133*, 86–92. [CrossRef]
25. Wang, J.; Deng, Y.; Qian, Y.; Qiu, X.; Ren, Y.; Yang, D. Reduction of lignin color via one-step UV irradiation. *Green Chem.* **2016**, *18*, 695–699. [CrossRef]
26. Ma, Y.-S.; Chang, C.-N.; Chiang, Y.-P.; Sung, H.-F.; Chao, A.C. Photocatalytic degradation of lignin using Pt/TiO2 as the catalyst. *Chemosphere* **2008**, *71*, 998–1004. [CrossRef]
27. Han, G.; Wang, X.; Hamel, J.; Zhu, H.; Sun, R. Lignin-AuNPs liquid marble for remotely-controllable detection of Pb2+. *Sci. Rep.* **2016**, *6*, 38164. [CrossRef]
28. Xue, Y.; Qiu, X.; Liu, Z.; Li, Y. Facile and Efficient Synthesis of Silver Nanoparticles Based on Biorefinery Wood Lignin and Its Application as the Optical Sensor. *ACS Sustain. Chem. Eng.* **2018**, *6*, 7695–7703. [CrossRef]
29. Li, Y.; Fu, Q.; Rojas, R.; Yan, M.; Lawoko, M.; Berglund, L. Lignin-Retaining Transparent Wood. *ChemSusChem* **2017**, *10*, 3445–3451. [CrossRef]
30. Li, Y.; Cheng, M.; Jungstedt, E.; Xu, B.; Sun, L.; Berglund, L.A. Optically Transparent Wood Substrate for Perovskite Solar Cells. *ACS Sustain. Chem. Eng.* **2019**, *7*, 6061–6067. [CrossRef]
31. Samanta, A.; Chen, H.; Samanta, P.; Popov, S.; Sychugov, I.; Berglund, L.A. Reversible Dual-Stimuli-Responsive Chromic Transparent Wood Biocomposites for Smart Window Applications. *ACS Appl. Mater. Interfaces* **2021**, *13*, 3270–3277. [CrossRef]
32. Li, T.; Zhai, Y.; He, S.; Gan, W.; Wei, Z.; Heidarinejad, M.; Dalgo, D.; Mi, R.; Zhao, X.; Song, J.; et al. A radiative cooling structural material. *Science* **2019**, *364*, 760–763. [CrossRef]
33. Wang, X.; Zhan, T.; Liu, Y.; Shi, J.; Pan, B.; Zhang, Y.; Cai, L.; Shi, S.Q. Large-Size Transparent Wood for Energy-Saving Building Applications. *ChemSusChem* **2018**, *11*, 4086–4093. [CrossRef]
34. Karla, V. Methodology of Research on Transparent Wood in Architectural Constructions. *Sel. Sci. Pap. J. Civ. Eng.* **2020**, *15*, 29–35. [CrossRef]
35. Katunský, D.; Kanócz, J.; Karľa, V. Structural elements with transparent wood in architecture. *Int. Rev. Appl. Sci. Eng.* **2018**, *9*, 101–106. [CrossRef]
36. Li, Y.; Vasileva, E.; Sychugov, I.; Popov, S.; Berglund, L. Optically Transparent Wood: Recent Progress, Opportunities, and Challenges. *Adv. Opt. Mater.* **2018**, *6*, 1800059. [CrossRef]
37. Bisht, P.; Pandey, K.K.; Barshilia, H.C. Photostable transparent wood composite functionalized with an UV-absorber. *Polym. Degrad. Stab.* **2021**, *189*, 109600. [CrossRef]

38. Bidsorkhi, H.C.; Soheilmoghaddam, M.; Pour, R.H.; Adelnia, H.; Mohamad, Z. Mechanical, thermal and flammability properties of ethylene-vinyl acetate (EVA)/sepiolite nanocomposites. *Polym. Test.* **2014**, *37*, 117–122. [CrossRef]
39. Mondal, S.; Tapon, D.A.S.; Ganguly, S.; Das, N. Revathy Ravindren a Others. Oxygen Permeability Properties of Ethylene Methyl Acrylate/Sepiolite Clay Composites with Enhanced Mechanical and Thermal Performance. *J. Polym. Sci. Appl.* **2017**, *2*, 2.
40. Ozdemir, E.; Arenas, D.R.; Kelly, N.L.; Hanna, J.V.; van Rijswijk, B.; Degirmenci, V.; McNally, T. Ethylene methyl acrylate copolymer (EMA) assisted dispersion of few-layer graphene nanoplatelets (GNP) in poly(ethylene terephthalate) (PET). *Polymer* **2020**, *205*, 122836. [CrossRef]
41. Lai, C.-F.; Li, J.-S. Self-assembly of colloidal Poly(St-MMA-AA) core/shell photonic crystals with tunable structural colors of the full visible spectrum. *Opt. Mater.* **2019**, *88*, 128–133. [CrossRef]
42. Kowalonek, J. Surface studies of UV-irradiated poly(vinyl chloride)/poly(methyl methacrylate) blends. *Polym. Degrad. Stab.* **2016**, *133*, 367–377. [CrossRef]
43. Hourston, D.J.; Satgurunathan, R.; Varma, H. Latex interpenetrating polymer networks based on acrylic polymers. III. Synthesis variations. *J. Appl. Polym. Sci.* **1987**, *33*, 215–225. [CrossRef]
44. Hourston, D.J.; Schäfer, F.-U. Poly(ether urethane)/poly(ethyl methacrylate) interpenetrating polymer networks: Morphology, phase continuity and mechanical properties as a function of composition. *Polymer* **1996**, *37*, 3521–3530. [CrossRef]
45. Poomali, S.; Suresha, B.; Lee, J.-H. Mechanical and three-body abrasive wear behaviour of PMMA/TPU blends. *Mater. Sci. Eng. A* **2008**, *492*, 486–490. [CrossRef]
46. Seeger, P.; Ratfisch, R.; Moneke, M.; Burkhart, T. Addition of thermo-plastic polyurethane (TPU) to poly(methyl methacrylate) (PMMA) to improve its impact strength and to change its scratch behavior. *Wear* **2018**, *406–407*, 68–74. [CrossRef]
47. Akinci, A.; Sen, S.; Sen, U. Friction and wear behavior of zirconium oxide reinforced PMMA composites. *Compos. Part B Eng.* **2014**, *56*, 42–47. [CrossRef]
48. De Castro Monsores, K.G.; Da Silva, A.O.; Oliveira, S.D.S.A.; Rodrigues, J.G.P.; Weber, R.P. Influence of ultraviolet radiation on polymethylmethacrylate (PMMA). *J. Mater. Res. Technol.* **2019**, *8*, 3713–3718. [CrossRef]
49. Tolvaj, L.; Mitsui, K. Light source dependence of the photodegradation of wood. *J. Wood Sci.* **2005**, *51*, 468–473. [CrossRef]
50. George, B.; Suttie, E.; Merlin, A.; Deglise, X. Photodegradation and photostabilisation of wood—The state of the art. *Polym. Degrad. Stab.* **2005**, *88*, 268–274. [CrossRef]
51. Argyropoulos, D.S. Wood and Cellulosic Chemistry. Second Edition, Revised and Expanded Edited by David, N.-S. Hon (Clemson University) and Nubuo Shiraishi (Kyoto University). Marcel Dekker: New York and Basel. 2001. vii + 914 pp. $250.00. ISBN 0-8247-0024-4. *J. Am. Chem. Soc.* **2001**, *123*, 8880–8881. [CrossRef]
52. Çaykara, T.; Güven, O. UV degradation of poly(methyl methacrylate) and its vinyltriethoxysilane containing copolymers. *Polym. Degrad. Stab.* **1999**, *65*, 225–229. [CrossRef]
53. Wochnowski, C.; Eldin, M.S.; Metev, S. UV-laser-assisted degradation of poly(methyl methacrylate). *Polym. Degrad. Stab.* **2005**, *89*, 252–264. [CrossRef]

MDPI
St. Alban-Anlage 66
4052 Basel
Switzerland
Tel. +41 61 683 77 34
Fax +41 61 302 89 18
www.mdpi.com

Polymers Editorial Office
E-mail: polymers@mdpi.com
www.mdpi.com/journal/polymers